Springer Series in Information Sciences 28

Editor: Thomas S. Huang

Springer Series in Information Sciences

Editors: Thomas S. Huang Teuvo Kohonen Manfred R. Schroeder
Managing Editor: H. K. V. Lotsch

1 **Content-Addressable Memories**
By T. Kohonen 2nd Edition

2 **Fast Fourier Transform and Convolution Algorithms**
By H. J. Nussbaumer 2nd Edition

3 **Pitch Determination of Speech Signals**
Algorithms and Devices By W. Hess

4 **Pattern Analysis and Understanding**
By H. Niemann 2nd Edition

5 **Image Sequence Analysis**
Editor: T. S. Huang

6 **Picture Engineering**
Editors: King-sun Fu and T. L. Kunii

7 **Number Theory in Science and Communication** With Applications in Cryptography, Physics, Digital Information, Computing, and Self-Similarity By M. R. Schroeder 2nd Edition

8 **Self-Organization and Associative Memory**
By T. Kohonen 3rd Edition

9 **Digital Picture Processing**
An Introduction By L. P. Yaroslavsky

10 **Probability, Statistical Optics, and Data Testing**
A Problem Solving Approach
By B. R. Frieden 2nd Edition

11 **Physical and Biological Processing of Images** Editors: O. J. Braddick and A. C. Sleigh

12 **Multiresolution Image Processing and Analysis** Editor: A. Rosenfeld

13 **VLSI for Pattern Recognition and Image Processing** Editor: King-sun Fu

14 **Mathematics of Kalman-Bucy Filtering**
By P. A. Ruymgaart and T. T. Soong 2nd Edition

15 **Fundamentals of Electronic Imaging Systems**
Some Aspects of Image Processing
By W. F. Schreiber 3rd Edition

16 **Radon and Projection Transform-Based Computer Vision**
Algorithms, A Pipeline Architecture, and Industrial Applications By J. L. C. Sanz, E. B. Hinkle, and A. K. Jain

17 **Kalman Filtering**
with Real-Time Applications
By C. K. Chui and G. Chen 2nd Edition

18 **Linear Systems and Optimal Control**
By C. K. Chui and G. Chen

19 **Harmony: A Psychoacoustical Approach**
By R. Parncutt

20 **Group-Theoretical Methods in Image Understanding**
By Ken-ichi Kanatani

21 **Linear Prediction Theory**
A Mathematical Basis for Adaptive Systems
By P. Strobach

22 **Psychoacoustics** Facts and Models
By E. Zwicker and H. Fastl

23 **Digital Image Restoration**
Editor: A. K. Katsaggelos

24 **Parallel Algorithms in Computational Science**
By D. W. Heermann and A. N. Burkitt

25 **Radar Array Processing**
Editors: S. Haykin, J. Litva, and T. J. Shepherd

26 **Signal Processing and Systems Theory**
Selected Topics
By C. K. Chui and G. Chen

27 **3D Dynamic Scene Analysis**
A Stereo Based Approach
By Z. Zhang and O. Faugeras

28 **Theory of Reconstruction from Image Motion**
By S. Maybank

29 **Motion and Structure from Image Sequences**
By J. Weng, T. S. Huang, and N. Ahuja

Stephen Maybank

Theory of Reconstruction from Image Motion

With 29 Figures

Springer-Verlag
Berlin Heidelberg New York
London Paris Tokyo
Hong Kong Barcelona
Budapest

Dr. Stephen Maybank
GEC-Marconi Limited, Hirst Research Centre,
East Lane, Wembley, Middlesex, HA9 7PP, UK

ISBN-13:978-3-642-77559-8 e-ISBN-13:978-3-642-77557-4
DOI: 10.1007/978-3-642-77557-4

Library of Congress Cataloging-in-Publication Data. Maybank, Stephen, 1954– . Theory of reconstruction from image motion / Stephen Maybank. p.cm. – (Springer series in information sciences; 28). Includes bibliographical references and index. ISBN-13:978-3-642-77559-8 1. Computer vision. 2. Image processing. I. Title. II. Series. TA1632.M3625 1993 006.3'7–dc 92-17979

Softcover reprint of the hardcover 1st edition 1993

Typesetting: Data conversion by Springer-Verlag
Production Editor: P. Treiber

54/3140-5 4 3 2 1 0 – Printed on acid-free paper

Preface

The image taken by a moving camera changes with time. These image motions contain information about the motion of the camera and about the shapes of the objects in the field of view. There are two main types of image motion, finite displacements and image velocities. Finite displacements are described by the point correspondences between two images of the same scene taken from different positions. Image velocities are the velocities of the points in the image as they move over the projection surface. Reconstruction is the task of obtaining from the image-motions information about the camera motion or about the shapes of objects in the field of view. In this book the theory underlying reconstruction is described.

Reconstruction from image motion is the subject matter of two different scientific disciplines, photogrammetry and computer vision. In photogrammetry the accuracy of reconstruction is emphasised; in computer vision the emphasis is on methods for obtaining information from images in real time in order to guide a mechanical device such as a robot arm or an automatic vehicle. This book arises from recent work carried out in computer vision. Computer vision is a young field but it is developing rapidly. The earliest papers on reconstruction in the computer vision literature date back only to the mid 1970s. As computer vision develops, the mathematical techniques applied to the analysis of reconstruction become more appropriate and more powerful. The advances in the theory are matched by the advances in the speed and capacity of the electronic equipment available for computer vision. It is now possible to implement in real time methods which were of only theoretical interest ten years ago.

The book is divided into six chapters. Chapter 1 is an introduction. Chapters 2 and 3 describe reconstruction from image correspondences, with particular emphasis on the ambiguous case, in which the same set of image correspondences yields two or more different reconstructions. Chapter 4 describes reconstruction from image velocities, again with an emphasis on the ambiguous case. Chapter 5 describes cases in which the data for reconstruction are just sufficient to ensure that only a finite number of reconstructions are possible. In Chap. 6 four algorithms for reconstruction from image motions are described. Two of the algorithms require image correspondences, and the remaining two require image velocities.

The chapters are divided into sections and the sections are divided into subsections. For example Sect. 2.1.1 is Subsection 1 of Section 1 of Chapter 2. Most of the results are stated either as theorems or as propositions. A proposition is usually less important or easier to prove than a theorem. The theorems, propositions and definitions are labelled in sequence, chapter by chapter. The numbering of the sections and subsections within a chapter is independent of the numbering of the theorems. The end of a proof of a theorem or a proposition is marked by a box, □, at the right-hand margin of the final line. The references are placed at the end of each chapter.

To return to the contents of the chapters in more detail: Chapter 1 describes the background to reconstruction. Certain useful mathematical concepts and notation are introduced. Two different but mathematically equivalent frameworks for reconstruction from point correspondences are described in Chap. 2, namely the Euclidean framework and the projective geometric framework. The Euclidean framework emphasises the description of the image in terms of coordinates, whilst the projective geometric framework emphasises the description of the image in terms of the lines in space passing through the optical centre of the camera. The ambiguous case of reconstruction is discussed within both frameworks. The properties of certain 3×3 matrices known as essential matrices are described. These matrices arise naturally in the Euclidean framework. They summarise the information about the camera displacement contained in a set of image correspondences. Reconstruction up to a collineation is discussed briefly at the end of Chap. 2. In this approach the full camera calibration is not required, but the reconstruction has a higher degree of ambiguity.

The geometry of the ambiguous case of reconstruction from image correspondences is developed in Chap. 3. Ambiguity is only possible if the points giving rise to the image correspondences lie on certain surfaces of degree two known as critical surfaces. There are at least two critical surfaces involved in each case of ambiguity, one for each possible reconstruction. The intersection of two critical surfaces contains a curve known as a horopter curve. Many properties of critical surfaces are directly related to the properties of horopter curves. The critical surfaces for the ambiguous case of reconstruction up to a collineation are discussed.

Chapter 4 describes the reconstruction of camera velocity from image velocities. It is shown that the ambiguous case is similar in many respects to the ambiguous case of reconstruction from image correspondences. The constraints on the camera velocity arising from four image velocity vectors are examined in detail. The translational velocity is subject to a quartic polynomial constraint. If the image velocity field is irregular then the leading order terms of the quartic split into two linear factors and a quadratic factor.

Chapter 5 discusses three cases of reconstruction in which the data obtained from the images are just sufficient to ensure that only a finite number of reconstructions are possible. The three cases are reconstruction from point correspondences, reconstruction from image velocities and reconstruction up to a collineation. In these minimal cases the algebraic nature of reconstruction is particularly apparent. Reconstruction is formulated as the problem of finding the common zeros of a particular set of multivariate polynomials. The degree of a reconstruction problem is defined to be the number of reconstructions obtained in the minimal case. If the degree is high then reconstruction is likely to be intrinsically difficult, regardless of the particular algorithm employed. The degree of reconstruction from point correspondences and from image velocities is ten in both cases. The degree of reconstruction up to a collineation is three.

In Chapter 6, four different reconstruction algorithms are described. Two are for reconstruction from image correspondences, and two are for reconstruction from image velocities. In all four algorithms it is assumed that the camera calibration is known. The algorithms for reconstruction from image velocities illustrate the importance of irregularities in the image velocity field. If the moving surface has a large change in depth or orientation over a small part of the field of view then reconstruction can be carried out using algorithms that are simpler than those designed for the general case. The results obtained are also more stable in the presence of noise.

The thesis on which part of this book is based began with a quotation from Machado suggested by Mari Carmen, 'Caminante no hay camino. Se hace camino al andar'. (Traveller there are no paths, Paths are made by walking.) Like a path, science is constructed, but at the same time it is not made arbitrarily. It has to *go* somewhere.

Sophia Antipolis S.J. Maybank
January 1992

Acknowledgements

Chapter 4 and Sect. 6.2 of this book are adapted from a Ph.D. thesis written whilst I was at Marconi Command and Control Systems Ltd., Frimley, UK (MCCS). The thesis is based on work carried out as part of a contract between the Royal Signals and Radar Establishment, Malvern, UK (RSRE) and MCCS. I thank MCCS and RSRE for the opportunity to write the thesis, and for permission to incorporate material into this book. The presentation, clarity and coherence of the thesis owe much to the untiring efforts of my supervisor, George Loizou, of Birkbeck College, University of London. Birkbeck is unique amongst British colleges in that the majority of the students are part time. I enrolled as a graduate student, and at the same time continued in employment at MCCS. The thesis was completed with the support of ESPRIT Project P940 (DMA). The revision of the thesis necessary for this book and the addition of new material was made possible by the support of ESPRIT Project P2502 (Voila). I thank the GEC-Marconi Hirst Research Centre, Wembley, for their support and encouragement during the writing of this book after my transfer there in April 1989. The final editing was carried out during a visit to INRIA, Sophia Antipolis, France.

I thank the many people that I have been fortunate enough to meet in the course of my work in computer vision. In particular, I am grateful to Bernard Buxton, David Castelow, David Corrall, Olivier Faugeras, Pavel Grossmann, Richard Johnson, Christopher Longuet-Higgins, George Loizou and Zhengyou Zhang. The book was typeset in LaTeX, using macros written by Zhengyou Zhang. I thank Robert Maybank for reading a preliminary draft and noting the many places where improvement was needed. Thanks are also due to Tom Huang for suggesting that the thesis should be published as a book, and to Dr. Lotsch of Springer-Verlag for making publication possible.

Contents

1 Introduction

A physically plausible reconstruction unique up to a single unknown scale factor can be obtained from just two different images of the same scene. The differences between the two images contain enough information about the camera position and about the location of the scene points relative to the camera to make the reconstruction possible. The scale factor cannot be recovered from the images alone. It is impossible to tell if the camera is near to a small object or far away from a large object. It is also possible to reconstruct a scene up to a single unknown scale factor using the velocities in an image taken by a moving camera.

In this introductory chapter reconstruction is placed in context as part of computer vision, the geometry underlying reconstruction is described and the conventions regarding image formation are given. The mathematical notation used in the remaining chapters is also described.

1.1 Background

Methods for reconstruction were first developed within the science of photogrammetry (Moffitt & Mikhail 1980; Wolf 1983). The task of photogrammetry is to measure accurately the lengths and angles in projected images, and to obtain useful information from these measurements. With the invention of the camera it became possible to obtain accurate and permanent records of projected images. The first experiments in reconstruction were carried out using photographs taken by two cameras on the ground separated by a baseline of known length. As early as 1840 the French geodesist Arago demonstrated the use of cameras for surveying the shape of terrain. This work was continued by Jordan, Stolze and Laussedat. The term 'photogrammetry' can be traced back to the 'photogrammetrie' of Laussedat. The first use of photogrammetry for the construction of maps from aerial photographs was in 1849 under the direction of Laussedat. Map making was the main application of photogrammetry until about 1960. More recently photogrammetry has found new applications in architecture, archaeology and engineering.

With the introduction of the aeroplane it became possible to take images in long sequences which systematically covered any terrain for which a map was required. The availability of such image sequences stimulated a rapid advance in the techniques of photogrammetry. In particular a stereoscopic plotting machine for combining two images of a scene was developed. The stereoscopic plotter gave an operator the ability to view a three dimensional reconstruction. Two slides made from consecutive images of a sequence are placed in projectors. The projectors are arranged such that they reproduce, up to scale, the relative positions of the camera from which the two images were obtained. When an operator views the two images his own stereoscopic vision system fuses the two images to create a three dimensional reconstruction of the original scene. The operator can then build up a map from the reconstruction as he views it. Stereo reconstruction by the operator is made easier using an anaglyphic method. One image is projected with red light and the other image is projected with green light. The operator wears spectacles in which one lens is red and the other lens is green. The red image cannot be seen through the green lens and the green image cannot be seen through the red lens. This ensures that each eye views only one of the projected images. The two images are then fused in the brain to give the three dimensional model.

The roots of computer vision lie not in photogrammetry but in biology and artificial intelligence (Gibson 1950; Marr 1982). The goal of computer vision is to obtain information from image sequences rapidly and reliably in order to guide the action of a machine. Computer vision became a practical possibility in the mid 1970s with the introduction of digital image storage and the development of computers capable of processing electronic images 'on line' as fast as they are obtained by the camera. Applications of computer vision to vehicle guidance, robotics and surveillance are being developed in laboratories around the world. Computer vision promises a combination of flexibility, adaptability and precision that is hard to achieve by other means.

It is apparent from the book by David Marr (1982) that in the 1970s there were close links between computer vision and the study of biological vision. In recent years these links have become weaker. Biological vision remains as a formidable example of what vision can achieve. However, the discipline of any particular application of computer vision does not allow the straightforward introduction of biological strategies. In addition it is not easy to make the correct inferences from a biological system. For example the optic nerve has a relatively low capacity for carrying information. Is this a defect which necessitates an increased amount of visual processing within the eye, or is it an advantage made possible by increasing the amount of visual processing carried out within the eye?

Vision systems require large amounts of computing power in order to process the stream of images emerging from an electronic camera. A typical electronic camera can deliver 25 images a second. Each image is a 512×512 grid of pixels (picture elements). Each pixel corresponds to a small region of the imaging surface of the camera. The pixel has associated with it a number or grey level which is a measure of the intensity of the light falling onto the image region. The grey level is typically between 0 and 127. A single such grey level thus contains one byte or equivalently eight binary digits of information. The total data rate of an electronic camera is thus $25 \times 512 \times 512 = 6.6$ Mbytes s^{-1}. The data rate is likely to increase as the resolution of electronic images increases. For example, high definition television may involve a 1024×1024 image, giving a data rate of at least 25 Mbytes s^{-1}. Many experimental visual systems for vehicle guidance require two or three cameras, each of which can have a data rate of 6.6 Mbytes s^{-1}. The important information is concealed within a vast stream of irrelevant information. The problem is to extract the important information rapidly and reliably.

An example is given to illustrate the degree of data compression that must be achieved by a computer vision system. In order to locate a line of length 100 pixels in an image it is necessary to process all the pixels within about ten pixels of the line. A reasonably accurate estimate of the velocity of the line across the image requires estimates of the position of the line in three consecutive frames. The total number of pixels involved is approximately $100 \times 20 \times 3 = 6000$. The total amount of data is thus 6000 bytes. Four numbers are required to describe the position and velocity of the line in the image. If two bytes are allowed for each number then this gives a total of 8 bytes. The amount of data is thus reduced by a factor of $6000/8 = 750$.

Reconstruction has attracted the attention of a large number of researchers in computer vision. One reason for this is the practical importance of reconstruction in the vision systems currently being developed. Another reason is that reconstruction can be specified cleanly in mathematical terms, as the problem of finding solutions to sets of equations. There are many important questions which naturally suggest themselves within the mathematical framework, for example how stable is the reconstruction when the data are subject to small perturbations? and what is the best way of obtaining an accurate reconstruction from a large amount of data of varying reliability?

1.2 Reconstruction

In this book two types of reconstruction are described, reconstruction from image correspondences, and reconstruction from image velocities. It is assumed that the image correspondences or the image velocities are known, but no assumptions are made about the camera velocity or about the shape of the scene other than that the scene is rigid. In applications such as map making the rigidity assumption clearly holds. In more complicated examples it is necessary to segment the image in order to separate out those parts that are due to the motion of a single rigid body.

Many variations on the basic reconstruction problem are omitted. These include reconstruction from the motions of straight lines relative to the camera (Liu & Huang 1988; Spetsakis & Aloimonos 1990), reconstruction from the motions of curves in the image (Faugeras 1990; Waxman et al. 1987) and reconstruction

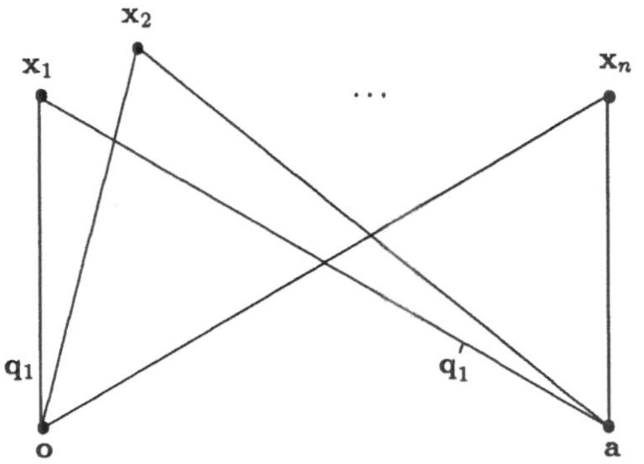

Fig. 1.1. Two views of a set of points in space

directly from grey level changes in the image (Negahdaripour & Horn 1987). The theory underlying reconstruction from moving lines or curves is not at present known in detail, although rapid progress is being made. Direct reconstruction from grey level changes is likely to be difficult because the link between rigid motion and changes in grey level is complicated.

Reconstruction from the point correspondences between two images is possible up to a single unknown scale factor, even when the relative position of the two viewpoints is unknown. The key assumption is that the objects in the field of view are rigid. The images are obtained from two different positions, as illustrated in Fig. 1.1. The point $\mathbf{o}$ is the optical centre of the camera when the first image is obtained, and $\mathbf{a}$ is the optical centre of the camera when the second image is obtained. The correspondences between some of the points in each image are assumed known. These points are usually easily identifiable surface markings or corners. If a point $\mathbf{x}$ in space projects to an easily identified point $\mathbf{q}$ in the first image, and to a point $\mathbf{q}'$ in the second image then it is assumed that $\mathbf{q}$ and $\mathbf{q}'$ can both be found, and that both points are known to be projections of the same point $\mathbf{x}$ in space. Given the two images and the correspondences between sufficiently many points in the images it is possible to reconstruct the relative position of the two cameras and the positions of the space points which project down to corresponding image points. The reconstruction is fixed up to a single unknown scale factor. There is no way of obtaining the scale factor without introducing additional information about the absolute sizes of the objects in the field of view. In addition to the unknown scale factor there is a 'twisted pair' ambiguity, as described in Sect. 2.1. The twisted pair ambiguity can usually be resolved without difficulty using the physically plausible condition that the reconstructed points $\mathbf{x}$ are in front of the camera.

Many different methods for establishing the image correspondences $\mathbf{q} \leftrightarrow \mathbf{q}'$ are described in the literature. A typical approach is to apply to the image a linear filter which has a high response in regions where the grey levels have a large local variance. The peak responses to the filter are located and then matched between the images. Each match yields a pair $\mathbf{q}$, $\mathbf{q}'$ of corresponding points.

Reconstruction is also possible using image velocities rather than image corrspondences. The velocity of the camera and the shape of objects in a scene are recovered up to a single unknown scale factor. Let $\mathbf{x}$ be a point in space with velocity $\dot{\mathbf{x}}$ relative to the camera, and let $\mathbf{x}$ project to a point $\mathbf{q}$ in the image. Then $\mathbf{q}$ has a velocity $\dot{\mathbf{q}}$ which is the projection of $\dot{\mathbf{x}}$. The reconstruction problem is to deduce the velocity of the camera and the positions of the points $\mathbf{x}$ relative to the camera using the information in the image velocities $\dot{\mathbf{q}}$.

Image velocities can be estimated using the correspondences $\mathbf{q} \leftrightarrow \mathbf{q}'$ between points in two images obtained by the camera at times separated by a small interval Δt. The estimate of $\dot{\mathbf{q}}$ is $\dot{\mathbf{q}} = (\mathbf{q}' - \mathbf{q})/\Delta t$. A more accurate estimate of $\dot{\mathbf{q}}$ can be obtained by using a sequence of three images (Murray & Buxton 1990).

It is also possible to estimate image velocities by differentiating the image grey levels. Let $I(x, y, t)$ be the grey level of the point (x, y) in the image at time t, and let $\mathbf{u}(x, y, t) = (u_1(x, y, t), u_2(x, y, t))$ be a velocity at (x, y) such that $\mathbf{u}(x, y, t)$ describes the rate of change of $I(x, y, t)$. In many cases it is a useful approximation to assume that $\mathbf{u}(x, y)$ is related to $I(x, y, t)$ by the motion constraint equation

$$\frac{\partial I}{\partial t} + u_1 \frac{\partial I}{\partial x} + u_2 \frac{\partial I}{\partial y} = 0 \tag{1.1}$$

The equation (1.1) is derived in Horn (1986) from the condition that the difference in grey levels

$$I(x + u_1 \Delta t, y + u_2 \Delta t, t + \Delta t) - I(x, y, t) \tag{1.2}$$

is of second order in Δt. The condition (1.2) ensures that the image grey levels are 'carried along' by the velocity field $\mathbf{u}$ without changing their values. Only one component of $\mathbf{u}$ at each point is determined by (1.1). The component of $\mathbf{u}$ normal to the grey level gradient $(\partial I/\partial x, \partial I/\partial y)^\top$ is unconstrained.

The drawback to the motion constraint equation is that the image grey levels may not be carried along by the image velocity field in such a way that (1.2) is second order in Δt. For example, if a white piece of paper is moved from an illuminated area into shadow then the brightness of the image of the paper decreases. Equation (1.1) then does not apply. The observed image velocity field $\mathbf{u}$ need not be the true projection of the motion relative to the camera. The motion constraint equation continues to be widely used in spite of this drawback. The equation has the advantage of simplicity. In addition there are theoretical arguments which show that the motion constraint equation is accurate in regions where the grey level gradients are large (Wu & Wohn 1991).

In the past the distinction between observable image velocity fields $\mathbf{u}$ and the true projection of three dimensional motion was not clearly made. More recently the term optical flow has been adopted for image velocity fields $\mathbf{u}$ which describe the observed grey level changes.

Reconstruction from image velocities is more important in computer vision than in photogrammetry, presumably because it is difficult to measure an image

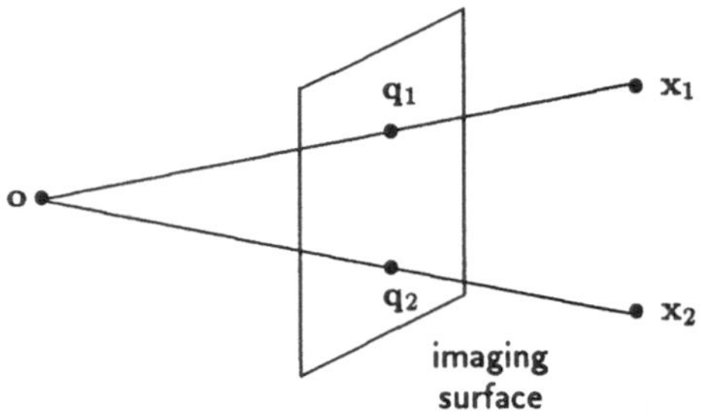

Fig. 1.2. Image formation

velocity field using conventional cameras. With the arrival of electronic cameras and powerful computers it is now possible to estimate image velocities accurately and to implement methods for reconstruction based on them.

1.3 Conventions About the Image

The image is assumed to arise by a linear projection from three-dimensional space to the imaging surface. It is often useful to think of image formation as a two step process. In the first step each point $\mathbf{x}$ in space is mapped to the line $\langle \mathbf{o}, \mathbf{x} \rangle$ joining $\mathbf{x}$ to the optical centre $\mathbf{o}$ of the camera. In the second step the lines through $\mathbf{o}$ are mapped linearly to the points of the image. The map is invertible, in that each line through $\mathbf{o}$ corresponds to a unique point in the image, and each point in the image corresponds to a unique line through $\mathbf{o}$. The image can be regarded as a parameterisation of the lines through $\mathbf{o}$, as illustrated in Fig. 1.2.

In the literature there are two standard methods of parameterising an image. The first method is to take the intersection of each line $\langle \mathbf{o}, \mathbf{x} \rangle$ with a plane placed near to $\mathbf{o}$, but not containing $\mathbf{o}$. The second method is to take the intersection of each line $\langle \mathbf{o}, \mathbf{x} \rangle$ with a sphere centred at $\mathbf{o}$. It is necessary either to restrict the image to a part of the sphere entirely contained within a single hemisphere, or to identify opposite points on the sphere and to work in the projective plane rather than on the surface of a sphere.

The image plane in Fig. 1.2 is placed between the optical centre and the object rather than behind the optical centre. This placement of the image plane is adopted almost universally in computer vision. The positions of the image points on a plane behind $\mathbf{o}$ are easy to deduce from the image on a plane in front of $\mathbf{o}$. The advantage of using a plane in front of $\mathbf{o}$ is that the image moves with the object. If the object moves to the left then the projected image, as shown in Fig. 1.2 also moves to the left. Similar remarks apply to the image formed on a sphere centred at $\mathbf{o}$.

In practice the process of image formation is more complicated than a linear transformation from space to the imaging surface (Wolf 1983). The image is subject to aberrations which degrade the sharpness, and to lens distortions, in

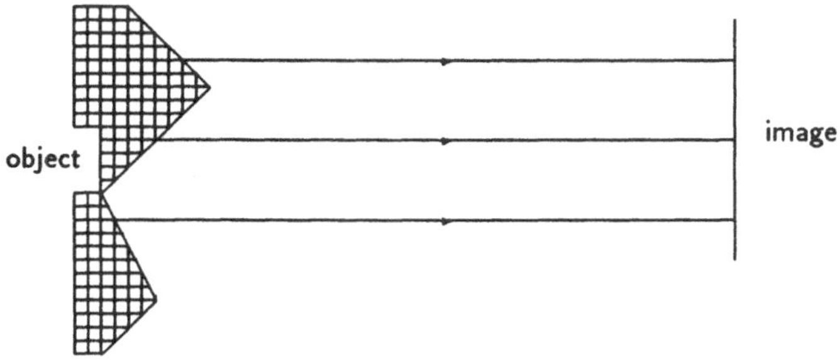

Fig. 1.3. Orthogonal projection

which the images of points are displaced slightly from their expected positions. It is assumed that the aberrations are slight, and that as a result feature points can be located accurately in the image. It is also assumed that the lens distortion can be corrected by a calibration procedure. Once the camera calibration is known the image can be transformed to yield the image that would have been obtained by a linear projection onto a plane or onto the unit sphere. In practice, a camera is calibrated by taking one or more images of an object of known dimensions, and then calculating the transformation from the points on the object to the corresponding points in the image. The values of the transformation at other points in space are obtained by interpolation. When discussing reconstruction from image correspondences or from image velocities it is assumed that the full camera calibration is known. If the transformation from space to the image is known to be linear, but no further information about the camera calibration is given then the term 'reconstruction up to a collineation' is used.

The case of image formation by orthogonal projection is frequently discussed in the computer vision literature. A good example is Harris (1990). In orthogonal projection the image is formed by a parallel beam of rays travelling from the surface of the object, as illustrated in Fig. 1.3. In certain circumstances, for example when the image is small and the range of depths is restricted, orthogonal projection is an adequate model for the process of image formation. In these circumstances reconstruction is often difficult because small errors in the data can have a large effect on the results. The additional errors which arise by assuming orthogonal projection are compensated by the greater stability of the equations underlying reconstruction from orthogonal projections (Harris 1990). In this book orthogonal projection is avoided because it is not an accurate model for the process of image formation.

1.4 Mathematical Background

Some knowledge of linear algebra and algebraic geometry is required. A good source of information on linear algebra is Sokolnikoff & Redheffer (1966). The authors also discuss rigid displacements and rigid velocities. The projective geometry necessary for Chap. 3 is covered by Semple & Kneebone (1952). Information about the singular value decomposition and about matrix algorithms can be found in Golub & Van Loan (1983). The algebraic geometry of curves is described by Fulton (1969), Semple & Roth (1949) and Walker (1962). Only a small part of the material in the last three books is required.

The terminology used to discuss image motions and the properties of cameras is introduced in Sect. 1.4.1. Background material on linear algebra and projective geometry is given in Sect. 1.4.2. The basic properties of algebraic curves are summarised in Sect. 1.4.3.

1.4.1 Terminology

An image motion can be either a finite change in an image arising from a camera displacement or it can be an instantaneous change arising from the camera velocity. Similarly, a camera motion is either a finite displacement of the camera, or a camera velocity. An image displacement field is a set of correspondences between the points in two images of the same scene taken from different viewpoints. Two points correspond if and only if they are both images of the same point in space. An image velocity field is a set of velocities of points across the image arising from camera velocity relative to the objects in the field of view. The terms 'image displacement field' or 'image velocity field' are employed even if only a small number of point correspondences or velocities are known. An image displacement field is said to be dense if it is defined for all points $\mathbf{q}$ in a non-empty open set in the image and if in addition the point correspondences $\mathbf{q} \leftrightarrow \mathbf{q}'$ define a homeomorphism from an open set in the first image onto an open set in the second image. A homeomorphism is a function that is continuous and that has a continuous inverse. A dense image velocity field is one that is known at all points in a non-empty open set of the image. An irregular image velocity field is one arising from a surface with significant variations in depth or orientation over a small part of the image. In this book derivatives of image velocity fields are always taken with respect to the image coordinates, never with respect to time.

The term 'reconstruction' refers to the problem of estimating the camera motion or the shape of a surface from image motions when the camera calibration is known. If the image is formed by a linear projection, but the details of the mapping from space to the image are otherwise unknown, then the term 'reconstruction up to a collineation' is used. The two images of the scene may be taken by the same camera or by different cameras. If the camera calibration is known then the same theory is applicable to both cases with only minor changes in the vocabulary. In reconstruction with a known camera calibration it is assumed that the images are obtained by a single moving camera.. In reconstruction up to a collineation it is assumed that the two images are taken by different cameras. The more restricted case in which both images are taken by a single uncalibrated camera lies outside the scope of this book.

1.4.2 Euclidean Space and Projective Space

Euclidean n-dimensional space $\mathbb{R}^n$ is the set of n-tuples $(x_1, \ldots, x_n)^\top$ of real numbers. Each n-tuple $(x_1, \ldots, x_n)^\top$ is a column vector. The corresponding row vector is $(x_1, \ldots, x_n)$. Whenever a vector is represented by a symbol such as $\mathbf{x}$ a column vector is always intended. The row vector corresponding to $\mathbf{x}$ is $\mathbf{x}^\top$. The origin of $\mathbb{R}^n$ is the point $(0, \ldots, 0)^\top$. The line joining two points $\mathbf{x}$, $\mathbf{y}$ of $\mathbb{R}^n$ is denoted by $\langle \mathbf{x}, \mathbf{y} \rangle$. The plane in $\mathbb{R}^n$ containing a line l and a point $\mathbf{x}$ not on l is denoted by $\langle l, \mathbf{x} \rangle$. A linear subspace of $\mathbb{R}^n$ of dimension $n-1$ is called a hyperplane. For example, the hyperplanes in $\mathbb{R}^2$ are lines, and the hyperplanes in $\mathbb{R}^3$ are planes.

Let $\mathbf{a}$, $\mathbf{b}$ be two vectors in $\mathbb{R}^3$. The vector product of $\mathbf{a}$ and $\mathbf{b}$ is the vector $\mathbf{c}$ defined by

$$\begin{aligned} c_1 &= a_2 b_3 - a_3 b_2 \\ c_2 &= a_3 b_1 - a_1 b_3 \\ c_3 &= a_1 b_2 - a_2 b_1 \end{aligned}$$

The vector $\mathbf{c}$ is normal to both $\mathbf{a}$ and $\mathbf{b}$. The length of $\mathbf{c}$ is equal to the area of the parallelogram with sides $\mathbf{a}$, $\mathbf{b}$ and the orientation of $\mathbf{c}$ is such that $\mathbf{a}$, $\mathbf{b}$, $\mathbf{c}$ form a right-handed triad of vectors. The notation $\mathbf{c} = \mathbf{a} \times \mathbf{b}$ is used. The symbol '$\times$' is also used occasionally for the multiplication of numbers or the multiplication of terms in an equation.

The space $\mathbb{C}^n$ is the set of n-tuples $(x_1, \ldots, x_n)^\top$ where the x_i are complex numbers. The usual symbol i is used for the square root of minus one. A scalar is an element of $\mathbb{R}$ or $\mathbb{C}$, depending on context.

The Euclidean distance $\|\mathbf{x}-\mathbf{y}\|$ between two vectors, $\mathbf{x}$ and $\mathbf{y}$ in $\mathbb{R}^n$ is defined by

$$\|\mathbf{x} - \mathbf{y}\| = \sqrt{\sum_{i=1}^{n} (x_i - y_i)^2}$$

The distance $\|\mathbf{x}\|$ between $\mathbf{x}$ and the origin is the Euclidean length of $\mathbf{x}$. It is also called the Euclidean norm or simply the norm of $\mathbf{x}$. The norm function has the following properties.

- $\|\mathbf{x}\| \geq 0$ and $\|\mathbf{x}\| = 0$ if and only if $\mathbf{x} = 0$.
- If λ is any real number then $\|\lambda \mathbf{x}\| = |\lambda| \, \|\mathbf{x}\|$.
- The triangle inequality holds, $\|\mathbf{x} + \mathbf{y}\| \leq \|\mathbf{x}\| + \|\mathbf{y}\|$.

A norm is any real valued function defined on $\mathbb{R}^n$ with the above three properties.

An $m \times n$ matrix is an array of numbers, real or complex, with m rows and n columns. Matrices have real entries unless stated otherwise. An $n \times n$ matrix A such that $A_{ij} = 0$ if $i \neq j$ is said to be diagonal. The tensor product $\mathbf{a} \otimes \mathbf{b}$ of two n-dimensional vectors $\mathbf{a}$, $\mathbf{b}$ is the 3×3 matrix with i, jth entry equal to $a_i b_j$.

Let A be an $m \times n$ matrix with real entries. The norm $\|A\|$ of A subordinate to the Euclidean vector norm is defined by

$$\|A\| = \max\{\|A\mathbf{x}\| \mid \|\mathbf{x}\| = 1\}$$

The space of $m \times n$ matrices is identical to $\mathbb{R}^{m \times n}$. The matrix norm $A \mapsto \|A\|$ has the usual properties of a norm on $\mathbb{R}^{m \times n}$. It has the additional property that for all $\mathbf{x}$ in $\mathbb{R}^n$

$$\|A\mathbf{x}\| \leq \|A\| \, \|\mathbf{x}\| \qquad \mathbf{x} \in \mathbb{R}^n$$

The trace, $\mathrm{tr}(A)$, of an $n \times n$ matrix A is defined by

$$\mathrm{tr}(A) = \sum_{i=1}^{n} A_{ii}$$

Let I be the $n \times n$ identity matrix and let r, s be any two scalars. The trace function has the following properties,

$$\begin{aligned} \mathrm{tr}(I) &= n \\ \mathrm{tr}(rA + sB) &= r\,\mathrm{tr}(A) + s\,\mathrm{tr}(B) \\ \mathrm{tr}(AB) &= \mathrm{tr}(BA) \\ \mathrm{tr}(A^\top A) &= \sum_{i,j=1}^{n} A_{ij}^2 \end{aligned}$$

The symmetric part, $\mathrm{sym}(A)$, of an $n \times n$ matrix A is defined by

$$\mathrm{sym}(A) = \frac{1}{2}(A + A^\top)$$

and the antisymmetric part, $\mathrm{asy}(A)$, of A is defined by

$$\mathrm{asy}(A) = \frac{1}{2}(A - A^\top)$$

It follows from the definitions of $\mathrm{sym}(A)$ and $\mathrm{asy}(A)$ that

$$A = \mathrm{sym}(A) + \mathrm{asy}(A)$$

Let A be an $m \times n$ matrix. The set of vectors $\mathbf{x}$ in $\mathbb{R}^n$ such that $A\mathbf{x} = 0$ is called the null space of A. The null space is a vector subspace of $\mathbb{R}^n$. An $n \times n$ matrix is invertible if and only if its null space contains only the zero vector. If A is invertible then the inverse A^{-1} of A is the unique matrix such that $A^{-1}A = AA^{-1} = I$. The set of vectors in $\mathbb{R}^m$ of the form $A\mathbf{x}$ for some vector $\mathbf{x}$ in $\mathbb{R}^n$ is called the range of A. The range is a vector subspace of $\mathbb{R}^m$. An $n \times n$ matrix is invertible if and only if the range of the matrix is the whole of $\mathbb{R}^n$.

Let A be an $n \times n$ matrix, let $\mathbf{e}$ be a non-zero vector and let λ be a scalar such that $A\mathbf{e} = \lambda\mathbf{e}$. Then $\mathbf{e}$ is an eigenvector of A, and λ is the corresponding eigenvalue of A. Each eigenvalue of A is a root of the characteristic equation

$$\det(A - \lambda I) = 0 \tag{1.3}$$

Conversely, if λ is a root of (1.3) then there exists a non-zero vector $\mathbf{e}$ such that $A\mathbf{e} = \lambda\mathbf{e}$. If (1.3) has n distinct roots then A has n linearly independent eigenvectors. If any of the roots of (1.3) are repeated then A may have less that n linearly independent eigenvectors. For example the complex 2×2 matrix

$$\begin{pmatrix} 2 & i \\ i & 0 \end{pmatrix}$$

has only one eigenvector, $(1, i)^\top$.

A 3×3 matrix R is said to be orthogonal if and only if $RR^\top = I$, where I is the identity matrix. The orthogonal matrices represent the rotations of $\mathbb{R}^3$. The rotations preserve the Euclidean lengths of vectors

$$\|R\mathbf{x}\|^2 = (R\mathbf{x})^\top.(R\mathbf{x}) = \mathbf{x}^\top R^\top R\mathbf{x} = \mathbf{x}.\mathbf{x} = \|\mathbf{x}\|^2$$

Each rotation is specified by giving the axis $\mathbf{n}$, and the angle of rotation θ about the axis. The vector $\mathbf{n}$ is usually normalised to ensure that $\mathbf{n}.\mathbf{n} = 1$. If $\theta = 180°$ then $R\mathbf{x}$ is given by

$$R\mathbf{x} = 2(\mathbf{x}.\mathbf{n})\mathbf{n} - \mathbf{x} \tag{1.4}$$

The points of projective space $\mathbb{P}^n$ are parameterised by $n+1$ tuples, $(x_1, \ldots, x_{n+1})^\top$, where at least one of the x_i is non-zero. Two $n+1$ tuples, $(x_1, \ldots, x_{n+1})^\top$ and $(y_1, \ldots, y_{n+1})^\top$, describe the same point of $\mathbb{P}^n$ if and only if there exists a non-zero scalar λ such that $x_i = \lambda y_i$ for $1 \le i \le n+1$. The same symbol $\mathbb{P}^n$ is used for a projective space defined over the complex numbers and for a projective space defined over the real numbers. If it is necessary to avoid the ambiguity then the notation $\mathbb{P}^n(\mathbb{C})$, $\mathbb{P}^n(\mathbb{R})$ is used. The space $\mathbb{R}^n$ is included in $\mathbb{P}^n$ by

$$(x_1, \ldots, x_n)^\top \mapsto (x_1, \ldots, x_n, 1)^\top$$

The set $\mathbb{P}^n(\mathbb{R}) \setminus \mathbb{R}^n$ is the hyperplane at infinity, H_∞. The hyperplane at infinity is defined by the equation $x_{n+1} = 0$. If $n = 3$ then H_∞ is called the plane at infinity. The hyperplane at infinity can be thought of as a set of directions. Let l be a line in $\mathbb{P}^n$ not contained in H_∞. The direction of l is the point $l \cap H_\infty$. Two lines l, l' are parallel if they intersect H_∞ at the same point.

In effect, $\mathbb{P}^n$ is obtained from $\mathbb{R}^n$ by adding additional points of intersection for parallel lines. This construction has profound and far reaching consequences. For example, in $\mathbb{P}^2(\mathbb{C})$ Bézout's Theorem holds for any two algebraic curves even though the points of $\mathbb{P}^2 \setminus \mathbb{C}^2$ are introduced only in order to provide additional points of intersection for lines. Bézout's Theorem states that two algebraic plane curves of respective degrees m, n either intersect in exactly mn points, counted with the correct multiplicities, or they have a common component.

A collineation is an invertible linear transformation from $\mathbb{P}^n$ to $\mathbb{P}^n$, $n \ge 2$. A collineation is so-called because it preserves collinearity: if $\mathbf{x}$, $\mathbf{y}$, $\mathbf{z}$ are collinear points in $\mathbb{P}^n$, and ω is a collineation of $\mathbb{P}^n$, then $\omega(\mathbf{x})$, $\omega(\mathbf{y})$, $\omega(\mathbf{z})$ are collinear points. If coordinates are chosen in $\mathbb{P}^n$, then a collineation is represented by a

$(n+1)\times(n+1)$ non-singular matrix. If A is a matrix representing a collineation ω, then any matrix λA, where λ is a non-zero scalar, also represents ω. An invertible linear transformation from $\mathbb{P}^1$ to $\mathbb{P}^1$ is called a homography. A homographic correspondence is often denoted by $\mathbf{x}\,\overline{\wedge}\,\mathbf{y}$. The advantage of the notation $\mathbf{x}\,\overline{\wedge}\,\mathbf{y}$ is that it is symmetrical in $\mathbf{x}$ and $\mathbf{y}$.

Let $\mathbf{x}_i$ for $i = 1, 2, 3$ be three distinct points in $\mathbb{P}^1$ and let $\mathbf{y}_i$ for $i = 1, 2, 3$ be a second set of three distinct points in $\mathbb{P}^1$. Then there exists a unique homography ω of $\mathbb{P}^1$ such that $\omega(\mathbf{x}_i) = \mathbf{y}_i$ for $i = 1, 2, 3$. If $\mathbf{x}_i = \mathbf{y}_i$ for $i = 1, 2, 3$ then ω is the identity. More generally, let $\mathbf{x}_i$ for $1 \le i \le n+2$ be $n+2$ distinct points in $\mathbb{P}^n$ such that no $n+1$ of the $\mathbf{x}_i$ lie in a hyperplane of $\mathbb{P}^n$. Let $\mathbf{y}_i$ for $1 \le i \le n+2$ be a second set of $n+2$ points of $\mathbb{P}^n$, again such that no $n+1$ of the $\mathbf{y}_i$ lie in a hyperplane. Then there is a unique collineation ω of $\mathbb{P}^n$ such that $\omega(\mathbf{x}_i) = \omega(\mathbf{y}_i)$ for $1 \le i \le n+2$. If $\mathbf{x}_i = \mathbf{y}_i$ for $1 \le i \le n+2$ then ω is the identity.

Certain special projective spaces are employed frequently. A pencil of lines is the set of coplanar lines which all pass through a single point. The point is known as the centre of the pencil. A star of lines is the set of lines in three dimensional space which all pass through a single point, known as the centre of the star. In three dimensional space a pencil of planes is a set of planes which all contain a single line, known as the axis of the pencil. In general, any projective space of dimension one may be referred to as a pencil, and any projective space of dimension two may be referred to as a star.

1.4.3 Algebraic Curves

Let coordinates x_1, x_2, x_3 be chosen in the projective plane $\mathbb{P}^2(\mathbb{C})$. An algebraic plane curve is the set of zeros of a non-trivial homogeneous polynomial f in x_1, x_2, x_3. The values of f are written as $f(x_1, x_2, x_3)$ or more simply as $f(\mathbf{x})$. Two homogeneous polynomials f, g define the same curve if and only if there exists a non-zero constant λ such that $f = \lambda g$. In the older literature a homogeneous polynomial is called a form. An algebraic plane curve is irreducible if and only if the defining form f does not split into non-trivial factors. If $f = f_1 f_2$, where f_1 and f_2 are each of degree one or greater then the curve defined by f is said to be reducible.

Let ∇ be the differential operator defined by

$$\nabla f = \left(\frac{\partial f}{\partial x_1}, \frac{\partial f}{\partial x_2}, \ldots, \frac{\partial f}{\partial x_n}\right)^\top$$

In the case of a plane curve $n = 3$. The tangent line at a point $\mathbf{u}$ on a plane curve f is the set of points $\mathbf{x}$ satisfying the equation

$$(\nabla f|_{\mathbf{u}}).\mathbf{x} = 0$$

Any point $\mathbf{u}$ on the algebraic plane curve f with the property $\nabla f|_{\mathbf{u}} = 0$ is a singular point of f. A non-singular curve is a curve without any singular points.

The tangent line at a point $\mathbf{u}$ on a curve is unique if and only if $\mathbf{u}$ is a non-singular point of a curve.

A conic is a curve in $\mathbb{P}^2$ defined by an equation of degree two. If coordinates are chosen in $\mathbb{P}^2$ then a general conic has an equation $\mathbf{x}^\top M\mathbf{x} = 0$, where M is a symmetric 3×3 matrix. The conic is non-singular if and only if the determinant $\det(M)$ is not zero. If $\det(M) = 0$ then the conic is singular. It then splits into a line pair. The two lines intersect at the point $\mathbf{p}$ defined by $M\mathbf{p} = 0$.

A space curve is a curve in $\mathbb{P}^3$ not contained in any plane of $\mathbb{P}^3$. The definition of an algebraic space curve is more complicated than the definition of an algebraic plane curve. For the present purposes it is sufficient to define an algebraic curve in $\mathbb{P}^3$ as the intersection of a set of algebraic surfaces. Let coordinates x_1, x_2, x_3, x_4 be chosen in $\mathbb{P}^3$. An algebraic surface is the set of zeros of a polynomial f homogeneous in the x_i. Let $f_1, \ldots, f_n$ be algebraic surfaces in $\mathbb{P}^3$. The set $f_1 \cap \ldots \cap f_n$ is an algebraic curve provided it is non-empty, it does not contain any isolated points and it has dimension one. If in addition the intersection is not contained in any plane of $\mathbb{P}^3$ then it is an algebraic space curve. If the curve is irreducible then it can be represented as the intersection of just three surfaces.

The singular points of surfaces in $\mathbb{P}^3$ are defined in a similar way to the singular points of a plane curve. Let coordinates x_1, x_2, x_3, x_4 be chosen in $\mathbb{P}^3$. Let f be an algebraic surface in $\mathbb{P}^3$. The tangent plane of f at a point $\mathbf{u}$ on f is the set of points $\mathbf{x}$ in $\mathbb{P}^3$ satisfying

$$(\nabla f|_{\mathbf{u}}).\mathbf{x} = 0$$

Any point $\mathbf{u}$ of the algebraic surface f with the property $\nabla f|_{\mathbf{u}} = 0$ is a singular point of f.

A real curve or a real surface is one which is defined by homogeneous polynomials with real coefficients. A real curve or a real surface may not have real points. For example the conic in $\mathbb{P}^2$ defined by

$$x_1^2 + x_2^2 + x_3^2 = 0$$

is real, but it does not have any real points. A line in $\mathbb{P}^2$ containing two real points $\mathbf{a}$, $\mathbf{b}$ is real, because it has an equation $(\mathbf{a} \times \mathbf{b}).\mathbf{x} = 0$.

The concept of a general position is very important in algebraic geometry. A set of points, curves or other geometrical figures is in general position if the figures do not have any special interrelations. For example if f is a curve in $\mathbb{P}^2$ and if $\mathbf{p}$ is a point of $\mathbb{P}^2$ in general position with respect to f then $\mathbf{p}$ is not a point of f. The point $\mathbf{p}$ is on f only if the coordinates of $\mathbf{p}$ satisfy $f(\mathbf{p}) = 0$. If two curves f and g in $\mathbb{P}^2$ are in general position then it can be safely assumed that they intersect in only a finite number of points, each point $\mathbf{p}$ of $f \cap g$ is a non-singular point of f and g and the tangents of f and g at $\mathbf{p}$ are distinct.

A general position requirement is included in the statement of many theorems in projective and algebraic geometry. For example three general planes in $\mathbb{P}^3$ intersect in a unique point. The use of the word 'general' avoids the listing of all the special cases in which the intersection of three planes is not a single point. With this convention the statement of the result is shorter and easier to understand. The disadvantage is than in any application it is necessary to check that the three planes actually encountered are in positions general enough to ensure that the result is true.

References

Faugeras O.D. 1990 On the motion of 3D curves and its relation to optical flow. *Proc. First European Conference on Computer Vision* (ed. O. Faugeras). Lecture Notes in Computer Science, vol. 427, 107-117. Berlin, Heidelberg and New York: Springer-Verlag.

Fulton W. 1969 *Algebraic Curves.* Reading, Massachusetts: W.A. Benjamin Inc., Mathematics Lecture Note Series (Reprinted 1974).

Gibson J.J. 1950 *The Perception of the Visual World.* Boston: Houghton Mifflin.

Golub G.H. & Van Loan C.F. 1983 *Matrix Computations.* Oxford: North Oxford Publishing Co. Ltd.

Harris C. 1990 Resolution of the bas-relief ambiguity in structure-from-motion under orthographic projection. *BMVC90 Proc. British Machine Vision Conference, Oxford*, 67-77.

Horn B.K.P. 1986 *Robot Vision.* Cambridge, Massachusetts: The MIT Press.

Liu Y. & Huang T.S. 1988 Estimation of rigid body motion using straight line correspondences. *Computer Vision, Graphics, and Image Processing* **44**, 35-57.

Marr D. 1982 *Vision.* San Francisco: W. H. Freeman & Co.

Moffitt F.H. & Mikhail E.M. 1980 *Photogrammetry* (3rd edition). New York: Harper and Row.

Murray D.W. & Buxton B.F. 1990 *Experiments in the Machine Interpretation of Visual Motion.* Cambridge, Massachusetts: The MIT Press.

Negahdaripour S. & Horn B.K.P. 1987 Direct passive navigation. *IEEE Trans. Pattern Analysis and Machine Intelligence* **9**, 168-176.

Semple J.G. & Kneebone G.T. 1953 *Algebraic Projective Geometry.* Oxford: Clarendon Press (reprinted 1979).

Semple J.G. & Roth R. 1949 *Introduction to Algebraic Geometry.* Oxford: Clarendon Press, reprinted 1985.

Sokolnikoff I.S. & Redheffer R.M. 1966 *Mathematics of Physics and Modern Engineering.* USA: McGraw-Hill.

Spetsakis M.E. & Aloimonos J. 1990 Structure from motion using line correspondences. *International J. Computer Vision* **4**, 171-183.

Walker R. 1962 *Algebraic Curves.* New York: Dover.

Waxman A.M., Kamgar-Parsi B. & Subbarao M. 1987 Closed form solutions to image flow equations for 3D structure and motion. *International J. Computer Vision* **1**, 239-258.

Wolf P.R. 1983 *Elements of Photogrammetry* (2nd edition). McGraw-Hill.

Wu J. & Wohn K. 1991 On the deformation of image intensity and zero-crossing contours under motion. *Computer Vision, Graphics, and Image Processing: Image Understanding* **53**, 66-75.

2 Reconstruction from Image Correspondences

The data for reconstruction from image correspondences consist of corresponding points from two images of the same scene taken from different viewpoints. A point in the first image corresponds to a point in the second image if and only if both points are projections of the same point in space. Reconstruction is the task of finding the relative positions of the two viewpoints compatible with the image correspondences. If sufficiently many image correspondences in general position are available then the relative position is determined up to a single unknown scale factor and a single 180° twist about the line joining the optical centres of the two cameras. There are many mathematically equivalent ways of formulating reconstruction. The imaging surface can be a plane or a sphere or some other more complicated shape. The rotation of the camera can be described using orthogonal matrices or quaternions and the geometry of reconstruction can be described in terms of Euclidean or projective geometry. The different formulations are equivalent, in that it is straightforward to translate from one formulation to another.

Two different frameworks for reconstruction are described; the first is Euclidean and the second is projective geometric. In the Euclidean framework the equations for reconstruction are expressed in vector notation, using the coordinates of points in the image. In the projective geometric framework the imaging surface is replaced by the star of lines through the optical centre of the camera. It is then no longer necessary to assume that the imaging surface has a particular form. In both the Euclidean and the projective frameworks it is assumed that the full camera calibration is known. Reconstruction up to a collineation, in which it is known only that the image is formed by a linear projection, is described in Sect. 2.4 at the end of this chapter.

Today the Euclidean framework is predominant in computer vision, partly because it leads quickly to algebraic equations which can form the basis of algorithms for reconstruction and partly because the properties of vectors and matrices are widely known. The projective geometric framework is older. It dates back at least to the first formulations of the reconstruction problem in the

19th century (Hesse 1863; Sturm 1869). The Euclidean and the projective geometric approaches to reconstruction have their roots in two different approaches to geometry, the analytic and the synthetic. In the analytic approach geometrical properties are described by algebraic equations and results are proved by manipulating the equations. The analytic approach is particularly appropriate for formulating algorithms and for implementing them on a computer. The synthetic approach is based directly on the geometric properties of figures such as lines, planes, conics, etc. The advantage of the synthetic approach is that gives the equations interpretations, without which they would be meaningless. The interpretations become more important as the number of equations increases and as each individual equation becomes larger. In the absence of a geometric interpretation, a large set of polynomial equations can be about as informative as a block of assembler code. The drawback of the synthetic approach is that it is difficult to make it completely rigorous. For example, it is easy to miss degenerate cases.

The Euclidean framework is described in Sect. 2.1, with particular emphasis on the ambiguous case in which two or more reconstructions are obtained from the same set of image correspondences. Certain 3×3 matrices known as essential matrices arise naturally in the Euclidean framework. The essential matrices are described in Sect. 2.2. The projective geometric framework is described in Sect. 2.3. The projective geometric treatment of ambiguity is briefly discussed, in preparation for a more extensive treatment in Chap. 3. Reconstruction up to a collineation is described in Sect. 2.4.

2.1 Euclidean Framework for Reconstruction

Two images of the same scene are taken by a camera from different positions. The optical centre of the camera for the first image is at a point $\mathbf{o}$ and the optical centre of the camera for the second image is at point $\mathbf{a}$. It is assumed that the camera calibration is known, and that the images are transformed into the images that would have been obtained by intersecting the light rays with the unit sphere centred at the optical centre of the camera. The advantage of using a sphere rather than a plane is that the distortions of the image that occur on the projection plane far from the optical centre of the camera are avoided.

Let coordinates be chosen in $\mathbb{R}^3$ with origin at $\mathbf{o}$. Let $\mathbf{x}$ be a point in space projecting to a point $\mathbf{q}$ in the first image. The line $\langle \mathbf{o}, \mathbf{x} \rangle$ cuts the unit sphere centred at $\mathbf{o}$ at two points, $\pm\mathbf{x}/\|\mathbf{x}\|$. The image $\mathbf{q}$ of $\mathbf{x}$ is defined by $\mathbf{q} = \mathbf{x}/\|\mathbf{x}\|$. The choice of the plus sign in the definition of $\mathbf{q}$ is arbitrary. No difficulties arise provided the image points are all contained within a single hemisphere of the projection sphere. The choice of sign can be avoided by using the projective plane rather than the projection sphere. The points $\mathbf{q}$ and $-\mathbf{q}$ are identified as a single point of the projective plane.

All points on the half-line drawn from $\mathbf{o}$ through $\mathbf{x}$ project down to $\mathbf{q}$. The coordinates of $\mathbf{q}$ are measured in a Cartesian coordinate frame attached to the

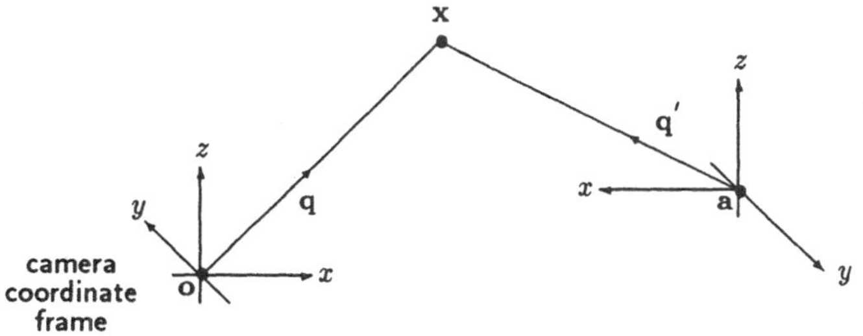

Fig. 2.1. A displacement of the camera

camera, with the origin of coordinates at the optical centre $\mathbf{o}$.

Let $\mathbf{q}'$ be the image of the same point $\mathbf{x}$ taken by the camera from the second position $\mathbf{a}$, as illustrated in Fig. 2.1. Then $\mathbf{q}$ and $\mathbf{q}'$ are corresponding image points, $\mathbf{q} \leftrightarrow \mathbf{q}'$. The lines $\langle \mathbf{o}, \mathbf{q} \rangle$ and $\langle \mathbf{a}, \mathbf{q}' \rangle$ intersect at $\mathbf{x}$. The coordinates of $\mathbf{q}'$ are measured in the Cartesian coordinate frame attached to the camera. This frame moves with the camera as it is taken from $\mathbf{o}$ to $\mathbf{a}$.

The displacement of the camera is described by the translation $\mathbf{a}$ which takes the optical centre of the camera from the first position to the second position and by the rotation $R^\top$ required to bring the two coordinate frames into alignment, after carrying out the translation. The notation $\{R, \mathbf{a}\}$ is used for the camera displacement. It is assumed that $\det(R) = 1$. The case $\det(R) = -1$ is excluded on physical grounds, because it involves a reflection of the camera coordinate frame.

In the first position of the camera coordinate frame $\mathbf{x} = p\mathbf{q}$, where $p = \|\mathbf{x}\|$. Let the camera be translated by $\mathbf{a}$ without rotation, and let $\mathbf{x}$ remain fixed in space. The coordinates of $\mathbf{x}$ in the translated camera coordinate frame are $p\mathbf{q} - \mathbf{a}$. A rotation $R^\top$ is now applied to bring the translated camera coordinate frame into alignment with the final position of the camera coordinate frame. The coordinates of $\mathbf{x}$ in the translated and rotated camera coordinate frame are $R(p\mathbf{q} - \mathbf{a})$. Let p' be the distance from $\mathbf{a}$ to $\mathbf{x}$. Then

$$p'\mathbf{q}' = R(p\mathbf{q} - \mathbf{a}) \tag{2.1}$$

Equation (2.1) is the basis of the following formulation of the reconstruction problem: given a set of n image correspondences $\mathbf{q}_i \leftrightarrow \mathbf{q}_i'$, find a camera displacement $\{R, \mathbf{a}\}$ and scalars p_i, p_i' such that (2.1) holds for each pair $\mathbf{q}_i$, $\mathbf{q}_i'$ of corresponding points.

In making this formulation of reconstruction from image correspondences certain physical aspects of the problem are omitted. For example there is no statistical description of the errors that arise in practice when the coordinates of $\mathbf{q}$ and $\mathbf{q}'$ are measured. The distances p_i and p_i' to points on the object surface are not required to be positive. In many applications p_i and p_i' are known to be positive because the imaging surface is a plane and the object is in front of the camera. In spite of these omissions the mathematical model is accurate enough to ensure that an analysis of the model can lead to physically meaningful results.

Reconstruction is always accompanied by a scaling ambiguity. The absolute size of an object cannot be determined, because the image of a small object near to the camera is identical to the image of a larger object further away from the camera. The scaling ambiguity is apparent from (2.1). If p, p' and $\mathbf{a}$ in (2.1) are scaled by the same non-zero constant λ then

$$\lambda p'\mathbf{q}' = R(\lambda p\mathbf{q} - \lambda\mathbf{a}) \tag{2.2}$$

It follows from (2.1) and (2.2) that $\{R, \mathbf{a}\}$ and $\{R, \lambda\mathbf{a}\}$ are both compatible with the same set of image correspondences. In view of this ambiguity in scale all the camera displacements $\{R, \lambda\mathbf{a}\}$ are counted as a single reconstruction. The scaling ambiguity can be resolved by imposing arbitrary conditions such as $\|\mathbf{a}\| = 1$, $a_3 \geq 0$.

There is a second ambiguity which is present in reconstruction whenever $\mathbf{o} \neq \mathbf{a}$. Let one reconstruction be given. Then a second reconstruction is obtained from it by rotating the position of the second camera by 180° about the line $\langle\mathbf{o}, \mathbf{a}\rangle$ joining the two optical centres. After carrying out the rotation corresponding rays still intersect in space, as shown in Fig. 2.2. Reconstruction is thus possible for the new orientation of the camera. The two reconstructed surfaces are different because the rotation changes the points at which corresponding rays intersect. In describing this second ambiguity it is convenient to scale $\mathbf{a}$ such that $\|\mathbf{a}\| = 1$. Let σ be a rotation through 180° with axis $\langle\mathbf{o}, \mathbf{a}\rangle$. It follows from the definition of σ that

$$\begin{aligned} \sigma(\mathbf{q}) &= 2(\mathbf{q}.\mathbf{a})\mathbf{a} - \mathbf{q} \\ \sigma(\mathbf{a}) &= \mathbf{a} \end{aligned} \tag{2.3}$$

Equations (2.1) and (2.3) yield

$$\begin{aligned} p'\mathbf{q}' &= R(p\mathbf{q} - \mathbf{a}) \\ &= R\sigma^2(p\mathbf{q} - \mathbf{a}) \\ &= R\sigma(2p(\mathbf{q}.\mathbf{a})\mathbf{a} - p\mathbf{q} - \mathbf{a}) \\ &= R\sigma(-p\mathbf{q} - (1 - 2p(\mathbf{q}.\mathbf{a}))\mathbf{a}) \end{aligned} \tag{2.4}$$

Let $\tilde{p}$, $\tilde{p}'$ be defined by

$$\begin{aligned} \tilde{p} &= -p(1 - 2p(\mathbf{q}.\mathbf{a}))^{-1} \\ \tilde{p}' &= p'(1 - 2p(\mathbf{q}.\mathbf{a}))^{-1} \end{aligned} \tag{2.5}$$

It follows from (2.4) and (2.5) that

$$\tilde{p}'\mathbf{q}' = R\sigma(\tilde{p}\mathbf{q} - \mathbf{a}) \tag{2.6}$$

On comparing (2.1) and (2.6) it is apparent that $\{R, \mathbf{a}\}$ and $\{R\sigma, \mathbf{a}\}$ are both compatible with the same set of image correspondences $\mathbf{q} \leftrightarrow \mathbf{q}'$. The camera displacements $\{R, \mathbf{a}\}$ and $\{R\sigma, \mathbf{a}\}$ are referred to as a twisted pair of solutions.

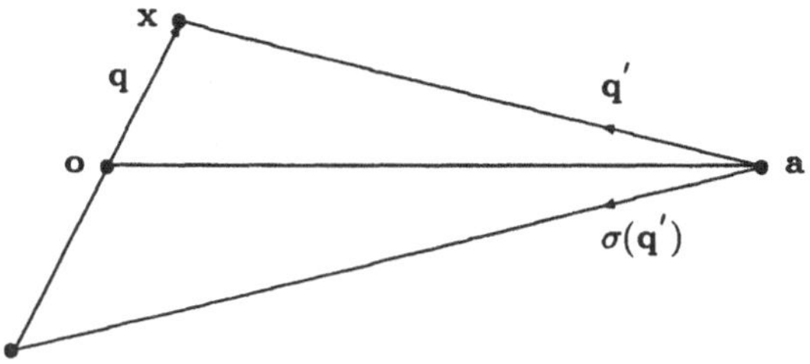

Fig. 2.2. A twisted pair of reconstructions

A twisted pair is treated as a single solution to the reconstruction problem. The convention as to whether a twisted pair counts as one solution or two solutions varies from author to author. For example, Horn (1990) counts each twisted pair as two solutions, but Negahdaripour (1990) counts each twisted pair as a single solution. Negahdaripour gives examples of surfaces reconstructed from twisted pairs of solutions. In most practical applications of reconstruction only one of a twisted pair of solutions is feasible in that it yields positive values for the distances p, p' of each scene point from the two optical centres $\mathbf{o}$, $\mathbf{a}$.

In obtaining (2.1) it is assumed that the point $\mathbf{x}$ projecting to $\mathbf{q}$ in the first image and to $\mathbf{q}'$ in the second image is in $\mathbb{R}^3$. If $\mathbf{x}$ is in the plane at infinity then the points $\mathbf{q}$ and $\mathbf{q}'$ are related by

$$\mathbf{q}' = R\mathbf{q} \tag{2.7}$$

Equation (2.7) is a limiting case of (2.1) as $p \to \infty$, $p' \to \infty$ and $p/p' \to 1$. The interpretation of (2.7) as a limit of (2.1) is that if $\mathbf{x}$ is very far away from the camera then the image of $\mathbf{x}$ is only slightly affected by the translation $\mathbf{a}$.

The distances p and p' are eliminated from (2.1). The scalar product of (2.1) with $R\mathbf{q} \times R\mathbf{a}$ yields

$$(R\mathbf{q} \times R\mathbf{a}).\mathbf{q}' = 0 \tag{2.8}$$

If (2.8) holds and if $\mathbf{q}' \times R\mathbf{q} \neq 0$ then p and p' can be found such that (2.1) holds. If $\mathbf{q}' \times R\mathbf{q} = 0$ then $\mathbf{q}' = R\mathbf{q}$.

Equation (2.8) is often written in a form involving antisymmetric matrices. Let T_a be the antisymmetric matrix associated with the vector $\mathbf{a}$. The entries of T_a are defined by

$$T_a = \begin{pmatrix} 0 & a_3 & -a_2 \\ -a_3 & 0 & a_1 \\ a_2 & -a_1 & 0 \end{pmatrix} \tag{2.9}$$

It follows from (2.9) that $T_a\mathbf{q} = \mathbf{q} \times \mathbf{a}$. Equation (2.8) is equivalent to

$$\mathbf{q}'^{\top} RT_a\mathbf{q} = 0 \tag{2.10}$$

Let E be the matrix defined by $E = RT_a$. It follows from (2.10) that

$$\mathbf{q}'^{\top} E\mathbf{q} = 0 \tag{2.11}$$

The matrix E is an example of an essential matrix (Longuet-Higgins 1981; Tsai & Huang 1984). The essential matrices are defined to be the non-zero 3×3 matrices which are the product of an orthogonal matrix and an antisymmetric matrix. Their properties are discussed in Sect. 2.2.

A second formulation of the reconstruction problem is obtained from (2.11). Given a set of n image correspondences $\mathbf{q}_i \leftrightarrow \mathbf{q}_i'$ find an essential matrix E such that (2.11) holds for each pair of corresponding points. The formulation involves the image correspondences and the camera displacement, but it does not refer explicitly to the distances from the two positions of the optical centre of the camera to points in the field of view. Once E is found the camera displacement and the relative positions of the points in space can be deduced, up to a single unknown scale factor and up to the twisted pair ambiguity.

The ever present scale ambiguity and the twisted pair ambiguity are easily handled using essential matrices. A scale change of λ in which the translation $\mathbf{a}$ is replaced by $\lambda\mathbf{a}$ is equivalent to a rescaling of E to give λE. If E_1 and E_2 are essential matrices obtained from a twisted pair of solutions then $E_1 = -E_2$, as shown below in Proposition 2.10. Both types of ambiguity can be removed by imposing conditions such as $\|E\| = 1$, $E_{33} \geq 0$.

2.1.1 Euclidean Treatment of Ambiguity

It has been shown that reconstruction is always accompanied by two types of ambiguity, the unknown scale, and the twisted pair of solutions. For a general set of image correspondences these are the only ambiguities that arise. However, certain special sets of image correspondences $\mathbf{q}_i \leftrightarrow \mathbf{q}_i'$ for $1 \leq i \leq n$, give rise to an additional ambiguity, in that there exist two linearly independent essential matrices, E, F such that

$$\mathbf{q}_i'^{\top} E \mathbf{q}_i = \mathbf{q}_i'^{\top} F \mathbf{q}_i = 0 \qquad (1 \leq i \leq n)$$

In this case the image correspondences are said to be ambiguous. Each essential matrix gives rise to a reconstruction compatible with the image correspondences. The two reconstructions are said to be essentially different, because they are obtained from linearly independent essential matrices. These observations are summarised in the following definitions.

Definition 2.1. *Two rigid displacements $\{R, \mathbf{a}\}$ and $\{S, \mathbf{b}\}$ are essentially different if the essential matrices RT_a, ST_b are linearly independent.*

Definition 2.2. *A set of image correspondences $\mathbf{q}_i \leftrightarrow \mathbf{q}_i'$, $1 \leq i \leq n$, is ambiguous if there exist two linearly independent essential matrices, E, F such that*

$$\mathbf{q}_i'^{\top} E \mathbf{q}_i = \mathbf{q}_i'^{\top} F \mathbf{q}_i = 0 \qquad (1 \leq i \leq n) \qquad (2.12)$$

The ambiguous case of reconstruction is described by many authors, including Faugeras & Maybank (1990), Horn (1990), Longuet-Higgins (1986, 1988),

Maybank (1990), Negahdaripour (1990) and Tsai & Huang (1981, 1984). The ambiguous case is described in the earlier photogrammetric literature by Hofmann (1950), Krames (1940) and Wunderlich (1942).

The first proposition describes two cases in which ambiguity cannot occur.

Proposition 2.3. *If $\{R, \mathbf{a}\}$, $\{R, \mathbf{b}\}$ are two camera displacements compatible with the same dense set of image correspondences $\mathbf{q} \leftrightarrow \mathbf{q}'$ then either* $\mathbf{a}$ *and* $\mathbf{b}$ *are zero, or* $\mathbf{a}$ *and* $\mathbf{b}$ *are non-zero and parallel. If $\{R, \mathbf{a}\}$, $\{S, \mathbf{a}\}$ are compatible with the same dense set of image correspondences then either $R = S$ or the two camera displacements are a twisted pair.*

Proof. Equation (2.1) is applied. If $\mathbf{a} = 0$ then $\mathbf{q}' = R\mathbf{q}$. On substituting $R\mathbf{q}$ for $\mathbf{q}'$ in (2.1) it follows that

$$p'R\mathbf{q} = R(p\mathbf{q} - \mathbf{b})$$

for all $\mathbf{q}$ in a non-empty open set in the image. On rearranging terms it follows that

$$(p' - p)R\mathbf{q} = \mathbf{b} \tag{2.13}$$

The vector $\mathbf{b}$ is fixed, thus (2.13) holds only if $p = p'$ and $\mathbf{b} = 0$. The vectors $\mathbf{a}$ and $\mathbf{b}$ are thus either both zero or both non-zero. It remains to show that if $\mathbf{a}$, $\mathbf{b}$ are both non-zero then they are parallel. Suppose if possible that $\mathbf{a} \neq \mathbf{b}$. It follows from (2.11) that

$$\begin{aligned} \mathbf{q}'^{\top} RT_a\mathbf{q} &= 0 \\ \mathbf{q}'^{\top} RT_b\mathbf{q} &= 0 \end{aligned} \tag{2.14}$$

It follows from (2.14) that

$$\begin{aligned} \mathbf{q}' &= RT_a\mathbf{q} \times RT_b\mathbf{q} \\ &= R((\mathbf{q} \times \mathbf{a}) \times (\mathbf{q} \times \mathbf{b})) \\ &= ((\mathbf{q} \times \mathbf{a}).\mathbf{b})R\mathbf{q} \end{aligned} \tag{2.15}$$

If $\mathbf{q}' = R\mathbf{q}$ than (2.13) holds, with the consequence $\mathbf{b} = 0$, contrary to the hypothesis $\mathbf{b} \neq 0$. If $\mathbf{q}' \neq R\mathbf{q}$ then $(\mathbf{q} \times \mathbf{a}).\mathbf{b} = 0$ for all $\mathbf{q}$ in a non-empty open set of the image. It follows that $\mathbf{a} = \mathbf{b}$, which contradicts the hypothesis $\mathbf{a} \neq \mathbf{b}$.

To prove the second part of the proposition it is assumed without loss of generality that $R = I$, where I is the 3×3 identity matrix. The camera displacement $\{I, \mathbf{a}\}$ is compatible with the image correspondences, thus $\mathbf{q}'^{\top} T_a\mathbf{q} = 0$. It follows that there exist scalars r, s depending on $\mathbf{q}'$ such that

$$\mathbf{q} = r\mathbf{q}' + s\mathbf{a} \tag{2.16}$$

The camera displacement $\{S, \mathbf{a}\}$ is compatible with the same set of image correspondences thus

$$\mathbf{q}'^{\top} ST_a\mathbf{q} = 0 \tag{2.17}$$

It follows from (2.16) and (2.17) that $\mathbf{q}'^{\top} S T_a \mathbf{q}' = 0$ for all $\mathbf{q}'$ in a non-empty open set of the image. The matrix ST_a is thus antisymmetric. The vector $\mathbf{a}$ is in the null space of ST_a. Thus there exists a non-zero scalar λ such that

$$ST_a = \lambda T_a \tag{2.18}$$

On taking norms of the left- and right-hand sides of (2.18) the equation $|\lambda| = 1$ is obtained. There are two cases to consider, $\lambda = 1$ and $\lambda = -1$. If $\lambda = 1$ then for any vector $\mathbf{x}$,

$$\begin{aligned} \mathbf{x} \times \mathbf{a} &= T_a \mathbf{x} \\ &= S T_a \mathbf{x} \\ &= S(\mathbf{x} \times \mathbf{a}) \end{aligned}$$

The rotation S thus acts as the identity on the plane of vectors normal to $\mathbf{a}$. The orthogonal matrix S has unit determinant, thus $S = I$. In the second case, $\lambda = -1$. The rotation S reverses the direction of each vector in the plane normal to $\mathbf{a}$. It follows that S is a rotation of 180° with axis $\mathbf{a}$. □

Corollary. Two camera displacements $\{R, \mathbf{a}\}$, $\{S, \mathbf{a}\}$ compatible with the same dense set of image correspondences are not essentially different. The corollary follows because $\{R, \mathbf{a}\}$ and $\{S, \mathbf{a}\}$ are either identical or a twisted pair. The essential matrices obtained from $\{R, \mathbf{a}\}$ and $\{S, \mathbf{a}\}$ are thus linearly dependent.

When discussing ambiguity certain conventions about the optical centres of the cameras are made. The initial position $\mathbf{o}$ of the optical centre is fixed at the origin of $\mathbb{R}^3$. In the ambiguous case there are at least two possible positions for the optical centre of the camera after the displacement such that the three optical centres are not collinear. The two positions are always denoted by $\mathbf{a}$, $\mathbf{b}$. It is always assumed that $\mathbf{o}$, $\mathbf{a}$, $\mathbf{b}$ are not collinear. This automatically excludes the case in which two of $\mathbf{o}$, $\mathbf{a}$, $\mathbf{b}$ are equal.

In the next theorem it is shown that ambiguity is only possible if the points in space projecting down to the two images lie on certain surfaces of degree two known as critical surfaces.

Theorem 2.4. *The image correspondences arising from the projections of a set of points $\mathbf{x}_i$, $1 \le i \le n$, in space are ambiguous only if the $\mathbf{x}_i$ lie on a surface of degree two. The surface contains $\mathbf{o}$ and both possible positions $\mathbf{a}$, $\mathbf{b}$ for the optical centre of the camera after the displacement.*

Proof. Let $\mathbf{q}_i \leftrightarrow \mathbf{q}_i'$, $1 \le i \le n$, be the image correspondences arising from the projections of the $\mathbf{x}_i$, and let E, F be linearly independent essential matrices such that (2.12) holds. It follows from (2.12) that $E\mathbf{q}_i \times F\mathbf{q}_i$ is parallel to $\mathbf{q}_i'$. In projective coordinates $\mathbf{q}_i' = E\mathbf{q}_i \times F\mathbf{q}_i$. Let $\{R, T_a\}$, $\{S, T_b\}$ be two rigid displacements such that $E = RT_a$, $F = ST_b$. In the rest of the proof the subscript i is omitted from $\mathbf{q}_i$ and $\mathbf{q}_i'$. The expansion of $E\mathbf{q} \times F\mathbf{q}$ yields

$$\mathbf{q}' = E\mathbf{q} \times F\mathbf{q}$$

$$\begin{aligned}
&= RT_a\mathbf{q} \times ST_b\mathbf{q} \\
&= R(\mathbf{q} \times \mathbf{a}) \times S(\mathbf{q} \times \mathbf{b}) \\
&= (R\mathbf{q} \times R\mathbf{a}) \times (S\mathbf{q} \times S\mathbf{b}) \\
&= [(S\mathbf{q} \times S\mathbf{b}).R\mathbf{q}]R\mathbf{a} - [(S\mathbf{q} \times S\mathbf{b}).R\mathbf{a}]R\mathbf{q} \\
&= R[((S\mathbf{b} \times S\mathbf{q}).R\mathbf{a})\mathbf{q} - ((S\mathbf{b} \times S\mathbf{q}).R\mathbf{q})\mathbf{a}] \qquad (2.19)
\end{aligned}$$

Let p be the distance from $\mathbf{o}$ to $\mathbf{x}$, and let p' be the distance from $\mathbf{a}$ to $\mathbf{x}$. Equation (2.1) states that

$$p'\mathbf{q}' = R(p\mathbf{q} - \mathbf{a}) \qquad (2.20)$$

On comparing (2.19) and (2.20) it is apparent that

$$p = \frac{(S\mathbf{b} \times S\mathbf{q}).R\mathbf{a}}{(S\mathbf{b} \times S\mathbf{q}).R\mathbf{q}} \qquad (2.21)$$

It follows from the definition of p that $\mathbf{x} = p\mathbf{q}$. On substituting $p^{-1}\mathbf{x}$ for $\mathbf{q}$ in (2.21) an equation in $\mathbf{x}$ is obtained,

$$(S\mathbf{b} \times S\mathbf{x}).R\mathbf{x} - (S\mathbf{b} \times S\mathbf{x}).R\mathbf{a} = 0 \qquad (2.22)$$

The surface defined by (2.22) is a quadric, because (2.22) is of degree two in $\mathbf{x}$. The surface contains $\mathbf{o}$, $\mathbf{a}$, $\mathbf{b}$ as required. □

In the German literature on photogrammetry critical surfaces are known as 'gefährliche Flächen' (Hofmann 1950), literally 'dangerous surfaces'. In the ambiguous case there are, by definition, two essentially different reconstructions, each of which yields a critical surface. Let ψ be the critical surface defined by (2.22). The second critical surface ϕ is obtained by interchanging $\{R, \mathbf{a}\}$ and $\{S, \mathbf{b}\}$ in the proof of Theorem 2.4. The equation of ϕ is

$$(R\mathbf{a} \times R\mathbf{x}).S\mathbf{x} - (R\mathbf{a} \times R\mathbf{x}).S\mathbf{b} = 0 \qquad (2.23)$$

The surfaces ψ, ϕ are together known as a critical surface pair. On setting $U = S^{\top}R$, (2.22) and (2.23) become

$$\begin{aligned}
(U\mathbf{x} \times \mathbf{x}).\mathbf{b} + (\mathbf{x} \times U\mathbf{a}).\mathbf{b} &= 0 \qquad (2.24) \\
(U^{\top}\mathbf{x} \times \mathbf{x}).\mathbf{a} + (\mathbf{x} \times U^{\top}\mathbf{b}).\mathbf{a} &= 0 \qquad (2.25)
\end{aligned}$$

Equations (2.24) and (2.25) are inhomogeneous in $\mathbf{x}$. The homogeneous versions of (2.24) and (2.25) appropriate for the projective space $\mathbb{P}^3$ are obtained by introducing a new coordinate x_4 and, in the case of (2.24) for example, writing

$$(U\mathbf{x} \times \mathbf{x}).\mathbf{b} + x_4(\mathbf{x} \times U\mathbf{a}).\mathbf{b} = 0 \qquad (2.26)$$

The plane $x_4 = 0$ is the plane at infinity. It follows from (2.26) that the intersection of ψ with the plane at infinity is the conic defined by the equations

$$x_4 = 0 \qquad\qquad (U\mathbf{x} \times \mathbf{x}).\mathbf{b} = 0$$

There are a several different classifications of the non-singular quadrics in $\mathbf{P}^3$ under the action of collineations of $\mathbf{P}^3$. The different classifications are obtained by restricting the collineations used to transform one quadric into another. If a general complex collineation of $\mathbf{P}^3$ is allowed then there is only one class of quadric. In other words if ψ_1 and ψ_2 are any two non-singular quadrics in $\mathbf{P}^3$ then there exists a collineation ω of $\mathbf{P}^3$ such that $\omega(\psi_1) = \psi_2$. If the allowable collineations are required to be real and affine then there are six classes of real non-singular quadrics (Mirsky 1961). A real affine collineation is a collineation defined by a real matrix that leaves the plane at infinity invariant. A real quadric is a quadric defined by a homogeneous polynomial equation of degree two with real coefficients.

The six classes of non-singular quadrics under the action of the real affine collineations of $\mathbf{P}^3$ are

- Virtual quadric
- Ellipsoid
- Hyperboloid of two sheets
- Hyperboloid of one sheet
- Elliptic paraboloid
- Hyperbolic paraboloid

The elliptic paraboloid and the hyperbolic paraboloid are tangent to the plane at infinity such that the point of tangency has real coordinates. Any tangent plane to a non-singular quadric intersects the quadric in two distinct lines, such that each line passes through the point of tangency. An elliptic paraboloid intersects the plane at infinity in two lines defined over the complex numbers. A hyperbolic paraboloid intersects the plane at infinity in two lines defined over the real numbers. The quadrics in the remaining four classes are not tangent to the plane at infinity. The four classes are distinguished as follows. A virtual quadric has no real points. An ellipsoid has real points, but it has no real points in the plane at infinity. A hyperboloid of two sheets and a hyperboloid of one sheet both intersect the plane at infinity in a non-singular conic containing an infinity of real points. The hyperboloid of two sheets contains no real lines, whilst the hyperboloid of one sheet does contain real lines. If the real singular quadrics are included then thirteen additional classes are obtained under the action of the real affine transformations (Mirsky 1961).

The critical surfaces are in general non-singular quadrics defined over the real numbers which intersect the plane at infinity in a non-singular conic containing an infinity of real points. The critical surfaces contain real lines, for example the line $\langle \mathbf{o}, \mathbf{b} \rangle$. It follows that the critical surfaces are, in general hyperboloids of one sheet. Degenerate hyperboloids of one sheet can also occur as critical surfaces in special cases. The degenerate critical surfaces include the hyperbolic paraboloids and certain types of cone.

2.1.2 Ambiguity and Instability

In practice it is rare to obtain image correspondences that are exactly ambiguous. It is more usual to find that reconstruction is unstable, in that a small change in the positions of the image points gives rise to a large change in the estimate of the camera displacement. It is shown that the unstable case is closely related to the ambiguous case in the following sense. If a set of image correspondences yields an unstable reconstruction then a small perturbation in the positions of the image points can be found such that the resulting set of image correspondences is ambiguous. This result suggests that the theory developed for the ambiguous case of reconstruction will be useful in the more common unstable case.

Theorem 2.5. *Let* $\mathbf{q}_i \leftrightarrow \mathbf{q}_i'$, $1 \le i \le n$, *be a set of image correspondences for which reconstruction is unstable. Let* E, F *be linearly independent essential matrices, normalised such that* $\|E\| = \|F\| = 1$ *and let* $\delta\mathbf{q}_i$, $\delta\mathbf{q}_i'$ *be perturbations of* $\mathbf{q}_i$, $\mathbf{q}_i'$ *respectively such that*

$$\begin{aligned} \mathbf{q}_i'^{\top} E \mathbf{q}_i &= 0 \\ (\mathbf{q}_i' + \delta\mathbf{q}_i')^{\top} F (\mathbf{q}_i + \delta\mathbf{q}_i) &= 0 \qquad (1 \le i \le n) \end{aligned} \tag{2.27}$$

The existence of E, F *and of the* $\delta\mathbf{q}_i$, $\delta\mathbf{q}_i'$ *follows from the hypothesis that reconstruction based on the* $\mathbf{q}_i \leftrightarrow \mathbf{q}_i'$ *is unstable. Let* Δ_i *be defined by*

$$\Delta_i = \max_i \{\|\delta\mathbf{q}_i\|, \|\delta\mathbf{q}_i'\|\} \tag{2.28}$$

Then there exist perturbations of each $\mathbf{q}_i'$ *of size at most* $3\sqrt{2}\Delta_i / \|E\mathbf{q}_i \times F\mathbf{q}_i\|$ *such that the resulting set of image correspondences is ambiguous.*

Proof. The subscript i is omitted from the notation. Let $\epsilon \pm 1$ and let $\mathbf{p}$ be the unit vector defined by

$$\mathbf{p} = \epsilon \|E\mathbf{q} \times F\mathbf{q}\|^{-1} (E\mathbf{q} \times F\mathbf{q})$$

The sign of ϵ is chosen such that $\mathbf{p}.\mathbf{q}' \ge 0$. The vector product of $\mathbf{p}$ with $\mathbf{q}'$ yields

$$\begin{aligned} \|E\mathbf{q} \times F\mathbf{q}\| (\mathbf{p} \times \mathbf{q}') &= (E\mathbf{q} \times F\mathbf{q}) \times \mathbf{q}' \\ &= (\mathbf{q}'^{\top} E\mathbf{q}) F\mathbf{q} - (\mathbf{q}'^{\top} F\mathbf{q}) E\mathbf{q} \\ &= -(\mathbf{q}'^{\top} F\mathbf{q}) E\mathbf{q} \end{aligned} \tag{2.29}$$

It follows from (2.27) and (2.28) that

$$\begin{aligned} \|\mathbf{q}'^{\top} F\mathbf{q}\| &= -\delta\mathbf{q}'^{\top} F\mathbf{q} - \mathbf{q}'^{\top} F \delta\mathbf{q} - \delta\mathbf{q}'^{\top} F \delta\mathbf{q} \\ &\le 3\Delta \end{aligned} \tag{2.30}$$

It follows from (2.29) and (2.30) that

$$\begin{aligned} \|\mathbf{p} \times \mathbf{q}'\| &\le 3\Delta \|E\mathbf{q}\| \, \|E\mathbf{q} \times F\mathbf{q}\|^{-1} \\ &\le 3\Delta \|E\mathbf{q} \times F\mathbf{q}\|^{-1} \end{aligned} \tag{2.31}$$

The vectors $\mathbf{p}$, $\mathbf{q}'$ are both unit vectors and $\mathbf{p}.\mathbf{q}' \geq 0$. Let θ be the angle between $\mathbf{p}$ and $\mathbf{q}'$. The condition $\mathbf{p}.\mathbf{q}' \geq 0$ ensures that $0 \leq \theta \leq \pi/2$. It follows that

$$\begin{aligned}
\frac{\|\mathbf{p}-\mathbf{q}'\|^2}{\|\mathbf{p}\times\mathbf{q}'\|^2} &= \frac{\|\mathbf{p}\|^2 - 2\mathbf{p}.\mathbf{q}' + \|\mathbf{q}'\|^2}{\|\mathbf{p}\times\mathbf{q}'\|^2} \\
&= 2\sin^{-2}(\theta)(1-\cos(\theta)) \\
&= 2(1+\cos(\theta))^{-1} \\
&\leq 2
\end{aligned} \tag{2.32}$$

It follows from (2.31) and (2.32) that

$$\|\mathbf{p}-\mathbf{q}'\| \leq 3\sqrt{2}\Delta\|E\mathbf{q}\times F\mathbf{q}\|^{-1}$$

The required perturbation is $\mathbf{q}' \mapsto \mathbf{p}$. □

Theorem 2.5 does not have a straightforward converse. If an ambiguous set of image correspondences is compatible with two linearly independent essential matrices E, F, normalised such that $\|E\| = \|F\| = 1$, then a small perturbation of the image points can be found such that the resulting correspondences are compatible with a unique essential matrix G_1 with unit norm such that $\|E - G_1\|$ is small, and a second small perturbation can be found such that the resulting image correspondences are compatible with a unique essential matrix G_2 with unit norm such that $\|F - G_2\|$ is small. This observation is a reason for regarding the ambiguous case of reconstruction as unstable. However, if the perturbed image correspondences are required to be ambiguous then there are in general two essential matrices G_1 and G_2 compatible with the perturbed image correspondences such that $\|E - G_1\|$ and $\|F - G_2\|$ are both small.

2.1.3 The Maximum Number of Reconstructions

It has been shown that certain sets of image correspondences are compatible with two essentially different reconstructions. The question arises of how many different reconstructions can be obtained in the worst case when a dense set of image correspondences is available. This question is answered by Hofmann (1950), Krames (1941), Longuet-Higgins (1988) and Negahdaripour (1990): there are at most three essentially different reconstructions compatible with a dense set of image correspondences.

The first step in obtaining this result is to count the number of essentially different reconstructions that arise when the two images are of a planar surface.

Theorem 2.6. *Let $\mathbf{q} \leftrightarrow \mathbf{q}'$ be a dense set of image correspondences arising from two images of a planar surface. Then any other surface also compatible with the $\mathbf{q} \leftrightarrow \mathbf{q}'$ is a plane.*

Proof. Let $\{R, \mathbf{a}\}$ and $\{S, \mathbf{b}\}$ be essentially different camera displacements compatible with the image correspondences. Let ψ, ϕ be the critical surface pair with equations (2.24) and (2.25) respectively. The hypothesis is that ψ contains a plane and it is required to prove that ϕ contains a plane.

As ψ contains a plane, the linear term $(\mathbf{x} \times U\mathbf{a}).\mathbf{b}$ of (2.24) divides the quadratic term $(U\mathbf{x} \times \mathbf{x}).\mathbf{b}$. Let $\mathbf{x}$ be a general point in the plane $(\mathbf{x} \times U\mathbf{a}).\mathbf{b} = 0$. Then there exist scalars r, s such that

$$\mathbf{x} = rU\mathbf{a} + s\mathbf{b} \tag{2.33}$$

On substituting the right-hand side of (2.33) for $\mathbf{x}$ in $(U\mathbf{x} \times \mathbf{x}).\mathbf{b}$ zero is obtained for all choices of r and s. The substitution is carried out in stages. It follows from (2.33) that

$$\mathbf{x} \times \mathbf{b} = (rU\mathbf{a} + s\mathbf{b}) \times \mathbf{b} = rU\mathbf{a} \times \mathbf{b} \tag{2.34}$$

It follows from (2.33) and (2.34) that

$$\begin{aligned} (U\mathbf{x} \times \mathbf{x}).\mathbf{b} &= (\mathbf{x} \times \mathbf{b}).U\mathbf{x} \\ &= r(U\mathbf{a} \times \mathbf{b}).U\mathbf{x} \\ &= r(\mathbf{a} \times U^{\top}\mathbf{b}).\mathbf{x} \\ &= r(\mathbf{a} \times U^{\top}\mathbf{b}).(rU\mathbf{a} + s\mathbf{b}) \end{aligned} \tag{2.35}$$

The right-hand side of (2.35) is zero for all choices of r and s if and only if

$$\begin{aligned} (\mathbf{a} \times U^{\top}\mathbf{b}).U\mathbf{a} &= 0 \\ (\mathbf{a} \times U^{\top}\mathbf{b}).\mathbf{b} &= 0 \end{aligned} \tag{2.36}$$

It follows from (2.36) that either $\mathbf{a} \times U^{\top}\mathbf{b} = 0$, or the four vectors $\mathbf{a}$, $U^{\top}\mathbf{b}$, $U\mathbf{a}$, $\mathbf{b}$ are coplanar.

In order to prove that ϕ contains a plane it is it is sufficient to show that $(\mathbf{x} \times U^{\top}\mathbf{b}).\mathbf{a}$ divides $(U^{\top}\mathbf{x} \times \mathbf{x}).\mathbf{a}$, and for this it is sufficient to show that any vector $\mathbf{x}$ satisfying the equation $(\mathbf{x} \times U^{\top}\mathbf{b}).\mathbf{a} = 0$ also satisfies $(U^{\top}\mathbf{x} \times \mathbf{x}).\mathbf{a} = 0$. A general vector $\mathbf{x}$ such that $(\mathbf{x} \times U^{\top}\mathbf{b}).\mathbf{a} = 0$ has the form

$$\mathbf{x} = rU^{\top}\mathbf{b} + s\mathbf{a} \tag{2.37}$$

for some choice of scalars r, s. It follows from (2.37) that

$$\mathbf{x} \times \mathbf{a} = rU^{\top}\mathbf{b} \times \mathbf{a} \tag{2.38}$$

It follows from (2.37) and (2.38) that

$$\begin{aligned} (U^{\top}\mathbf{x} \times \mathbf{x}).\mathbf{a} &= (\mathbf{x} \times \mathbf{a}).U^{\top}\mathbf{x} \\ &= r(U^{\top}\mathbf{b} \times \mathbf{a}).U^{\top}\mathbf{x} \\ &= r(\mathbf{b} \times U\mathbf{a}).\mathbf{x} \end{aligned} \tag{2.39}$$

Equation (2.37) is used to substitute for $\mathbf{x}$ on the right-hand side of (2.39) to yield

$$\begin{aligned} (U^{\top}\mathbf{x} \times \mathbf{x}).\mathbf{a} &= r(\mathbf{b} \times U\mathbf{a}).(rU^{\top}\mathbf{b} + s\mathbf{a}) \\ &= r^2(\mathbf{b} \times U\mathbf{a}).U^{\top}\mathbf{b} + r\,s(\mathbf{b} \times U\mathbf{a}).\mathbf{a} \end{aligned}$$

If $\mathbf{a} \times U^\top \mathbf{b} = 0$ then $\mathbf{b} \times U\mathbf{a} = 0$. It then follows that $(U^\top \mathbf{x} \times \mathbf{x}).\mathbf{a} = 0$ for all choices of r, s in (2.37). If $\mathbf{a} \times U^\top \mathbf{b} \neq 0$ then $\mathbf{a}$, $U^\top \mathbf{b}$, $U\mathbf{a}$, $\mathbf{b}$ are coplanar. Again it follows that $(U^\top \mathbf{x} \times \mathbf{x}).\mathbf{a} = 0$ for all choices of r, s. □

If a critical surface contains a plane then it splits into a plane pair. One of the planes contains the optical centres $\mathbf{o}$, $\mathbf{a}$, $\mathbf{b}$. It is fixed by the camera positions independently of the positions of the objects in the field of view. For this reason it is usual to omit the plane from the discussion.

It follows from Theorem 2.6 that planar and non-planar surfaces can be considered separately when discussing ambiguity. There is no danger that a set of image correspondences arising from two images of a planar surface can also arise from two views of a non-planar surface. The planar case turns up frequently in practice, and it has its own special properties, the most notable of which is that there is almost always a two way ambiguity.

Theorem 2.7. *Let $\mathbf{q} \leftrightarrow \mathbf{q}'$ be a dense set of image correspondences arising from two images of the plane surface $\mathbf{x}.\mathbf{u} = 1$, and let $\{R, \mathbf{a}\}$ be the associated rigid displacement. If $\mathbf{a} \times \mathbf{u} \neq 0$ then there is up to scale exactly one other plane surface which can also give rise to the image correspondences $\mathbf{q} \leftrightarrow \mathbf{q}'$.*

Proof. Let $p = \|\mathbf{x}\|$. The equation $\mathbf{x}.\mathbf{u} = 1$ is equivalent to $p\,\mathbf{q}.\mathbf{u} = 1$. The proof is based on a trick which dates back at least as far as Hay (1966). On multiplying the term $-R\mathbf{a}$ on the right-hand side of (2.1) by $1 = p\,\mathbf{q}.\mathbf{u}$ it follows that

$$\begin{aligned} p'\mathbf{q}' &= pR\mathbf{q} - p(\mathbf{q}.\mathbf{u})R\mathbf{a} \\ &= pR(I - \mathbf{a} \otimes \mathbf{u})\mathbf{q} \end{aligned} \tag{2.40}$$

It is convenient to think of the two images as copies of the projective plane $\mathbf{P}^2$. It follows from (2.40) that the image correspondences $\mathbf{q} \leftrightarrow \mathbf{q}'$ define a collineation between the two images. The matrix W of this collineation is determined by the image correspondences up to a single scalar multiple. It follows from (2.40) that there exists a scalar λ such that

$$\lambda W = R(I - \mathbf{a} \otimes \mathbf{u}) \tag{2.41}$$

The orthogonal matrix R is eliminated from (2.41),

$$\begin{aligned} \lambda^2 W^\top W &= (I - \mathbf{u} \otimes \mathbf{a})(I - \mathbf{a} \otimes \mathbf{u}) \\ &= I - \mathbf{u} \otimes \mathbf{a} - \mathbf{a} \otimes \mathbf{u} + (\mathbf{a}.\mathbf{a})\mathbf{u} \otimes \mathbf{u} \end{aligned} \tag{2.42}$$

It is assumed without loss of generality that the translation $\mathbf{a}$ is normalised such that $\mathbf{a}.\mathbf{a} = 1$. Let $\mathbf{p}$ be the vector defined by $\mathbf{p} = \mathbf{u} - \mathbf{a}$. The tensor product $\mathbf{p} \otimes \mathbf{p}$ is given by

$$\mathbf{p} \otimes \mathbf{p} = \mathbf{u} \otimes \mathbf{u} + \mathbf{a} \otimes \mathbf{a} - \mathbf{a} \otimes \mathbf{u} - \mathbf{u} \otimes \mathbf{a} \tag{2.43}$$

Equations (2.42), (2.43) and the assumption $\mathbf{a}.\mathbf{a} = 1$ yield

$$\lambda^2 W^\top W = I + \mathbf{p} \otimes \mathbf{p} - \mathbf{a} \otimes \mathbf{a} \tag{2.44}$$

The hypothesis $\mathbf{a} \times \mathbf{u} \neq 0$ ensures that $\mathbf{a} \times \mathbf{p} \neq 0$. It follows from (2.44) that

$$\lambda^2 W^\top W(\mathbf{a} \times \mathbf{p}) = \mathbf{a} \times \mathbf{p} \tag{2.45}$$

One of the eigenvalues of $\lambda^2 W^\top W$ is thus equal to 1. Let $\mathbf{y}$ be any unit vector such that $\mathbf{y}.\mathbf{p} \neq 0$, $\mathbf{y}.\mathbf{a} = 0$. It follows that

$$\lambda^2 \mathbf{y}^\top W^\top W \mathbf{y} = 1 + (\mathbf{y}.\mathbf{p})^2 > 1$$

The matrix $\lambda^2 W^\top W$ thus has an eigenvalue strictly greater than 1. Similarly, let $\mathbf{y}$ now be any unit vector such that $\mathbf{y}.\mathbf{p} = 0$, $\mathbf{y}.\mathbf{a} \neq 0$. It follows that

$$\lambda^2 \mathbf{y}^\top W^\top W \mathbf{y} = 1 - (\mathbf{y}.\mathbf{a})^2 < 1$$

The matrix $\lambda^2 W^\top W$ thus has an eigenvalue strictly less than 1.

The matrix $W^\top W$ is diagonalised. Let $\lambda_1 > \lambda_2 > \lambda_3$ be the eigenvalues of $\lambda^2 W^\top W$, and let U be a known orthogonal matrix such that

$$\lambda^2 U^\top W^\top W U = \begin{pmatrix} \lambda_1 & 0 & 0 \\ 0 & \lambda_2 & 0 \\ 0 & 0 & \lambda_3 \end{pmatrix} \tag{2.46}$$

The scale factor λ is determined up to sign by the condition $\lambda_2 = 1$. It follows from (2.45) and (2.46) that $U^\top(\mathbf{p} \times \mathbf{a})$ is parallel to $(0,1,0)^\top$. Let $\mathbf{p}'$ and $\mathbf{a}'$ be the vectors defined by $\mathbf{p}' = U^\top \mathbf{p}$ and $\mathbf{a}' = U^\top \mathbf{a}$. It follows from (2.44) and (2.46) that

$$\begin{aligned} \lambda^2 U^\top W^\top W U - I &= U^\top \mathbf{p} \otimes \mathbf{p} U - U^\top \mathbf{a} \otimes \mathbf{a} U \\ &= \mathbf{p}' \otimes \mathbf{p}' - \mathbf{a}' \otimes \mathbf{a}' \end{aligned} \tag{2.47}$$

Let the coordinates of $\mathbf{p}'$ and $\mathbf{a}'$ be defined by $\mathbf{p}' = (p_1', p_2', p_3')^\top$, $\mathbf{a}' = (a_1', a_2', a_3')^\top$. The values of p_2' and a_2' are zero because

$$\begin{aligned} p_2' &= \mathbf{p}'.U^\top(\mathbf{p} \times \mathbf{u}) = \mathbf{p}'.(\mathbf{p}' \times \mathbf{u}') = 0 \\ a_2' &= \mathbf{a}'.U^\top(\mathbf{p} \times \mathbf{a}) = \mathbf{a}'.(\mathbf{p}' \times \mathbf{a}') = 0 \end{aligned}$$

The matrix equation (2.47) reduces to

$$\begin{pmatrix} \lambda_1 - 1 & 0 & 0 \\ 0 & 0 & 0 \\ 0 & 0 & \lambda_3 - 1 \end{pmatrix} = \begin{pmatrix} p_1'^2 - a_1'^2 & 0 & p_1'p_3' - a_1'a_3' \\ 0 & 0 & 0 \\ p_1'p_3' - a_1'a_3' & 0 & p_3'^2 - a_3'^2 \end{pmatrix}$$

from which it follows that

$$\begin{aligned} \lambda_1 &= 1 + p_1'^2 - a_1'^2 \\ \lambda_3 &= 1 + p_3'^2 - a_3'^2 \\ 0 &= p_1'p_3' - a_1'a_3' \end{aligned} \tag{2.48}$$

It follows from the equation $\mathbf{a}' = U\mathbf{a}$ that $\mathbf{a}'.\mathbf{a}' = \mathbf{a}.\mathbf{a}$. The condition $\mathbf{a}.\mathbf{a} = 1$ thus yields an additional equation

$$\|\mathbf{a}'\|^2 = a_1'^2 + a_3'^2 = 1 \tag{2.49}$$

Equations (2.48) and (2.49) are solved for $\mathbf{p}'$, $\mathbf{a}'$. It follows from (2.49) that there exists an angle θ such that

$$\begin{aligned} a_1' &= \cos(\theta) \\ a_3' &= \sin(\theta) \end{aligned}$$

The terms $\cos(\theta)$ and $\sin(\theta)$ are substituted for a_1' and a_3' respectively in (2.48) to yield

$$\begin{aligned} \lambda_1 &= p_1'^2 + \sin^2(\theta) \\ \lambda_3 &= p_3'^2 + \cos^2(\theta) \\ 0 &= p_1'p_3' - \cos(\theta)\sin(\theta) \end{aligned} \tag{2.50}$$

The terms p_1' and p_3' are eliminated from (2.50) to yield

$$(\lambda_1 - \sin^2(\theta))(\lambda_3 - \cos^2(\theta)) = \cos^2(\theta)\sin^2(\theta)$$

from which it follows that

$$\lambda_1\lambda_3 - \sin^2(\theta)\lambda_3 - \cos^2(\theta)\lambda_1 = 0 \tag{2.51}$$

The solutions to (2.51) are

$$\begin{aligned} \cos(\theta) &= \pm\sqrt{\frac{\lambda_3(\lambda_1 - 1)}{\lambda_1 - \lambda_3}} \\ \sin(\theta) &= \pm\sqrt{\frac{\lambda_1(1 - \lambda_3)}{\lambda_1 - \lambda_3}} \end{aligned} \tag{2.52}$$

In view of the scaling ambiguity in reconstruction, the vector $\mathbf{a}'$ can be multiplied by $\pm\sqrt{\lambda_1 - \lambda_3}$ without changing the form of the solution in any significant way. After this scaling there remain from (2.52) two distinct solutions for $\mathbf{a}'$,

$$(\sqrt{\lambda_3(\lambda_1 - 1)}, 0, \sqrt{\lambda_1(1 - \lambda_3)})^\top \quad \text{and} \quad (\sqrt{\lambda_3(\lambda_1 - 1)}, 0, -\sqrt{\lambda_1(1 - \lambda_3)})^\top \tag{2.53}$$

The solutions (2.53) are distinct because $\lambda_1 > 1 > \lambda_3$. One of the solutions (2.53) yields the known value of $\mathbf{a}$ via the equation $\mathbf{a} = U\mathbf{a}'$. The remaining solution yields a second planar surface which gives rise to the same set of image correspondences. □

The proof of Theorem 2.7 is due to Longuet-Higgins (1986). In the case $\mathbf{a} \times \mathbf{u} = 0$ excluded by the hypotheses of Theorem 2.7 the two reconstructions merge to give a single reconstruction, in which the translation $\mathbf{a}$ of the camera is parallel to the vector $\mathbf{u}$ defining the normal of the plane giving rise to the image correspondences.

The intermediate results obtained in the proof of Theorem 2.7 are used to show that any real collineation from the first image to the second image can be defined by the image correspondences arising from two images of a planar surface.

Theorem 2.8. *Any real collineation from the first image to the second image can be realised from two views of a plane surface.*

Proof. Let W be the matrix of a collineation from the first image to the second image. In view of the statement and proof of Theorem 2.7 it is sufficient to show that there exists a scale factor λ, an orthogonal matrix R and vectors $\mathbf{a}$, $\mathbf{u}$ such that

$$\lambda W = R(I - \mathbf{a} \otimes \mathbf{u}) \tag{2.54}$$

If W is a scalar multiple of an orthogonal matrix then it suffices to choose $\mathbf{a} = 0$. Thus it is assumed that W is not a scalar multiple of an orthogonal matrix, and that $\mathbf{a} \neq 0$. It is assumed without any loss of generality that $\mathbf{a}.\mathbf{a} = 1$. By taking a product WU where U is an orthogonal matrix, it is sufficient to consider the case in which $W^\top W$ is diagonal with decreasing entries on the leading diagonal. The scalar λ is chosen such that

$$\lambda^2 W^\top W = \begin{pmatrix} \lambda_1 & 0 & 0 \\ 0 & 1 & 0 \\ 0 & 0 & \lambda_3 \end{pmatrix} \tag{2.55}$$

It follows that $\lambda_1 \geq 1 \geq \lambda_3$. Let $\mathbf{p} = (p_1, p_2, p_3)^\top$ and $\mathbf{a} = (a_1, a_2, a_3)^\top$ be vectors such that $\mathbf{a}.\mathbf{a} = 1$ and

$$\begin{pmatrix} \lambda_1 & 0 & 0 \\ 0 & 1 & 0 \\ 0 & 0 & \lambda_3 \end{pmatrix} = I + \mathbf{p} \otimes \mathbf{p} - \mathbf{a} \otimes \mathbf{a} \tag{2.56}$$

If $\lambda_1 = 1$ or if $\lambda_3 = 1$ then a solution to (2.56) exists with $\mathbf{p}$ parallel to $\mathbf{a}$. If $\lambda_1 \neq 1$ and $\lambda_3 \neq 1$ then the solution to (2.56) is obtained by following the proof of Theorem 2.7. Let $a_2 = p_2 = 0$ and let $\mathbf{a} = (\cos(\theta), 0, \sin(\theta))^\top$. A solution to (2.55) can be found such that

$$\cos(\theta) = \pm\sqrt{\frac{\lambda_3(\lambda_1 - 1)}{\lambda_1 - \lambda_3}} \tag{2.57}$$

The conditions $\lambda_1 \geq 1 \geq \lambda_3$ ensure that

$$0 \leq \frac{\lambda_3(1 - \lambda_1)}{\lambda_3 - \lambda_1} \leq 1$$

It follows that a real value of θ can always be found such that (2.57) holds.

Let $\mathbf{u}$ be defined by $\mathbf{u} = \mathbf{p} + \mathbf{a}$. It follows from (2.55) and (2.56) that

$$\lambda^2 W^\top W = (I - \mathbf{a} \otimes \mathbf{u})^\top (I - \mathbf{a} \otimes \mathbf{u})$$

The condition $\mathbf{p} \neq 0$ is sufficient to ensure that $I - \mathbf{a} \otimes \mathbf{u}$ is invertible. The matrix $\lambda W(I - \mathbf{a} \otimes \mathbf{u})^{-1}$ is orthogonal. Equation (2.54) follows if

$$R = \lambda W(I - \mathbf{a} \otimes \mathbf{u})^{-1}$$

□

Theorem 2.9. *Let* $\mathbf{q} \leftrightarrow \mathbf{q}'$ *be a dense set of image correspondences arising from two images of a rigid surface. Then there are at most three essentially different camera displacements compatible with the* $\mathbf{q} \leftrightarrow \mathbf{q}'$.

Proof. Suppose, if possible, that the image correspondences are compatible with three essentially different camera displacements $\{R_1, \mathbf{a}_1\}$, $\{R_2, \mathbf{a}_2\}$, $\{R_3, \mathbf{a}_3\}$. It follows from Proposition 2.3 that no two of $\mathbf{a}_1$, $\mathbf{a}_2$, $\mathbf{a}_3$ are parallel, and no two of R_1, R_2, R_3 are equal. To prove the result it suffices to show that $\{R_3, \mathbf{a}_3\}$ is uniquely determined by $\{R_1, \mathbf{a}_1\}$ and $\{R_2, \mathbf{a}_2\}$.

The camera displacements $\{R_1, \mathbf{a}_1\}$, $\{R_2, \mathbf{a}_2\}$ are used to construct a critical surface ψ such that the image correspondences arise from two images of ψ taken by a camera which undergoes a displacement $\{R_1, \mathbf{a}_1\}$ between taking the first and second images. Let U_2 be the orthogonal matrix defined by $U_2 = R_2^\top R_1$. It follows from (2.24) that the equation of ψ is

$$(U_2\mathbf{x} \times \mathbf{x}).\mathbf{a}_2 + (\mathbf{x} \times U_2\mathbf{a}_1).\mathbf{a}_2 = 0 \tag{2.58}$$

The same surface ψ is constructed from the rigid displacements $\{R_1, \mathbf{a}_1\}$, $\{R_3, \mathbf{a}_3\}$. Let U_3 be the orthogonal matrix defined by $U_3 = R_3^\top R_1$. Equation (2.24) yields a second expression for ψ,

$$(U_3\mathbf{x} \times \mathbf{x}).\mathbf{a}_3 + (\mathbf{x} \times U_3\mathbf{a}_1).\mathbf{a}_3 = 0 \tag{2.59}$$

Equations (2.58) and (2.59) describe the same surface, thus they are identical up to a scalar multiple. It is assumed without loss of generality that this scalar multiple is equal to one. The parts of (2.58) and (2.59) of degree one in $\mathbf{x}$ are equated and the parts of degree two in $\mathbf{x}$ are equated to obtain the two equations

$$(U_2\mathbf{a}_1 \times \mathbf{x}).\mathbf{a}_2 = (U_3\mathbf{a}_1 \times \mathbf{x}).\mathbf{a}_3 \tag{2.60}$$

$$(U_2\mathbf{x} \times \mathbf{x}).\mathbf{a}_2 = (U_3\mathbf{x} \times \mathbf{x}).\mathbf{a}_3 \tag{2.61}$$

Let E_2, E_3 be the essential matrices defined by $E_2 = U_2^\top T_{a_2}$, $E_3 = U_3^\top T_{a_3}$, and let M be the symmetric matrix defined by $M = \mathrm{sym}(E_2)$. It follows from (2.61) that $\mathbf{x}^\top E_2 \mathbf{x} = \mathbf{x}^\top E_3 \mathbf{x}$ for all $\mathbf{x}$, thus

$$M = \mathrm{sym}(E_2) = \mathrm{sym}(E_3)$$

The matrix $E_3 - E_2$ is thus antisymmetric. Let $\mathbf{d}$ be the vector such that

$$E_3 - E_2 = T_d \tag{2.62}$$

It follows from (2.60) that $\mathbf{a}_2 \times U_2\mathbf{a}_1 = \mathbf{a}_3 \times U_3\mathbf{a}_1$, thus

$$(E_2^\top - E_3^\top)\mathbf{a}_1 = 0 \tag{2.63}$$

It follows from (2.63) and (2.62) that $\mathbf{d} \times \mathbf{a}_1 = 0$. Hence $\mathbf{d}$ is parallel to $\mathbf{a}_1$. Let μ be a scalar such that $\mathbf{d} = \mu\mathbf{a}_1$. Equation (2.62) yields

$$E_3 = E_2 + \mu T_{a_1} \tag{2.64}$$

The right-hand side of (2.64) is homogenised by introducing a new scalar unknown λ,

$$E_3 = \lambda E_2 + \mu T_{a_1} \tag{2.65}$$

The determinant of E_3 is equal to zero, thus (2.65) yields

$$\det(\lambda E_2 + \mu T_{a_1}) = 0 \tag{2.66}$$

Equation (2.66) is a homogeneous polynomial constraint on (λ, μ) of degree three. Any two solutions of (2.66) that differ only by a non-zero scalar multiple yield essentially the same reconstruction. If (2.66) does not hold identically in λ and μ then there are up to scale three solutions for (λ, μ), namely $(1,0)$, $(0,1)$ and $(1,1)$. If $(\lambda, \mu) = (1,0)$ then $E_3 = E_2$. If $(\lambda, \mu) = (0,1)$ then $E_3 = T_{a_1}$, and $\mathbf{a}_1$ is parallel to $\mathbf{a}_3$. This possibility is ruled out by Proposition 2.3. There remains only the solution $(\lambda, \mu) = (1,1)$. There is thus at most one additional camera displacement $\{R_3, \mathbf{a}_3\}$ compatible with the image correspondences, but essentially different from $\{R_1, \mathbf{a}_1\}$ and $\{R_3, \mathbf{a}_3\}$.

There remains the case in which (2.66) holds identically in λ and μ. In this case there exists a vector $\mathbf{y}$ depending on (λ, μ) such that for all (λ, μ),

$$\lambda E_2\mathbf{y} + \mu T_{a_1}\mathbf{y} = 0 \tag{2.67}$$

It follows that

$$\begin{aligned} \mathbf{y}^\top E_2\mathbf{y} &= 0 \\ \mathbf{a}_1^\top E_2\mathbf{y} &= 0 \end{aligned} \tag{2.68}$$

If the line $\mathbf{a}_1^\top E_2\mathbf{y}$ is not a component of the conic $\mathbf{y}^\top E_2\mathbf{y}$ then there are at most two solutions of (2.68) for $\mathbf{y}$. Hence $\mathbf{y}$ is in general independent of (λ, μ). If the limit $(\lambda, \mu) \to (0,1)$ is taken then it follows that this constant value of $\mathbf{y}$ is $\mathbf{a}_1$. If the limit $(\lambda, \mu) \to (1,0)$ is taken then it follows that this constant value of $\mathbf{y}$ is $\mathbf{a}_2$. It follows that $\mathbf{y} = \mathbf{a}_1 = \mathbf{a}_2$. This contradicts the hypothesis $\mathbf{a}_1 \neq \mathbf{a}_2$.

If the line $\mathbf{a}_1^\top E_2\mathbf{y}$ is a component of the conic $\mathbf{y}^\top E_2\mathbf{y}$ then it follows from (2.58) that ψ splits into a plane pair. This contradicts the hypothesis that the image correspondences are compatible with three essentially different camera displacements. Finally, there remains the possibility that $\mathbf{y}^\top E_2\mathbf{y} = 0$ for all $\mathbf{y}$. The matrix E_2 is then antisymmetric. It follows that U_2 is either the identity, or a rotation of 180° with axis $\mathbf{a}_2$. An application of Proposition 2.3 and the accompanying Corollary excludes both cases. □

2.2 Essential Matrices

The essential matrices arise naturally in the Euclidean formulation of reconstruction from point correspondences. An essential matrix is, by definition, a 3×3 matrix which is the product of an orthogonal matrix and a non-zero antisymmetric matrix. Essential matrices have already been applied several times, most notably in Theorem 2.9 where they are used to show that there are at most three essentially different reconstructions compatible with a dense set of image correspondences. In this section the essential matrices are studied independently of their application to reconstruction. Among the properties of essential matrices that are obtained, the most important is that they can be defined as the zeros of a set of nine homogeneous polynomial equations of degree three in the coefficients of 3×3 matrices. A second result of practical importance concerns the singular values of the essential matrices with real entries. A real 3×3 matrix is an essential matrix if and only if one singular value is zero and the remaining two singular values are equal and non-zero. Most of the results deal with essential matrices with real entries, as the essential matrices that arise in reconstruction are of this form. However in Sect. 2.2.1, the essential matrices are allowed to have complex entries.

The first result of this section deals with essential matrices and twisted pairs $\{R, \mathbf{a}\}$, $\{R\sigma, \mathbf{a}\}$ of camera displacements.

Proposition 2.10. *Let E be an essential matrix, $E = RT_a$. Let S be the orthogonal matrix defined by $S = R\sigma$, where σ is a rotation of $180°$ about the axis $\mathbf{a}$. Then E is given by*

$$E = RT_a = -ST_a$$

If $\{U, \mathbf{b}\}$ is any rigid displacement such that $E = UT_b$ and $\det(U) = 1$, then either $\{U, \mathbf{b}\} = \{R, \mathbf{a}\}$ or $\{U, \mathbf{b}\} = \{S, -\mathbf{a}\}$.

Proof. The matrix $-ST_a$ is applied to an arbitrary vector $\mathbf{x}$ of $\mathbb{R}^3$ to yield

$$\begin{aligned} -ST_a\mathbf{x} &= -R\sigma(\mathbf{x} \times \mathbf{a}) \\ &= -R(\sigma\mathbf{x} \times \mathbf{a}) \\ &= R(\mathbf{x} \times \mathbf{a}) \\ &= E\mathbf{x} \end{aligned}$$

It follows that $E = -ST_a$.

To prove the second part of the proposition, let $\{U, \mathbf{b}\}$ be any rigid displacement such that $E = UT_b$. It follows that $\mathbf{a} \times \mathbf{b} = 0$ because

$$0 = E\mathbf{a} = UT_b\mathbf{a} = U(\mathbf{a} \times \mathbf{b})$$

Let λ be a scalar such that $\mathbf{b} = \lambda\mathbf{a}$. It follows from the hypotheses that

$$E = RT_a = \lambda UT_a \tag{2.69}$$

The application of the matrix norm $\|.\|$ to (2.69) yields

$$\begin{aligned} \|E\| &= \|RT_a\| = \|T_a\| \\ \|E\| &= |\lambda|\,\|UT_a\| = |\lambda|\,\|T_a\| \end{aligned}$$

from which it follows that $\lambda = \pm 1$. In the case $\lambda = 1$, $RT_a = UT_a$. Let $\mathbf{e}$ be any vector normal to $\mathbf{a}$. Then there exists a vector $\mathbf{f}$ normal to $\mathbf{a}$ such that $T_a\mathbf{f} = \mathbf{e}$. It follows that

$$R\mathbf{e} = RT_a\mathbf{f} = UT_a\mathbf{f} = U\mathbf{e}$$

The matrix $U^\top R$ thus acts as the identity on the plane of vectors normal to $\mathbf{a}$. The matrices U, R are both orthogonal, and both of determinant $+1$. It follows that $U^\top R$ is equal to the identity, hence $U = R$, and $\mathbf{b} = \mathbf{a}$. A similar argument shows that in the case $\lambda = -1$, $U^\top R$ reverses the directions of all the vectors normal to $\mathbf{a}$. It follows that $U^\top R = \sigma$, and $\{U, \mathbf{a}\} = \{R\sigma, -\mathbf{a}\}$. □

It is shown next that the set of essential matrices is invariant under rotations and scaling. These results, although simple, are exceedingly useful in proofs involving essential matrices.

Proposition 2.11. *Let E be an essential matrix with real or complex entries, let U, V be orthogonal matrices and let λ be a non-zero scalar. Then λE, UEV and $E^\top$ are all essential matrices. If $E = RT_a$, and if* $\mathbf{b}$ *is the vector defined by* $\mathbf{b} = R\mathbf{a}$, *then $E = T_bR$.*

Proof. To prove that λE is an essential matrix it suffices to observe that

$$\lambda E = R(\lambda T_a) = RT_{\lambda a}$$

To prove that UEV is an essential matrix it suffices to observe that

$$UEV = URT_aV = URVV^\top T_aV \tag{2.70}$$

The matrix $V^\top T_aV$ is antisymmetric, and URV is orthogonal, thus it follows from (2.70) that UEV is an essential matrix. To prove that $E^\top$ is an essential matrix it suffices to observe that

$$E^\top = T_a^\top R^\top = -R^\top RT_aR^\top \tag{2.71}$$

The matrix $RT_aR^\top$ is antisymmetric, thus it follows from (2.71) that $E^\top$ is an essential matrix. Finally, to show that $E = T_bR$ it suffices to observe that

$$E = RT_a = RT_aR^\top R$$

It remains to show that $RT_aR^\top = T_b$, where $\mathbf{b} = R\mathbf{a}$. If $\mathbf{x}$ is any vector in $\mathbb{R}^3$ then

$$RT_aR^\top\mathbf{x} = R(R^\top\mathbf{x} \times \mathbf{a}) = \mathbf{x} \times R\mathbf{a} = \mathbf{x} \times \mathbf{b} = T_b\mathbf{x} \tag{2.72}$$

It follows from (2.72) that $RT_aR^\top = T_b$, thus $E = T_bR$, as required. □

2.2.1 A Characterisation of Essential Matrices

In this subsection it is shown that the coefficients of an essential matrix satisfy a set of nine homogeneous polynomial equations of degree three. The equations come close to defining the set of essential matrices. Any matrix of rank two that satisfies the equations is an essential matrix. These results hold for essential matrices with both real and complex entries. In the complex case the matrix RT_a is essential if and only if $R^\top R = I$ and T_a is antisymmetric. It should be noted that the condition $R^\top R = I$ is different from the Hermitian condition $\overline{R}^\top R = I$. The main consequence of these results is that the essential matrices form an algebraic variety in $\mathbb{P}^8$. This characterisation of essential matrices leads to a formulation of reconstruction within the framework of algebraic geometry, as described in detail in Sect. 5.2.

The set of essential matrices is referred to as an algebraic variety rather than as a manifold for two reasons. The first reason is that the terminology emphasises the algebraic nature of the essential matrices. The second reason is that the closure in $\mathbb{P}^8$ of the set of essential matrices is not a manifold. The closure of the set of essential matrices is by definition the smallest closed set in $\mathbb{P}^8$ which contains the essential matrices. A point E of $\mathbb{P}^8$ is in the closure if and only if there exists a sequence E_i of essential matrices such that $\lim_{i\to\infty} E_i = E$. The closure of the set of essential matrices is an algebraic variety but it is not a manifold because it contains singular points.

The inclusion of the essential matrices with complex entries in the discussion is not in accord with practical applications of reconstruction. The essential matrices arising in practice invariably have real entries. The advantage of including the complex case is that the algebra becomes much simpler.

Some properties of orthogonal matrices are recalled. If $\mathbf{x}$, $\mathbf{y}$ are two non-zero vectors with complex coordinates such that $\|\mathbf{x}\| = \|\mathbf{y}\|$, then there exists an orthogonal matrix R such that $R\mathbf{x} = \mathbf{y}$. The case $\|\mathbf{x}\| \neq 0$ is straightforward. A proof is given here for the more unusual case $\|\mathbf{x}\| = 0$. It suffices to show that if $\|\mathbf{x}\| = 0$, $\mathbf{x} \neq 0$ then there exists an orthogonal matrix R such that $R\mathbf{x} = (1, i, 0)^\top$. If $x_3 \neq 0$ then let R_1 be the orthogonal matrix defined by

$$R_1 = \begin{pmatrix} 0 & 0 & 1 \\ i\,x_3^{-1}x_1 & i\,x_3^{-1}x_2 & 0 \\ i\,x_3^{-1}x_2 & -i\,x_3^{-1}x_1 & 0 \end{pmatrix}$$

If $x_3 = 0$ then let R_1 be the identity matrix. It follows from the definition of R_1 that

$$R_1\mathbf{x} = (a, b, 0)^\top \tag{2.73}$$

where $a^2 + b^2 = 0$, $a \neq 0$, $b \neq 0$. Next, let $\epsilon = \pm 1$ and let R_2 be the orthogonal matrix defined by

$$R_2 = \begin{pmatrix} (2a)^{-1} & (2b)^{-1} & -1 \\ -\epsilon(b + (2b)^{-1}) & \epsilon(a + (2a)^{-1}) & \epsilon(a(2b)^{-1} - b(2a)^{-1}) \\ \epsilon a & \epsilon b & \epsilon \end{pmatrix}$$

The sign of ϵ is chosen to ensure that $R_2(a,b,0)^\top = (1,i,0)^\top$. The matrix R is defined by $R = R_2R_1$.

Rodrigues' formula is a useful way of expressing an orthogonal matrix R in terms of the axis $\mathbf{n}$ and angle of rotation θ, provided $\mathbf{n}.\mathbf{n} \neq 1$.

$$R = \mathbf{n} \otimes \mathbf{n} + \cos(\theta)(I - \mathbf{n} \otimes \mathbf{n}) - \sin(\theta)T_n \tag{2.74}$$

The matrix $R^\top$ is obtained by substituting $-\theta$ for θ on the right-hand side of (2.74). It follows from (2.74) that

$$R^\top - R = 2\sin(\theta)T_n \tag{2.75}$$

If $\mathbf{n}.\mathbf{n} = 0$ then the analogue of Rodrigues' formula is

$$R = I + tT_n + \frac{t^2}{2}\mathbf{n} \otimes \mathbf{n}$$

where t is a scalar depending on R.

Theorem 2.12. *If E is an essential matrix with real or complex entries then*

$$EE^\top E = \frac{1}{2}\mathrm{tr}(EE^\top)E \tag{2.76}$$

Proof. It follows from Proposition 2.11 that E can be written in the form $E = T_aR$, where T_a is an antisymmetric matrix and R is an orthogonal matrix. The left-hand side of (2.76) reduces to

$$EE^\top E = T_aRR^\top T_a^\top T_aR = T_aT_a^\top T_aR \tag{2.77}$$

It follows from the properties of the trace function that

$$\mathrm{tr}(EE^\top)E = \mathrm{tr}(T_aT_a^\top)T_aR \tag{2.78}$$

In order to obtain (2.76) from (2.77) and (2.78) it suffices to prove that

$$T_aT_a^\top T_a = \frac{1}{2}\mathrm{tr}(T_aT_a^\top)T_a \tag{2.79}$$

Let $\mathbf{x}$ be an arbitrary vector in $\mathbb{C}^3$. The application of $T_aT_a^\top$ to $\mathbf{x}$ yields

$$\begin{aligned} T_aT_a^\top\mathbf{x} &= T_a(\mathbf{a} \times \mathbf{x}) \\ &= (\mathbf{a} \times \mathbf{x}) \times \mathbf{a} \\ &= (\mathbf{a}.\mathbf{a})\mathbf{x} - (\mathbf{a}.\mathbf{x})\mathbf{a} \end{aligned}$$

It follows that

$$T_aT_a^\top = (\mathbf{a}.\mathbf{a})I - \mathbf{a} \otimes \mathbf{a} \tag{2.80}$$

The application of T_a to the right-hand side of (2.80) yields

$$T_a T_a^\top T_a = (\mathbf{a}.\mathbf{a})T_a \tag{2.81}$$

It is observed that

$$\frac{1}{2}\mathrm{tr}(T_a T_a^\top) = \frac{1}{2}\mathrm{tr}((\mathbf{a}.\mathbf{a})I - \mathbf{a}\otimes\mathbf{a}) = \mathbf{a}.\mathbf{a}$$

Equation (2.79) now follows from (2.81). □

The converse to Theorem 2.12 is more difficult. It is obtained from the following three results.

Proposition 2.13 *Let E be any real or complex matrix satisfying (2.76). Then it follows that* $\det(E) = 0$.

Proof. Let E satisfy (2.76) and suppose if possible that $\det(E) \neq 0$. A contradiction is obtained. By hypothesis, E has a non-zero determinant, thus E is invertible. On postmultiplying (2.76) by $2E^{-1}$ the following equation is obtained

$$2EE^\top = \mathrm{tr}(EE^\top)I \tag{2.82}$$

The trace of (2.82) yields $\mathrm{tr}(EE^\top) = 0$. It then follows from (2.76) that $EE^\top = 0$. This yields $\det(E) = 0$, which contradicts the hypothesis $\det(E) \neq 0$. □

Proposition 2.14. *Let E be a real or complex matrix of rank two with rows $\mathbf{e}_1^\top$, $\mathbf{e}_2^\top$, $\mathbf{e}_3^\top$, and let E satisfy (2.76). Then at least one of the scalar products $\mathbf{e}_i.\mathbf{e}_i$, $i = 1,2,3$, is non-zero.*

Proof. The proof is by contradiction. Suppose if possible that

$$\mathbf{e}_1.\mathbf{e}_1 = \mathbf{e}_2.\mathbf{e}_2 = \mathbf{e}_3.\mathbf{e}_3 = 0 \tag{2.83}$$

It follows from (2.83) that $\mathrm{tr}(EE^\top) = 0$. The substitution of $\mathrm{tr}(EE^\top) = 0$ into (2.76) yields

$$EE^\top E = 0 \tag{2.84}$$

The matrix E has rank two, thus it follows from (2.84) that $EE^\top$ has rank one or less. The case $EE^\top = 0$ is ruled out because E has rank two. It follows that $EE^\top$ has rank one. In addition, $EE^\top$ is symmetric with zero entries on the leading diagonal. It follows that $EE^\top = \mathbf{u}\otimes\mathbf{u}$ and that the components u_1, u_2, u_3 of $\mathbf{u}$ satisfy $u_1^2 = u_2^2 = u_3^2 = 0$. This is a contradiction because $EE^\top \neq 0$. □

The converse to Theorem 2.12 can now be obtained. The proof uses repeatedly the fact that a 3×3 matrix E satisfies (2.76) if and only if UEV satisfies (2.76), where U, V are orthogonal matrices.

Theorem 2.15. *Let E be any 3×3 matrix of rank two or three such that (2.76) holds. Then E is an essential matrix.*

Proof. It follows from Proposition 2.13 that $\det(E) = 0$, thus E does not have rank three. Let $\mathbf{e}_1^\top$, $\mathbf{e}_2^\top$, $\mathbf{e}_3^\top$ be the rows of E, and let $e_{ij} = (\mathbf{e}_i)_j$ be the entries of E. The matrix E has rank two, thus there exists a non-zero vector $\mathbf{a}$ such that $\mathbf{a}^\top E = 0$. The proof divides into two cases.

Case 1. $\mathbf{a}.\mathbf{a} \neq 0$. On scaling $\mathbf{a}$ and on premultiplying E by an appropriate orthogonal matrix it is supposed without loss of generality that $\mathbf{a} = (0,0,1)^\top$. It follows that $\mathbf{e}_3 = 0$. It follows from Proposition 2.14 that either $\mathbf{e}_1.\mathbf{e}_1 \neq 0$ or $\mathbf{e}_2.\mathbf{e}_2 \neq 0$. It is assumed without loss of generality that that $\mathbf{e}_2.\mathbf{e}_2 \neq 0$. On dividing the entries of E by $\mathbf{e}_2.\mathbf{e}_2$, and then post-multiplying E by an appropriate orthogonal matrix, it is assumed without loss of generality that $\mathbf{e}_2 = (0,1,0)^\top$, $e_{13} = 0$. The substitution of E into (2.76) yields

$$\begin{pmatrix} e_{11}(\mathbf{e}_1.\mathbf{e}_1) & e_{12}(\mathbf{e}_1.\mathbf{e}_1 + 1) & 0 \\ e_{11}e_{12} & 1 + e_{12}^2 & 0 \\ 0 & 0 & 0 \end{pmatrix} = \frac{1}{2}(1 + \mathbf{e}_1.\mathbf{e}_1) \begin{pmatrix} e_{11} & e_{12} & 0 \\ 0 & 1 & 0 \\ 0 & 0 & 0 \end{pmatrix} \tag{2.85}$$

It follows from the second row of (2.85) that $e_{11}e_{12} = 0$. The matrix E has rank two thus $e_{11} \neq 0$. It follows that $e_{12} = 0$. The top left-hand entry in the first row of (2.85) yields, after cancelling e_{11},

$$\mathbf{e}_1.\mathbf{e}_1 = \frac{1}{2}(1 + \mathbf{e}_1.\mathbf{e}_1)$$

It follows that $\mathbf{e}_1.\mathbf{e}_1 = e_{11}^2 = 1$. As a result $e_{11} = \epsilon$ where $\epsilon = \pm 1$. The matrix E is written as the product of an antisymmetric matrix and an orthogonal matrix as follows:

$$E = \begin{pmatrix} \epsilon & 0 & 0 \\ 0 & 1 & 0 \\ 0 & 0 & 0 \end{pmatrix} = \begin{pmatrix} 0 & 1 & 0 \\ -1 & 0 & 0 \\ 0 & 0 & 0 \end{pmatrix} \begin{pmatrix} 0 & -1 & 0 \\ \epsilon & 0 & 0 \\ 0 & 0 & \epsilon \end{pmatrix}$$

Case 2. $\mathbf{a}.\mathbf{a} = 0$. This case can arise only if E has complex entries. On pre-multiplying E by an orthogonal matrix it is supposed that $\mathbf{a} = (0,1,i)^\top$. It follows from the equation $\mathbf{a}^\top E = 0$ that $\mathbf{e}_2 + i\mathbf{e}_3 = 0$, thus $\mathbf{e}_3 = i\mathbf{e}_2$. The substitution of $i\mathbf{e}_2$ for $\mathbf{e}_3$ and the evaluation of $EE^\top E$ yields

$$EE^\top E = \begin{pmatrix} (\mathbf{e}_1.\mathbf{e}_1)\mathbf{e}_1^\top \\ (\mathbf{e}_1.\mathbf{e}_2)\mathbf{e}_1^\top \\ i(\mathbf{e}_1.\mathbf{e}_2)\mathbf{e}_1^\top \end{pmatrix} \tag{2.86}$$

It follows from (2.86) that $EE^\top E$ has rank one or less. The matrix E has rank two, thus (2.76) is satisfied only if $EE^\top E = 0$, and $\mathrm{tr}(EE^\top) = 0$. The expansion of $\mathrm{tr}(EE^\top)$ yields

$$\begin{aligned} 0 &= \mathrm{tr}(EE^\top) \\ &= \mathbf{e}_1.\mathbf{e}_1 + \mathbf{e}_2.\mathbf{e}_2 + \mathbf{e}_3.\mathbf{e}_3 \\ &= \mathbf{e}_1.\mathbf{e}_1 \end{aligned} \tag{2.87}$$

It follows from Proposition 2.14 and (2.87) that either $\mathbf{e}_2.\mathbf{e}_2 \neq 0$, or $\mathbf{e}_3.\mathbf{e}_3 \neq 0$. The equation $\mathbf{e}_2 + i\mathbf{e}_3 = 0$ holds, thus $\mathbf{e}_2.\mathbf{e}_2 \neq 0$ and $\mathbf{e}_3.\mathbf{e}_3 \neq 0$. After post-multiplying E by an orthogonal matrix and scaling E, it is supposed that $\mathbf{e}_2 =$

$(1,0,0)^\top$. If $\mathbf{e}_1 = 0$ then E has rank one, contrary to hypothesis, thus $\mathbf{e}_1 \neq 0$. The equation (2.86) yields $\mathbf{e}_1.\mathbf{e}_2 = 0$. It follows that $e_{11} = 0$. The application of the condition $\mathbf{e}_1.\mathbf{e}_1 = 0$ yields $\mathbf{e}_1 = (0, e_{12}, \epsilon i e_{12})^\top$ where $\epsilon = \pm 1$. Let θ be the angle such that

$$-e_{12} = \cos(\theta) + i\sin(\theta)$$

In fact, $\theta = -i\log(-e_{12})$. The matrix E is an essential matrix because

$$E = \begin{pmatrix} 0 & e_{12} & \epsilon i e_{12} \\ 1 & 0 & 0 \\ i & 0 & 0 \end{pmatrix} = \begin{pmatrix} 0 & -\epsilon & -i\epsilon \\ \epsilon & 0 & 0 \\ i\epsilon & 0 & 0 \end{pmatrix} \begin{pmatrix} \epsilon & 0 & 0 \\ 0 & \epsilon\cos(\theta) & -\sin(\theta) \\ 0 & \epsilon\sin(\theta) & \cos(\theta) \end{pmatrix}$$

□

Theorems 2.12 and 2.15 are due to Demazure (1988). A consequence of the two theorems is that reconstruction from point correspondences is algebraic, in that it is equivalent to the problem of finding a common zero of a set of polynomial equations. If the point correspondences are $\mathbf{q}_i \leftrightarrow \mathbf{q}'_i$ for $1 \leq i \leq n$ then the equations are

$$\mathbf{q}_i'^\top E \mathbf{q}_i = 0 \qquad (1 \leq i \leq n) \tag{2.88}$$

$$E E^\top E = \frac{1}{2}\mathrm{tr}(E E^\top) E \tag{2.89}$$

In the case of reconstruction up to a collineation (2.89) is replaced by the single equation $\det(E) = 0$. The algebraic properties of reconstruction stemming from (2.88) and (2.89) are explored further in Sect. 5.2.

The equations (2.76) are not the only polynomial constraints on the coefficients of the essential matrices. They are notable firstly because of their symmetry and secondly because any rank two matrix which satisfies them is an essential matrix. It is shown in Sect. 5.2 that there are rank one matrices which satisfy (2.76). Such matrices are not the product of an orthogonal matrix and a non-zero antisymmetric matrix, because any such product has rank two. It is not possible to find a homogeneous polynomial in the entries of 3×3 matrices which vanishes for every essential matrix but which does not vanish on the rank one matrices satisfying (2.76), because there exist rank one matrices which are limits of sequences of essential matrices.

The polynomial constraints on E in the next theorem were suggested by H.C. Longuet-Higgins (Huang & Faugeras 1989).

Theorem 2.16. *Let E be a real or complex 3×3 matrix with rows $\mathbf{e}_1^\top$, $\mathbf{e}_2^\top$, $\mathbf{e}_3^\top$ such that the scalar products $\mathbf{e}_i.\mathbf{e}_j$ are all non-zero. Then E is an essential matrix if and only if E has rank two and*

$$\frac{\mathbf{e}_1}{\mathbf{e}_2.\mathbf{e}_3} + \frac{\mathbf{e}_2}{\mathbf{e}_3.\mathbf{e}_1} + \frac{\mathbf{e}_3}{\mathbf{e}_1.\mathbf{e}_2} = 0 \tag{2.90}$$

Proof. Let E be an essential matrix, and let $E = T_a R$ where T_a is an antisymmetric matrix and R is an orthogonal matrix. Then E satisfies (2.90) if and only if $ER^\top$ satisfies (2.90). It is thus assumed that $E = T_a$. The hypothesis $\mathbf{e}_i.\mathbf{e}_j \neq 0$ for all i, j ensures that the three components of $\mathbf{a}$ are non-zero. Let $\mathbf{a} = (a_1, a_2, a_3)^\top$. It follows from $E = T_a$ that

$$\begin{aligned}
(\mathbf{e}_2.\mathbf{e}_3)^{-1}\mathbf{e}_1 &= (0, -a_2^{-1}, a_3^{-1})^\top \\
(\mathbf{e}_3.\mathbf{e}_1)^{-1}\mathbf{e}_2 &= (a_1^{-1}, 0, -a_3^{-1})^\top \\
(\mathbf{e}_1.\mathbf{e}_2)^{-1}\mathbf{e}_3 &= (-a_1^{-1}, a_2^{-1}, 0)^\top
\end{aligned} \tag{2.91}$$

Equation (2.90) follows from (2.91).

To obtain the converse result, let E be any matrix satisfying (2.90), and let a, b, c be the scalars defined by

$$\begin{aligned}
a^{-1} &= \mathbf{e}_1.\mathbf{e}_2 \\
b^{-1} &= \mathbf{e}_2.\mathbf{e}_3 \\
c^{-1} &= \mathbf{e}_3.\mathbf{e}_1
\end{aligned} \tag{2.92}$$

It is assumed that E is scaled such that $abc = -1$. On taking the scalar product of (2.90) with $\mathbf{e}_1$, $\mathbf{e}_2$ and $\mathbf{e}_3$ in turn and then applying (2.92) the following equations are obtained

$$\begin{aligned}
\mathbf{e}_1.\mathbf{e}_1 &= -b^{-1}\left(\frac{a}{c} + \frac{c}{a}\right) \\
\mathbf{e}_2.\mathbf{e}_2 &= -c^{-1}\left(\frac{b}{a} + \frac{a}{b}\right) \\
\mathbf{e}_3.\mathbf{e}_3 &= -a^{-1}\left(\frac{c}{b} + \frac{b}{c}\right)
\end{aligned} \tag{2.93}$$

It follows from (2.92), (2.93) and the assumption $abc = -1$ that

$$EE^\top = \begin{pmatrix} c^2 + a^2 & a^{-1} & c^{-1} \\ a^{-1} & b^2 + a^2 & b^{-1} \\ c^{-1} & b^{-1} & c^2 + b^2 \end{pmatrix} \tag{2.94}$$

Let $\mathbf{s}$ be the vector defined by $\mathbf{s} = (b, c, a)^\top$. The expansion of $T_s T_s^\top$ yields

$$T_s T_s^\top = \begin{pmatrix} c^2 + a^2 & -bc & -ab \\ -bc & b^2 + a^2 & -ac \\ -ab & -ac & c^2 + b^2 \end{pmatrix} \tag{2.95}$$

It follows from (2.94) and (2.95) that

$$EE^\top = T_s T_s^\top \tag{2.96}$$

The proof is completed by showing that (2.76) holds and then applying Theorem 2.15. The vector $\mathbf{s}^\top E$ is equal to zero because

$$\begin{aligned} \mathbf{s}^\top E &= a\,\mathbf{e}_1^\top + b\,\mathbf{e}_2^\top + c\,\mathbf{e}_3^\top \\ &= (\mathbf{e}_2.\mathbf{e}_3)^{-1}\mathbf{e}_1^\top + (\mathbf{e}_3.\mathbf{e}_1)^{-1}\mathbf{e}_2^\top + (\mathbf{e}_3.\mathbf{e}_1)^{-1}\mathbf{e}_3^\top \\ &= 0 \end{aligned} \tag{2.97}$$

It follows from (2.96) and (2.97) that

$$\begin{aligned} EE^\top E &= T_s T_s^\top E \\ &= ((\mathbf{s}.\mathbf{s})I - \mathbf{s}\otimes\mathbf{s})E \\ &= (\mathbf{s}.\mathbf{s})E \\ &= \frac{1}{2}\mathrm{tr}(EE^\top)E \end{aligned}$$

The result now follows from Theorem 2.15. □

2.2.2 The Singular Value Decomposition

The singular value decomposition (SVD) is an exceedingly useful technique developed in numerical analysis for studying the stability of solutions to matrix equations (Golub & Van Loan 1983). The SVD of a $m \times n$ real matrix A is given by $A = U^\top \Sigma V$, where U is an $m \times m$ orthogonal matrix, V is an $n \times n$ orthogonal matrix, and Σ is an $m \times n$ diagonal matrix. The matrices U, V are chosen such that the entries Σ_{ii} of Σ satisfy

$$\Sigma_{11} \geq \Sigma_{22} \geq \ldots \geq \Sigma_{kk} \geq 0$$

where $k = \min\{m, n\}$. The singular values σ_i of A are defined by $\sigma_i = \Sigma_{ii}$. The orthogonal matrices U, V are unique if and only if the singular values of A are distinct.

The first result is due to Huang & Faugeras (1989).

Theorem 2.17. *A 3×3 matrix E is an essential matrix if and only if the singular values $\sigma_1 \geq \sigma_2 \geq \sigma_3$ of E satisfy $\sigma_1 = \sigma_2 > 0$, $\sigma_3 = 0$.*

Proof. Let E be an essential matrix and let $E = T_a R$, where T_a is an antisymmetric matrix and R is an orthogonal matrix. It follows from the properties of the singular value decomposition that the singular values of E are equal to the singular values of T_a. The singular values of T_a are the square roots of the eigenvalues of the symmetric matrix $T_a T_a^\top$. The expansion of $T_a T_a^\top$ yields

$$T_a T_a^\top = (\mathbf{a}.\mathbf{a})I - \mathbf{a}\otimes\mathbf{a} \tag{2.98}$$

It follows from (2.98) that two of the eigenvalues of $T_a T_a^\top$ are 0 and $\mathbf{a}.\mathbf{a}$. The sum of the eigenvalues of $T_a T_a^\top$ is equal to $\mathrm{tr}(T_a T_a^\top) = 2\,\mathbf{a}.\mathbf{a}$, thus the remaining eigenvalue of $T_a T_a^\top$ is equal to $\mathbf{a}.\mathbf{a}$. It follows that the singular values of T_a and hence of E are $\|\mathbf{a}\|$, $\|\mathbf{a}\|$ and 0.

To prove the converse, let $\sigma_1 = \sigma_2 = \sigma > \sigma_3 = 0$ be the singular values of E. It follows from the properties of the singular value decomposition that there exist orthogonal matrices U, V such that

$$U^\top EV = \begin{pmatrix} \sigma & 0 & 0 \\ 0 & \sigma & 0 \\ 0 & 0 & 0 \end{pmatrix} = \begin{pmatrix} 0 & 1 & 0 \\ 1 & 0 & 0 \\ 0 & 0 & -1 \end{pmatrix} \begin{pmatrix} 0 & \sigma & 0 \\ -\sigma & 0 & 0 \\ 0 & 0 & 0 \end{pmatrix} \tag{2.99}$$

It follows from (2.99) that $U^\top EV$ is an essential matrix. An application of Proposition 2.11 shows that E is an essential matrix. □

Proposition 2.18. *Let E be a 3×3 matrix. Then E is an essential matrix if and only if there exists a vector* $\mathbf{a}$ *such that $EE^\top = T_a T_a^\top$.*

Proof. If E is an essential matrix then there exist an orthogonal matrix R and an antisymmetric matrix T_a such that $E = T_a R$. It follows that $EE^\top = T_a T_a^\top$ as required.

Conversely, let $EE^\top = T_a T_a^\top$ for some vector $\mathbf{a}$. The eigenvalues of $EE^\top$ are $\|\mathbf{a}\|^2$, $\|\mathbf{a}\|^2$, 0. It follows that the singular values of E are $\|\mathbf{a}\|$, $\|\mathbf{a}\|$, 0. The result follows from Theorem 2.17. □

2.2.3 Symmetric and Antisymmetric Parts

Two main results concerning the symmetric and the antisymmetric parts of an essential matrix are obtained. The first result is that one of the eigenvalues of the symmetric part of an essential matrix is equal to the sum of the other two eigenvalues. The second result is that one of the eigenvectors of the symmetric part of an essential matrix is normal to the vector associated with the antisymmetric part of the matrix. These results are proved for essential matrices with real entries. The proofs in the real case are simplified because of the fact that a real symmetric matrix is diagonalisable by an orthogonal transformation. It is not always possible to diagonalise a symmetric complex matrix using an orthogonal transformation. For example let M be a symmetric matrix and let $\mathbf{e}$ be an eigenvector of M such that $\|\mathbf{e}\| = 0$, $M\mathbf{e} = \mathbf{e}$. If $\mathbf{e}$ is up to a scale factor the only eigenvector of M with eigenvalue one then there is no orthogonal matrix U such that $U^\top MU$ is diagonal.

The eigenvalues of a real symmetric matrix are real and the eigenvectors associated with different eigenvalues are orthogonal. To see this, let λ be an eigenvalue of an $n \times n$ real symmetric matrix A and let $\mathbf{e}$ be the associated eigenvector. An overline denotes complex conjugation. Thus $\overline{\mathbf{e}}$ is the vector obtained by taking the complex conjugates of the components of $\mathbf{e}$. It follows from the definitions of λ and $\mathbf{e}$ that

$$\lambda \mathbf{e}.\overline{\mathbf{e}} = (A\mathbf{e})^\top \overline{\mathbf{e}}$$

$$\begin{aligned} &= \mathbf{e}^\top A^\top \overline{\mathbf{e}} \\ &= \mathbf{e} A \overline{\mathbf{e}} \\ &= \mathbf{e} \overline{A \mathbf{e}} \\ &= \overline{\lambda} \mathbf{e}.\overline{\mathbf{e}} \end{aligned}$$

The eigenvector $\mathbf{e}$ is by definition non-zero, thus $\mathbf{e}.\overline{\mathbf{e}} \neq 0$ and hence $\lambda = \overline{\lambda}$. Let $\mathbf{e}_i$, $\mathbf{e}_j$ be two eigenvectors of A, and let λ_1, λ_2 be the associated eigenvalues. It follows that

$$\begin{aligned} \lambda_i \mathbf{e}_i.\mathbf{e}_j &= (A\mathbf{e}_i).\mathbf{e}_j \\ &= \mathbf{e}_i^\top A^\top \mathbf{e}_j \\ &= \mathbf{e}_i^\top A \mathbf{e}_j \\ &= \lambda_j \mathbf{e}_i.\mathbf{e}_j \end{aligned}$$

If $\lambda_i \neq \lambda_j$, then $\mathbf{e}_i.\mathbf{e}_j = 0$.

The first result is an expression for one of the eigenvalues of the symmetric part of an essential matrix.

Theorem 2.19. *Let $E = RT_a$, where R is an orthogonal matrix and T_a is antisymmetric. Let* $\mathbf{n}$ *be the axis of R, and let θ be the angle of rotation of R. Then $\mathbf{a} - R\mathbf{a}$ is an eigenvector of* $\mathrm{sym}(E)$ *with eigenvalue* $(\mathbf{a}.\mathbf{n})\sin(\theta)$.

Proof. If $\mathbf{a} = R\mathbf{a}$ then R is either the identity or a rotation with axis $\mathbf{a}$. In both cases $\mathrm{sym}(E)$ is singular, and the result holds trivially. Thus it is assumed that $\mathbf{a} \neq R\mathbf{a}$. Let $\mathbf{b}$ be the vector defined by $\mathbf{b} = R\mathbf{a}$. It follows from the definition of $\mathrm{sym}(E)$ that

$$\begin{aligned} 2\,\mathrm{sym}(E) &= E + E^\top \\ &= RT_a + T_a^\top R^\top \\ &= RT_a - R^\top T_b \end{aligned} \tag{2.100}$$

It follows from (2.100) that

$$\begin{aligned} 2\,\mathrm{sym}(E)(\mathbf{a} - \mathbf{b}) &= -R^\top T_b \mathbf{a} - RT_a \mathbf{b} \\ &= (R - R^\top)(\mathbf{a} \times \mathbf{b}) \end{aligned} \tag{2.101}$$

Equation (2.75) is used to replace $R - R^\top$ by $-2\sin(\theta)T_n$ in (2.101). This substitution yields

$$\begin{aligned} \mathrm{sym}(E)(\mathbf{a} - \mathbf{b}) &= -\sin(\theta)T_n(\mathbf{a} \times \mathbf{b}) \\ &= \sin(\theta)[\mathbf{n} \times (\mathbf{a} \times \mathbf{b})] \\ &= \sin(\theta)[(\mathbf{n}.\mathbf{b})\mathbf{a} - (\mathbf{n}.\mathbf{a})\mathbf{b}] \end{aligned} \tag{2.102}$$

It follows from the definitions of $\mathbf{b}$ and $\mathbf{n}$ that

$$\mathbf{b}.\mathbf{n} = \mathbf{a}^\top R^\top \mathbf{n} = \mathbf{a}.\mathbf{n} \tag{2.103}$$

The required result

$$\text{sym}(E)(\mathbf{a}-\mathbf{b}) = (\mathbf{a}.\mathbf{n})\sin(\theta)(\mathbf{a}-\mathbf{b})$$

follows from (2.103) and (2.102). □

Proposition 2.20. *Let $E = RT_a$ where R is an orthogonal matrix, and T_a is an antisymmetric matrix. Let $\mathbf{n}$ be the axis of R and let θ be the angle of rotation of R. Then the trace of E is given by*

$$\frac{1}{2}\text{tr}(E) = (\mathbf{a}.\mathbf{n})\sin(\theta) \tag{2.104}$$

Proof. It follows from the properties of the trace function that

$$\begin{aligned} 2\text{tr}(E) &= 2\,\text{tr}(\text{sym}(E)) \\ &= \text{tr}(RT_a + T_a^\top R^\top) \\ &= \text{tr}((R - R^\top)T_a) \end{aligned} \tag{2.105}$$

It follows from (2.75) that

$$(R - R^\top)T_a = -2\,\sin(\theta)T_nT_a \tag{2.106}$$

The trace of (2.106) yields

$$\begin{aligned} \text{tr}((R - R^\top)T_a) &= -2\,\sin(\theta)\text{tr}(T_nT_a) \\ &= 4(\mathbf{n}.\mathbf{a})\sin(\theta) \end{aligned} \tag{2.107}$$

Equation (2.104) follows from (2.105) and (2.107). □

Corollary. One of the eigenvalues of sym(E) is equal to the sum of the other two eigenvalues. To prove the corollary note first that it follows from Theorem 2.19 and (2.104) that $(1/2)\text{tr}(E)$ is an eigenvalue of sym(E). Let λ_1, λ_2, λ_3 be the eigenvalues of sym(E), indexed such that $\lambda_3 = (1/2)\text{tr}(E)$. It follows from the properties of the trace function that

$$\text{tr}(E) = \text{tr}(\text{sym}(E)) = \lambda_1 + \lambda_2 + \lambda_3 \tag{2.108}$$

The substitution of $2\,\lambda_3$ for tr(E) in (2.108) yields $\lambda_3 = \lambda_1 + \lambda_2$.

The antisymmetric part of an essential matrix is considered next.

Proposition 2.21. *Let $E = RT_a$, where R is an orthogonal matrix, and T_a is an antisymmetric matrix. Let $\mathbf{c}$ be the vector such that* asy$(E) = T_c$. *Then*

$$R(\mathbf{c}\times\mathbf{a}) + R^\top\mathbf{c}\times\mathbf{a} = 0 \tag{2.109}$$

Proof. It follows from the definition of asy(E) that

$$\begin{aligned} 0 &= 2\,\mathrm{asy}(E)\mathbf{c} \\ &= (RT_a - T_a^\top R^\top)\mathbf{c} \\ &= R(\mathbf{c}\times\mathbf{a}) + R^\top\mathbf{c}\times\mathbf{a} \end{aligned}$$

The result follows. □

Propositions 2.20 and 2.21 are applied to obtain a connection between the symmetric and the antisymmetric parts of an essential matrix.

Theorem 2.22. *Let* $E = RT_a$, *and let* $\mathbf{c}$ *be the vector such that* asy(E) $= T_c$. *Then* $\mathbf{c}$ *is normal to an eigenvector of* sym(E).

Proof. It is shown that $\mathbf{c}$ is normal to the eigenvector $\mathbf{a} - R\mathbf{a}$ of sym(E) obtained in Theorem 2.19,

$$\mathbf{c}.(\mathbf{a} - R\mathbf{a}) = 0 \tag{2.110}$$

If $\mathbf{a} = R\mathbf{a}$ then the result holds trivially. Thus it is assumed that $\mathbf{a} \neq R\mathbf{a}$. Let $\mathbf{n}$ be the axis of the orthogonal matrix R. The scalar product of (2.109) with $\mathbf{n}$ yields

$$\begin{aligned} 0 &= \mathbf{n}.R(\mathbf{c}\times\mathbf{a}) + \mathbf{n}.(R^\top\mathbf{c}\times\mathbf{a}) \\ &= \mathbf{n}.(\mathbf{c}\times\mathbf{a}) + \mathbf{n}.(\mathbf{c}\times R\mathbf{a}) \\ &= \mathbf{c}.(\mathbf{a}\times\mathbf{n} + R\mathbf{a}\times\mathbf{n}) \\ &= \mathbf{c}.((\mathbf{a} + R\mathbf{a})\times\mathbf{n}) \end{aligned} \tag{2.111}$$

The proof divides into two cases. Suppose first that $(\mathbf{a} + R\mathbf{a})\times\mathbf{n} \neq 0$. To obtain (2.110) from (2.111) it suffices to show that $(\mathbf{a} + R\mathbf{a})\times\mathbf{n}$ is parallel to $\mathbf{a} - R\mathbf{a}$. To do this, it suffices to show that $\mathbf{n}$ and $\mathbf{a} + R\mathbf{a}$ are both normal to $\mathbf{a} - R\mathbf{a}$. It follows from the definition of $\mathbf{n}$ that $\mathbf{n}.(\mathbf{a} - R\mathbf{a}) = 0$. The vector $\mathbf{a} + R\mathbf{a}$ is normal to $\mathbf{a} - R\mathbf{a}$ because

$$\begin{aligned} (\mathbf{a} + R\mathbf{a}).(\mathbf{a} - R\mathbf{a}) &= \mathbf{a}^\top(I + R^\top)(I - R)\mathbf{a} \\ &= \mathbf{a}^\top(R^\top - R)\mathbf{a} \\ &= 0 \end{aligned}$$

The case $(\mathbf{a} + R\mathbf{a})\times\mathbf{n} = 0$ is dealt with as follows.

$$\begin{aligned} 2T_c\mathbf{n} &= 2\,\mathrm{asy}(E)\mathbf{n} \\ &= (RT_a - T_a^\top R^\top)\mathbf{n} \\ &= RT_a\mathbf{n} + T_a\mathbf{n} \\ &= -T_a\mathbf{n} + T_a\mathbf{n} \\ &= 0 \end{aligned}$$

It follows that $\mathbf{c}$ is parallel to $\mathbf{n}$, and that $\mathbf{c}.(\mathbf{a} - R\mathbf{a}) = 0$, as required. □

A partial converse to Theorem 2.22 is obtained.

Theorem 2.23. *Let E be a 3×3 matrix with zero determinant. Let λ_1, λ_2, λ_3 be the eigenvalues of* $\mathrm{sym}(E)$, *and let* $\mathbf{c}$ *be the vector such that* $\mathrm{asy}(E) = T_c$. *If $\lambda_1 + \lambda_2 = \lambda_3$, and if* $\mathbf{c}$ *is normal to the eigenvector associated with λ_3 then E is an essential matrix.*

Proof. It follows from Proposition 2.11 that for any orthogonal matrix U, E is an essential matrix if and only if $U^\top EU$ is an essential matrix. It is thus supposed, without loss of generality, that the matrix $M = \mathrm{sym}(E)$ is diagonal. Let $M_{ii} = \lambda_i$ $(i = 1,2,3)$. It follows that $E = M + T_c$, subject to the condition $\mathbf{c}.(0,0,1)^\top = c_3 = 0$. The entries of E are

$$E = \begin{pmatrix} \lambda_1 & 0 & 0 \\ 0 & \lambda_2 & 0 \\ 0 & 0 & \lambda_3 \end{pmatrix} + \begin{pmatrix} 0 & 0 & c_2 \\ 0 & 0 & -c_1 \\ -c_2 & c_1 & 0 \end{pmatrix} \tag{2.112}$$

It follows from (2.112) and the hypothesis $\det(E) = 0$ that

$$\lambda_1\lambda_2(\lambda_1 + \lambda_2) + \lambda_1 c_1^2 + \lambda_2 c_2^2 = 0$$

It also follows from (2.112) that

$$EE^\top = \begin{pmatrix} \lambda_1^2 + c_2^2 & -c_1c_2 & \lambda_2 c_2 \\ -c_1c_2 & \lambda_2^2 + c_1^2 & -\lambda_1 c_1 \\ \lambda_2 c_2 & -\lambda_1 c_1 & c_1^2 + c_2^2 + \lambda_3^2 \end{pmatrix} \tag{2.113}$$

Let τ be defined by $\tau = (1/2)\mathrm{tr}(EE^\top)$. It follows from (2.113) and the hypothesis $\lambda_3 = \lambda_1 + \lambda_2$ that

$$\begin{aligned} \tau &= \frac{1}{2}(\lambda_1^2 + \lambda_2^2 + \lambda_3^2) + c_1^2 + c_2^2 \\ &= \lambda_1^2 + \lambda_2^2 + \lambda_1\lambda_2 + c_1^2 + c_2^2 \end{aligned}$$

A straightforward but lengthy calculation yields

$$EE^\top E = \frac{1}{2}\mathrm{tr}(EE^\top)E \tag{2.114}$$

The full calculation leading to (2.114) is omitted. An example is given to demonstrate the method. The top left-hand entry of $EE^\top E$ is given by

$$\begin{aligned} (EE^\top E)_{11} &= \lambda_1(\lambda_1^2 + c_2^2) - \lambda_2 c_2^2 v \\ &= \lambda_1(\lambda_1^2 + c_2^2) + \lambda_1 c_1^2 + \lambda_1\lambda_2(\lambda_1 + \lambda_2) \\ &= \lambda_1(\lambda_1^2 + c_1^2 + c_2^2 + \lambda_1\lambda_2 + \lambda_2^2) \\ &= \tau\lambda_1 \end{aligned}$$

in agreement with (2.114). The calculations for the other entries of $EE^\top E$ are similar. The result follows on applying Theorem 2.15. □

2.2.4 Ambiguity

In Theorem 2.9 the detailed algebraic properties of essential matrices are used to show that there are at most three essentially different reconstructions compatible with a dense set of image correspondences. In this subsection a second proof of the same result is given. The proof also uses essential matrices, but it places more emphasis on the properties of the algebraic variety of all essential matrices.

It is useful to begin with some remarks on rational maps and quadratic transformations. The rational maps are polynomial transformations between projective spaces. A quadratic transformation from $\mathbb{P}^2$ to $\mathbb{P}^2$ is the simplest type of non-linear rational map, in which the defining polynomials have degree two. Rational maps are not defined here in their full generality. A complete account can be found in Hartshorne (1977).

A rational map is a map from $\mathbb{P}^m$ to $\mathbb{P}^n$ defined by polynomials. For almost all $\mathbf{x}$ in $\mathbb{P}^m$ the value of the rational map is given by

$$\mathbf{x} \mapsto (f_1(\mathbf{x}), \ldots, f_{n+1}(\mathbf{x}))^\top \tag{2.115}$$

where the f_i for $1 \le i \le n+1$ are homogeneous polynomials, all of the same degree. The conditions that the f_i are homogeneous and of the same degree are necessary to ensure that the rational map is defined on the points $\mathbf{x}$ of $\mathbb{P}^m$ and that it takes values in $\mathbb{P}^n$. The rational map (2.115) is not defined at those points $\mathbf{x}$ which are common zeros of the f_i for $1 \le i \le n+1$. A rational map Φ from $\mathbb{P}^n$ to $\mathbb{P}^n$ is invertible if there exists a rational map Ψ from $\mathbb{P}^n$ to $\mathbb{P}^n$ such that $\Phi(\Psi(\mathbf{x})) = \mathbf{x}$ and $\Psi(\Phi(\mathbf{x})) = \mathbf{x}$ for almost all $\mathbf{x}$ in $\mathbb{P}^n$.

The quadratic transformations are the simplest examples of non-linear rational maps. Let ϕ_1, ϕ_2, ϕ_3 be homogeneous polynomials of degree two in $\mathbf{x} = (x_1, x_2, x_3)^\top$ such that the ϕ_i do not possess a common component. Let Φ be the rational map from $\mathbb{P}^2$ to $\mathbb{P}^2$ defined by

$$\Phi(\mathbf{x}) = (\phi_1(\mathbf{x}), \phi_2(\mathbf{x}), \phi_3(\mathbf{x}))^\top \tag{2.116}$$

The rational map Φ is a quadratic transformation if and only if it is invertible and non-linear. It is shown that Φ is invertible and non-linear if and only if the ϕ_i have exactly three common zeros. To see this, let $(a, b, 1)^\top$ be a point of $\mathbb{P}^2$. Then $(a, b, 1)^\top$ is in the range of Φ if and only if the following two equations have a solution for $\mathbf{x}$ that varies with a, b,

$$\begin{aligned} \phi_1(\mathbf{x}) - a\phi_3(\mathbf{x}) &= 0 \\ \phi_2(\mathbf{x}) - b\phi_3(\mathbf{x}) &= 0 \end{aligned} \tag{2.117}$$

The conics defined by (2.117) intersect in four points. If ϕ_1, ϕ_2, ϕ_3 have exactly three common zeros then only one of these four intersections will vary with $(a, b, 1)^\top$. The variable intersection is the point $\mathbf{x}$ for which $\phi(\mathbf{x}) = (a, b, 1)^\top$. If the ϕ_i have strictly less than three common zeros then there will be two or more points $\mathbf{x}$ such that $\Phi(\mathbf{x}) = (a, b, 1)^\top$. In this case Φ is not a quadratic transformation because it fails to be one to one at a general point in the range. If ϕ_1, ϕ_2, ϕ_3 have four or more common zeros then either Φ maps $\mathbb{P}^3$ to a single line, or Φ is a collineation. In both cases Φ is not a quadratic transformation.

If the three common zeros of the ϕ_i, lie on a single line l, then l is a component of each of the ϕ_i. This case is excluded because the ϕ_i are required not to have a common component.

The common zeros $\mathbf{a}_1$, $\mathbf{a}_2$, $\mathbf{a}_3$ of the ϕ_i are called the fundamental points of Φ, and the three lines $\langle \mathbf{a}_1, \mathbf{a}_2 \rangle$, $\langle \mathbf{a}_2, \mathbf{a}_3 \rangle$, $\langle \mathbf{a}_3, \mathbf{a}_1 \rangle$ are called the fundamental lines of Φ. The image of a fundamental line under Φ is a single point, and this point is a fundamental point of the inverse quadratic transformation Φ^{-1}. Let coordinates be chosen in $\mathbb{P}^2$ such that

$$\begin{aligned} \mathbf{a}_1 &= (1,0,0)^\top \\ \mathbf{a}_2 &= (0,1,0)^\top \\ \mathbf{a}_3 &= (0,0,1)^\top \end{aligned}$$

The quadratic transformation Φ is then given by

$$\Phi(\mathbf{x}) = A \begin{pmatrix} x_2 x_3 \\ x_3 x_1 \\ x_1 x_2 \end{pmatrix}$$

where A is an invertible 3×3 matrix with coefficients depending on Φ and on the choice of coordinates. Coordinates are now be chosen in the range of Φ such that

$$\Phi(\mathbf{x}) = (x_2 x_3, x_3 x_1, x_1 x_2)^\top \tag{2.118}$$

Equation (2.118) is the standard quadratic transformation. Any other quadratic transformation can be reduced to (2.118) by suitable choices of coordinates in the domain space and the range space.

Let $\mathbf{y} = (y_1, y_2, y_3)^\top$ be a general point in the range of Φ and let Ψ be the quadratic transformation from the range of Φ to the domain of Φ defined by

$$\Psi(\mathbf{y}) = (y_2 y_3, y_3 y_1, y_1 y_2)^\top \tag{2.119}$$

It follows from (2.118) and (2.119) that

$$\begin{aligned} \Psi(\Phi(\mathbf{x})) &= (x_1^2 x_2 x_3, x_1 x_2^2 x_3, x_1 x_2 x_3^2)^\top &= \mathbf{x} \\ \Phi(\Psi(\mathbf{y})) &= (y_1^2 y_2 y_3, y_1 y_2^2 y_3, y_1 y_2 y_3^2)^\top &= \mathbf{y} \end{aligned}$$

It follows that $\Psi = \Phi^{-1}$. It is noted that a composition $\Psi(\Phi(\mathbf{x}))$ is defined even at the points $\mathbf{x}$ for which $\Phi(\mathbf{x})$ is not defined.

Inversion in the unit circle is a good example of a quadratic transformation. Let Φ be the quadratic transformation defined on $\mathbb{P}^2$ by

$$\Phi(\mathbf{x}) = (x_1 x_3, x_2 x_3, x_1^2 + x_2^2)^\top$$

The fundamental points of Φ are $(1,i,0)^\top$, $(1,-i,0)^\top$, $(0,0,1)^\top$. The Euclidean space $\mathbb{R}^2$ is included in the projective plane by $(x_1,x_2)^\top \mapsto (x_1,x_2,1)^\top$. Under this inclusion the action of Φ on the points of $\mathbb{R}^2$ is given by

$$\Phi(x_1,x_2) = \left(\frac{x_1}{x_1^2+x_2^2}, \frac{x_2}{x_1^2+x_2^2}\right)^\top$$

In polar coordinates, $x_1 = r\cos(\theta)$, $x_2 = r\sin(\theta)$, the quadratic transformation Φ is defined on $\mathbb{R}^2 \subset \mathbb{P}^2$ by

$$\Phi(r\cos(\theta), r\sin(\theta)) = (r^{-1}\cos(\theta), r^{-1}\sin(\theta))^\top$$

The unit circle $r = 1$ is invariant under Φ and each line through the origin $(0,0)^\top$ is also invariant under Φ. A point outside the unit circle and distance r from the origin is mapped by Φ to a point inside the circle at a distance $1/r$ from the origin. The origin is one of the fundamental points of Φ.

Under the inversion Φ each neighbourhood of the origin is mapped to a neighbourhood of the line at infinity $x_3 = 0$. To see this, let ξ be a line segment of length δ based at the origin and with direction θ. The points of ξ are defined in $\mathbb{P}^2$ by

$$\xi = \{(t\cos(\theta), t\sin(\theta), 1)^\top \mid 0 < t < \delta\}$$

The image of ξ under Φ is the line segment of $\mathbb{P}^2$ defined by

$$\begin{aligned}\Phi(\xi) &= (t^{-1}\cos(\theta), t^{-1}\sin(\theta), 1)^\top \\ &= (\cos(\theta), \sin(\theta), t)^\top \qquad (0 < t < \delta)\end{aligned}$$

The segment $\Phi(\xi)$ is thus directed towards the point $(\cos(\theta), \sin(\theta), 0)^\top$ in the line at infinity. Under the action of Φ each point in the line at infinity corresponds to the direction of a line segment at the origin. If c is an algebraic plane curve passing through the origin then $\Phi(c)$ passes through the point in the line at infinity corresponding to the direction of the tangent of c at the origin. If c has a singularity at the origin and if l_1, l_2 are distinct lines through the origin tangent to c then $\Phi(c)$ intersects the line at infinity at two points, corresponding to the directions of l_1 and l_2 at the origin.

Quadratic transformations are used in algebraic geometry to change or resolve the singular points of plane curves. If c is an algebraic plane curve then there exists a sequence of quadratic transformations Φ_i, $1 \le i \le n$, such that the curve $\Phi_n(\ldots\Phi_1(c)\ldots)$ has only ordinary singular points. An ordinary singular point is a point at which the curve has distinct tangent lines. An example is given of the resolution of a singularity of a cubic plane curve by the application of a quadratic transformation. Let a be a small real number and let c be the cubic plane curve defined by

$$x_1^3 - (x_2^2 - a^2x_1^2)(x_3 - x_1) = 0 \qquad (2.120)$$

If $a \neq 0$ then c has a node at $x_1 = 0$, $x_2 = 0$. A node is by definition a singular point of order two such that the two tangents at the singular point are distinct. The two tangents of c at the node $(0,0,1)^\top$ are defined by $x_2 = \pm a x_1$. If $a = 0$ then the two tangent lines to c at $(0,0,1)^\top$ coincide. The curve c is then a cuspidal cubic, with a cusp at $(0,0,1)^\top$.

Let $a = 0$. There exists a quadratic transformation Φ such that each singular point of $\Phi(c)$ is an ordinary singular point in that it has distinct tangent lines. In fact in this example it is possible to find Φ such that $\Phi(c)$ has no singular points at all. The cusp is chosen as one of the fundamental points of Φ, and the remaining two fundamental points of Φ are chosen on c. Inversion in the unit circle is a suitable choice for Φ. Under inversion the curve c with $a = 0$ transforms to the curve $\Phi(c)$ given by

$$x_2^2 x_3^2 (x_1^2 + x_2^2) = x_1^3 x_3^3 + x_1 x_2^2 x_3^3$$

On cancelling the product of lines $x_3^2(x_1^2 + x_2^2)$, the equation for $\Phi(c)$ reduces to $x_2^2 = x_1 x_3$. It follows that $\Phi(c)$ is a non-singular conic which is tangent to the line at infinity at the point $(1,0,0)^\top$. The cusp singularity of c has been resolved by the quadratic transformation Φ. The tangency or 'double intersection' of $\Phi(c)$ with the line at infinity arises because the curve c has a double tangent line at $(0,0,1)^\top$.

After this digression on quadratic transformations it is back to the properties of essential matrices.

Proposition 2.24. *Let E be an essential matrix and let B be an arbitrary 3×3 matrix. Then $E\mathbf{q} \times B\mathbf{q} = 0$ for all vectors $\mathbf{q}$ if and only if there exists a scalar λ such that $B = \lambda E$.*

Proof. If $B = \lambda E$ then $E\mathbf{q} \times B\mathbf{q} = 0$ for all $\mathbf{q}$. Conversely, let $E\mathbf{q} \times B\mathbf{q} = 0$ for all $\mathbf{q}$. Let $E = RT_a$, where R is an orthogonal matrix and T_a is an antisymmetric matrix. The expansion of $E\mathbf{q} \times B\mathbf{q}$ yields

$$\begin{aligned} 0 &= E\mathbf{q} \times B\mathbf{q} \\ &= R(\mathbf{q} \times \mathbf{a}) \times B\mathbf{q} \\ &= (\mathbf{q}^\top B^\top R\mathbf{q})R\mathbf{a} - (\mathbf{q}^\top B^\top R\mathbf{a})R\mathbf{q} \end{aligned} \tag{2.121}$$

The vectors $R\mathbf{q}$ and $\mathbf{a}$ are, in general linearly independent; thus (2.121) holds if and only if, for all $\mathbf{q}$,

$$\begin{aligned} \mathbf{q}^\top B^\top R\mathbf{a} &= 0 \\ \mathbf{q}^\top B^\top R\mathbf{q} &= 0 \end{aligned} \tag{2.122}$$

It follows from (2.122) that $B^\top R\mathbf{a} = 0$, and that $B^\top R$ is antisymmetric. Let T_b be the antisymmetric matrix defined by $T_b = B^\top R$. The condition $B^\top R\mathbf{a} = 0$ yields

$$0 = B^\top R\mathbf{a} = T_b\mathbf{a} = \mathbf{a} \times \mathbf{b} \tag{2.123}$$

It follows from (2.123) that there exists a scalar λ such that $T_b = -\lambda T_a$. Hence, $B^\top R = -\lambda T_a$, from which it follows that $B = \lambda R T_a = \lambda E$. □

In the next theorem Proposition 2.24 is applied to show that the essential matrices compatible with a dense set of image correspondences are linearly dependent. The theorem holds trivially if there is only a two way ambiguity. The interesting case is three way ambiguity.

Theorem 2.25. *Let E_1, E_2, E_3 be essential matrices such that*

$$\mathbf{q}'^\top E_i \mathbf{q} = 0 \qquad (1 \le i \le 3) \tag{2.124}$$

for a dense set of image correspondences $\mathbf{q} \leftrightarrow \mathbf{q}'$ and let the correspondences $\mathbf{q} \mapsto \mathbf{q}'$ define a non-linear map from the first image to the second image. Then E_1, E_2, E_3 are linearly dependent.

Proof. It is assumed that any two of the E_i are linearly independent, because otherwise the result follows immediately. If $E_1\mathbf{q} \times E_2\mathbf{q} = 0$ for all $\mathbf{q}$ then it follows from Proposition 2.24 that E_1 and E_2 are linearly dependent, contrary to assumption. It follows that $E_1\mathbf{q} \times E_2\mathbf{q} \neq 0$ for almost all $\mathbf{q}$. Thus (2.124) yields

$$\mathbf{q}' = E_1\mathbf{q} \times E_2\mathbf{q} \tag{2.125}$$

A similar argument establishes that $\mathbf{q}' = E_1\mathbf{q} \times E_3\mathbf{q}$. Let Φ be the transformation $\Phi : \mathbf{q} \mapsto \mathbf{q}'$. It follows from (2.125) that Φ is defined by polynomials of degree two in the components of $\mathbf{q}$. The symmetry between $\mathbf{q}$ and $\mathbf{q}'$ ensures that Φ is invertible, and the hypotheses of the theorem ensure that Φ is non-linear. It follows that Φ is a quadratic transformation.

Let $\mathbf{n}$ be a point of $\mathbb{P}^2$ such that Φ is defined at $\mathbf{n}$. Then $E_1\mathbf{n}$, $E_2\mathbf{n}$, $E_3\mathbf{n}$ are linearly dependent, because $\Phi(\mathbf{n}).E_i\mathbf{n} = 0$, $i = 1, 2, 3$. Let a_1, a_2, a_3, be coefficients, possibly depending on $\mathbf{n}$, such that

$$a_1 E_1 \mathbf{n} + a_2 E_2 \mathbf{n} + a_3 E_3 \mathbf{n} = 0$$

Let G be the matrix defined by

$$G = a_1 E_1 + a_2 E_2 + a_3 E_3$$

and let Ψ be the transformation of $\mathbb{P}^2$ defined by

$$\Psi(\mathbf{q}) = E_1\mathbf{q} \times G\mathbf{q} = E_1\mathbf{q} \times (a_2 E_2 + a_3 E_3)\mathbf{q}$$

The vectors $E_1\mathbf{q} \times E_2\mathbf{q}$ and $E_1\mathbf{q} \times E_3\mathbf{q}$ are parallel for all $\mathbf{q}$, thus either $\Psi = \Phi$ or $\Psi = 0$. The case $\Psi = \Phi$ does not apply because $\Psi(\mathbf{n}) = 0$, $\Phi(\mathbf{n}) \neq 0$. It follows that $\Psi = 0$, and that as a consequence $E_1\mathbf{q}$ and $G\mathbf{q} = (a_2 E_2 + a_3 E_3)\mathbf{q}$ are parallel for all $\mathbf{q}$. The application of Proposition 2.24 shows that E_1 and G are linearly dependent. □

Theorem 2.26. *Let* $\mathbf{q} \leftrightarrow \mathbf{q}'$ *be a dense set of image correspondences. Then there are at most three essential matrices* E *such that*

$$\mathbf{q}'^{\top} E \mathbf{q} = 0 \tag{2.126}$$

Proof. It is assumed first that the function $\mathbf{q} \mapsto \mathbf{q}'$ is non-linear. It follows from Theorem 2.25 that the essential matrices E satisfying (2.126) form a line l in the projective space $\mathbb{P}^8$ constructed from the non-zero 3×3 matrices. Let E_1, E_2 be two essential matrices contained in l. Each matrix in l is a linear combination of E_1 and E_2,

$$\lambda_1 E_1 + \lambda_2 E_2 \tag{2.127}$$

It is shown in the proof of Theorem 2.9 that the determinant of (2.127) yields a non-trivial homogeneous cubic polynomial constraint on λ_1, λ_2. Up to scale there are at most three pairs (λ_1, λ_2) satisfying the constraint. The result follows for the case in which $\mathbf{q} \mapsto \mathbf{q}'$ is non-linear.

There remains the case in which the function defined by $\mathbf{q} \mapsto \mathbf{q}'$ is linear. It follows from Theorem 2.8 that there exists a plane Π in space such that each point of Π projects to corresponding points $\mathbf{q}$, $\mathbf{q}'$ in the first and second images respectively. It follows from Theorem 2.7 that there are in general exactly two planes Π and it follows from Theorem 2.6 that the image correspondences cannot also arise from the projections of points on a non-planar surface. □

2.3 Projective Framework for Reconstruction

The first mathematical descriptions of reconstruction were developed in the 19th century using projective geometry rather than Euclidean geometry (Sturm 1869). The projective geometric approach emphasises the star of projection lines through the optical centre of the camera, rather than the intersections of these lines with a particular imaging surface.

The epipoles and the epipolar transformation are fundamental concepts in the projective geometric framework. They replace the pair $\{R, \mathbf{a}\}$ used to describe the displacement of the camera in the Euclidean framework. There are two epipoles, one in each image. The epipole $\mathbf{p}$ in the first image is the projection of the optical centre $\mathbf{a}$ of the camera after the displacement and the epipole $\mathbf{p}'$ in the second image is the projection of the optical centre $\mathbf{o}$ before the displacement. The epipoles are illustrated in Fig. 2.3. The epipolar transformation is a particular homography from the lines through $\mathbf{p}$ in the first image to the lines through $\mathbf{p}'$ in the second image.

There are many properties of the ambiguous case of reconstruction that can be described very cleanly in the projective geometric framework. An introduction to the geometry of ambiguity is given in this chapter. Further information is given in Chap. 3.

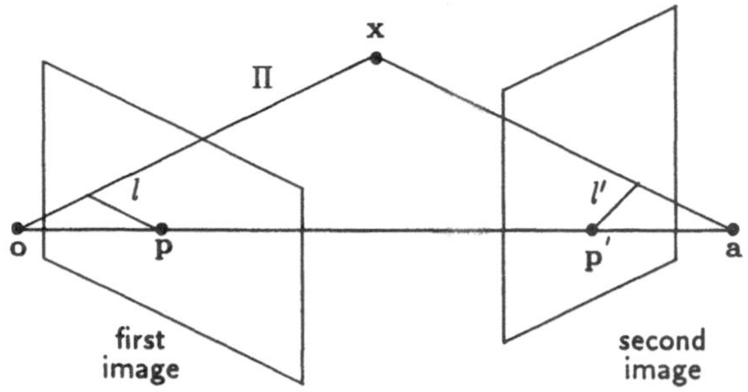

Fig. 2.3. The epipolar transformation

The epipolar transformation is described in in Sect. 2.3.1. The projective geometric approach to ambiguity is introduced in Sect. 2.3.2 and some of the properties of critical surfaces are obtained in Sect. 2.3.3.

2.3.1 The Epipolar Transformation

The epipolar transformation from the lines through $\mathbf{p}$ to the lines through $\mathbf{p}'$ is constructed as illustrated in Fig. 2.3. The epipole $\mathbf{p}$ in the first image is the projection of $\langle \mathbf{o}, \mathbf{a} \rangle$ from $\mathbf{o}$, and the epipole $\mathbf{p}'$ in the second image is the projection of $\langle \mathbf{o}, \mathbf{a} \rangle$ from $\mathbf{a}$. Let l be any line through $\mathbf{p}$ in the first image. Then l is the projection of a plane Π in space containing $\langle \mathbf{o}, \mathbf{a} \rangle$. It follows from the definition of $\mathbf{p}'$ that the projection of Π into the second image is a line l' which contains $\mathbf{p}'$. The projection $\Pi \mapsto l$ is a homography from the pencil of planes containing $\langle \mathbf{o}, \mathbf{a} \rangle$ to the pencil of lines in the first image containing $\mathbf{p}$. Similarly the projection $\Pi \mapsto l'$ is a homography to the pencil of lines in the second image with centre $\mathbf{p}'$. The two homographies $\Pi \mapsto l$ and $\Pi \mapsto l'$ define the epipolar transformation $l \mapsto l'$. The notation $l \barwedge l'$ is used if l and l' correspond under the epipolar transformation. The planes Π containing $\langle \mathbf{o}, \mathbf{a} \rangle$ are called the epipolar planes.

The epipolar transformation is of central importance for matching points between the first and second images. Let $\mathbf{x}$ be any point of $\mathbb{P}^3$, let $\mathbf{q}$ be the projection of $\mathbf{x}$ in the first image, and let l_x be the epipolar line containing $\mathbf{q}$. Then the point $\mathbf{q}'$ in the second image that matches $\mathbf{q}$ lies on the epipolar line l'_x corresponding to l_x under the epipolar transformation. Once the epipolar transformation is known, the problem of finding $\mathbf{q}'$ is simplified because it is only necessary to search in the one dimensional space l'_x rather than in a two dimensional region of the second image.

The epipolar transformation can be calculated from the corresponding points in two different images of the same scene. It is not necessary to know the full camera calibration. The only requirement is that the projection from space to the image is linear. If the full camera calibration is known then the images can be parameterised such that the epipolar transformation preserves angles in the following sense. If l_1, l_2 are epipolar lines in the first image, and if l'_1, l'_2 are the corresponding epipolar lines in the second image, then the angle between l_1 and l_2 is equal to the angle between l'_1 and l'_2. The angle between two epipolar lines can be obtained by a geometric construction described in Sect. 3.1. It is not necessary to choose a special parameterisation of the image.

In the projective geometric framework reconstruction is formulated in terms of the epipoles and the epipolar transformation. Let $\mathbf{q}_i \leftrightarrow \mathbf{q}'_i$ for $1 \leq i \leq n$, be a set of image correspondences. Then the reconstruction problem is to find points $\mathbf{p}$ and $\mathbf{p}'$ in the first and second images respectively, such that there is an angle preserving homography,

$$\langle \mathbf{p}, \mathbf{q}_i \rangle \barwedge \langle \mathbf{p}' \mathbf{q}'_i \rangle \qquad (1 \leq i \leq n) \qquad (2.128)$$

Once the epipoles $\mathbf{p}$, $\mathbf{p}'$ have been found the relative position of the two cameras and the positions of the points $\mathbf{x}_i$ in space projecting down to the points $\mathbf{q}_i$, $\mathbf{q}'_i$ can be calculated, up to a single unknown scale factor and up to a twisted pair of camera displacements.

In the next proposition it is shown that the formulation of reconstruction based on epipoles is equivalent to the formulation in terms of essential matrices given in Sect. 2.1.

Proposition 2.27. *Let $\mathbf{q}_i \leftrightarrow \mathbf{q}'_i$, $1 \leq i \leq n$, be a set of $n \geq 3$ image correspondences. Then to each essential matrix E such that*

$$\mathbf{q}'^{\top}_i E \mathbf{q}_i = 0 \qquad (1 \leq i \leq n) \qquad (2.129)$$

there corresponds a unique pair of epipoles $\mathbf{p}$, $\mathbf{p}'$ such that the homography (2.128) preserves angles. Conversely, each pair of epipoles $\mathbf{p}$, $\mathbf{p}'$ such that the homography (2.128) preserves angles yields an essential matrix E such that (2.129) holds.

Proof. Let E be an essential matrix such that (2.129) holds. The matrix E has rank two. Let $\mathbf{p}$, $\mathbf{p}'$ be defined by $\mathbf{p}'^{\top} E = 0$, $E\mathbf{p} = 0$. It follows from the hypothesis that E is an essential matrix that there exists an orthogonal matrix R such that $E = RT_p$. The equations

$$0 = \mathbf{p}'^{\top} E = \mathbf{p}'^{\top} R T_p$$

yield $R^{\top}\mathbf{p}' = \mathbf{p}$. It follows that R induces an angle preserving homography from the lines through $\mathbf{p}$ to the lines through $\mathbf{p}'$. On substituting RT_p for E in (2.129) the equation $\mathbf{q}'^{\top}_i RT_p\mathbf{q}_i = 0$ is obtained. It follows that $\mathbf{q}'^{\top}_i R(\mathbf{q}_i \times \mathbf{p}) = 0$. The point $R\mathbf{q}_i$ is thus on the line $\langle R\mathbf{p}, \mathbf{q}'_i \rangle = \langle \mathbf{p}', \mathbf{q}'_i \rangle$. It follows that $\mathbf{p}$, $\mathbf{p}'$ are the epipoles, and that R induces the epipolar transformation.

Conversely, let $\mathbf{p}$, $\mathbf{p}'$ be points such that (2.128) is an angle preserving homography. Let R be a rotation such that $R\mathbf{p} = \mathbf{p}'$, and such that for each i the line $R\langle \mathbf{p}, \mathbf{q}_i \rangle$ contains $\mathbf{q}'_i$. By definition R induces the epipolar transformation.

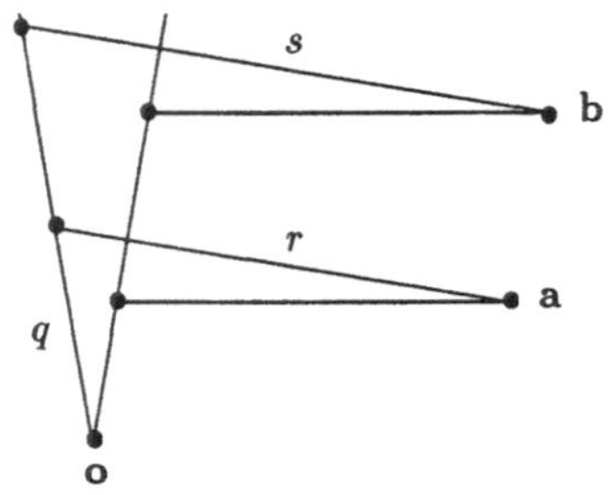

Fig. 2.4. The geometry of ambiguity

Let E be the essential matrix defined by $E = RT_p$. It follows that for each i,

$$\begin{aligned} \mathbf{q}_i'^{\top} E \mathbf{q}_i &= \mathbf{q}_i'^{\top} R T_p \mathbf{q}_i \\ &= \mathbf{q}_i'^{\top} R(\mathbf{q}_i \times \mathbf{p}) \\ &= \mathbf{q}_i'.(R\mathbf{q}_i \times R\mathbf{p}) \\ &= (\mathbf{p}' \times \mathbf{q}_i').R\mathbf{q}_i \\ &= 0 \end{aligned}$$

as required. □

The epipoles $\mathbf{p}$, $\mathbf{p}'$ found in the proof of Proposition 2.27 do not uniquely determine E. There exists a one parameter family of essential matrices E such that $\mathbf{p}'^{\top} E = 0$ and $E\mathbf{p} = 0$.

2.3.2 Ambiguity

The conventions established in Sect. 2.1.1 for the ambiguous case of reconstruction are adopted. The optical centre of the camera before the displacement is fixed at $\mathbf{o}$, and the two possible positions for the optical centre of the camera after the displacement are at $\mathbf{a}$ and $\mathbf{b}$. It is assumed that $\mathbf{o}$, $\mathbf{a}$, $\mathbf{b}$ are not collinear. Let q be the line of points in space projecting to the point $\mathbf{q}$ in the first image, and let r be the line of points in space projecting to the point $\mathbf{q}'$ corresponding to $\mathbf{q}$ when the optical centre of the camera is at $\mathbf{a}$. Similarly, let s be the line of points in space projecting to $\mathbf{q}'$ in the second image when the optical centre of the camera is at $\mathbf{b}$. The assumption that reconstruction is possible either when the optical centre of the second camera is at $\mathbf{a}$ or when it is at $\mathbf{b}$ ensures that q intersects both r and s, as shown in Fig. 2.4. In general, q is the unique line through $\mathbf{o}$ that intersects both r and s. The line q is the intersection of the two planes $\langle \mathbf{o}, r\rangle$ and $\langle \mathbf{o}, s\rangle$. As the line q through $\mathbf{o}$ varies, the intersection of the lines q and r traces out a surface ψ. Similarly the intersection of the lines q and s traces out a surface ϕ. The surfaces ψ and ϕ are the critical surface pair described in Sect. 2.1.1.

It is convenient to suppose that the parameterisations of the stars of lines through **o**, **a**, **b** are chosen such that the images are formed by taking the intersections of the lines with the surface of the unit sphere centred at the optical centre of the camera. The image taken by the second camera has the same appearance for both positions **a**, **b** of the optical centre. It follows that the angle between two lines r_i, r_j through **a** is equal to the angle between the corresponding pair of lines s_i, s_j through **b**. The correspondence ω defined by $r \mapsto \omega(r) = s$ from the lines through **a** to the lines through **b** is thus angle preserving. It follows that ω is linear and orthogonal.

The choice of a particular parameterisation of the image can be avoided by specifying the camera calibration in terms of the image of the absolute conic. The absolute conic and the relation between the absolute conic and the camera calibration are described in Sect. 3.1. The requirement that ω is orthogonal is equivalent to the requirement that if r is any line through **a** which intersects the absolute conic then $\omega(r)$ is a line through **b** which also intersects the absolute conic. It is often convenient to parameterise the stars of lines through **a** and **b** with the points of the plane at infinity Π_∞. The collineation ω from the star of lines through **a** to the star of lines through **b** defines a collineation of Π_∞, also denoted by ω. If **r** is any point of Π_∞, then $\omega(\mathbf{r})$ is defined by $\omega\langle \mathbf{a}, \mathbf{r}\rangle = \langle \mathbf{b}, \omega(\mathbf{r})\rangle$. The induced collineation ω of Π_∞ leaves invariant the absolute conic, i.e. if Ω is the absolute conic then $\omega(\Omega) = \Omega$.

A geometrical construction of the critical surface pair ψ and ϕ is illustrated in Fig. 2.5. In the figure, **o**, **a**, **b** are, as usual, the three different positions of the optical centre of the camera. The line h through **a** is defined by $h = \omega^{-1}\langle \mathbf{o}, \mathbf{b}\rangle$. The plane Π is an arbitrary plane containing h. The collineation ω from the lines through **a** to the lines through **b** induces a collineation from the planes through **a** to the planes through **b**, also denoted by ω. The plane $\omega(\Pi)$ contains $\omega(h) = \langle \mathbf{o}, \mathbf{b}\rangle$. As Π varies through the pencil of planes containing h, $\omega(\Pi)$ varies through the pencil of planes containing $\langle \mathbf{o}, \mathbf{b}\rangle$. It is shown that as Π varies the line $l = \Pi \cap \omega(\Pi)$ sweeps out the critical surface ψ.

Let **x** be any point on l, let $r = \langle \mathbf{a}, \mathbf{x}\rangle$, and let $s = \omega(r)$. Then s and $\langle \mathbf{o}, \mathbf{x}\rangle$ are both contained in the same plane $\omega(\Pi)$. It follows that s intersects $\langle \mathbf{o}, \mathbf{x}\rangle$. Thus $\langle \mathbf{o}, \mathbf{x}\rangle$ is the unique common transversal of r and s which passes through **o**. It follows from the definition of ψ that **x** is a point of ψ. The point **x** is an arbitrary point of l, hence the entire line l is contained in ψ. Thus, as Π varies, l sweeps out ψ.

It is shown geometrically that the degree of ψ is two. For a general choice of ω, the lines $\langle \mathbf{o}, \mathbf{b}\rangle$ and h are not coplanar. The surface ψ contains both $\langle \mathbf{o}, \mathbf{b}\rangle$ and h, thus ψ is not a plane. It follows that the degree of ψ is greater than or equal to two. Let k be an arbitrary fixed line in space. As Π varies through the space of planes containing h, the linear transformation $\Pi \mapsto \omega(\Pi)$ induces a linear transformation ρ of k defined by $\rho(\Pi \cap k) = \omega(\Pi) \cap k$. For a general choice of k, ρ is not the identity. The fixed points of ρ are precisely the points at which k intersects ψ. A non-trivial projective linear transformation of a line

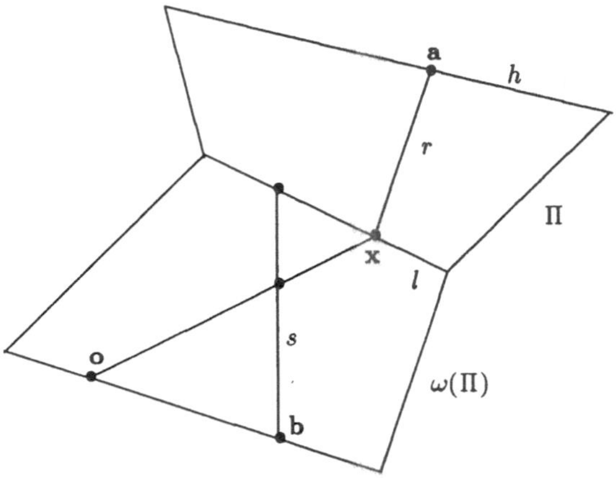

Fig. 2.5. Geometric construction of a critical surface

has at most two fixed points, thus k meets ψ in at most two points. It follows that the degree of ψ is at most two. The degree of ψ is not one, thus the degree of ψ is exactly two.

The keys to the geometry of a non-singular quadric are two families $\mathcal{F}_1$, $\mathcal{F}_2$ of straight lines contained in the quadric. Let ψ be a non-singular quadric in $\mathbf{P}^3$ and let $\mathbf{p}$ be any point of ψ. Then there are exactly two lines $g_1(\mathbf{p})$, $g_2(\mathbf{p})$ through $\mathbf{p}$ that are contained in ψ. The line $g_1(\mathbf{p})$ is from $\mathcal{F}_1$ and the line $g_2(\mathbf{p})$ is from $\mathcal{F}_2$. Any two lines from the same family $\mathcal{F}_i$ are skew and any two lines from different families intersect. The lines are called the generators of ψ. In the construction of ψ described above in this subsection the lines $l = \Pi \cap \omega(\Pi)$ sweep out one of the families of generators of ψ as Π varies through the pencil of planes containing $\langle \mathbf{o}, \mathbf{b} \rangle$.

There are many different ways of constructing a quadric surface. For example, let h, k be two skew lines in $\mathbf{P}^3$ and let ρ be a homography from the points of h to the points of k. Then the lines $\langle \mathbf{p}, \rho(\mathbf{p}) \rangle$ sweep out a non-singular quadric surface as $\mathbf{p}$ ranges over the points of h. Conversely, let h, k be two lines in the same family of generators $\mathcal{F}_1$ of a non-singular quadric surface. Each line l in $\mathcal{F}_2$ intersects h and k. The correspondence $l \cap h \mapsto l \cap k$ is a homography between the points of h and the points of k.

It has already been noted that a quadric surface can be constructed from a homography between two pencils of planes. The converse is also true. Let h, k be two lines in the same family of generators $\mathcal{F}_1$ of a non-singular quadric surface, and let l be a general line in $\mathcal{F}_2$. The mapping $\langle h, l \rangle \mapsto \langle k, l \rangle$ defines a homography between the pencil of planes through h and the pencil of planes through k.

A singular quadric surface is a cone. If the cone has more that one singular point then the cone splits into a plane pair, and every point in the intersection of the two planes is a singular point. If the cone has exactly one singular point then this point is called the vertex of the cone. A cone that does not split into a plane pair has only one family of generators. Every line in the family of generators passes through the vertex.

2.3.3 The Intersection of a Critical Surface Pair

The space curve formed by the intersection of the critical surface pair ψ, ϕ is examined. It is clear that $\psi \cap \phi$ is a curve of degree four because the surfaces ψ, ϕ are each of degree two. The most important property of the curve $\psi \cap \phi$ is that it splits into two components.

Proposition 2.28. *The space curve of degree four formed by the intersection of a critical surface pair splits into a straight line and a curve of degree three. The line contains the optical centre* $\mathbf{o}$ *of the camera prior to the displacement.*

Proof. The notation of Fig. 2.5 is used. Let $\mathbf{\Phi}$ be the plane $\mathbf{\Phi} = \langle h, \mathbf{o} \rangle$. It follows from the construction of the critical surface ψ that the line $g = \mathbf{\Phi} \cap \omega(\mathbf{\Phi})$ is contained in ψ. The line g contains $\mathbf{o}$, because $\mathbf{\Phi}$ and $\omega(\mathbf{\Phi})$ both contain $\mathbf{o}$. It follows from the symmetry between ψ and ϕ that ϕ is swept out by the lines $\Pi \cap \omega(\Pi)$ as Π ranges through the pencil of planes containing $\langle \mathbf{o}, \mathbf{a} \rangle$. The plane $\mathbf{\Phi}$ contains $\langle \mathbf{o}, \mathbf{a} \rangle$, thus ϕ contains $g = \mathbf{\Phi} \cap \omega(\mathbf{\Phi})$. Thus g is contained in $\psi \cap \phi$. It follows that $\psi \cap \phi$ splits into the line g and a curve c of degree $3 = 4 - 1$. □

A space curve is, by definition, a curve in $\mathbf{P}^3$ that is not contained in a plane. If $\psi \cap \phi$ is contained in a plane then $\psi \cap \phi$ is a line pair with each line counted twice in the intersection. This highly degenerate case does not occur in general, thus the curve c obtained in Proposition 2.28 is indeed a space curve. The curve c is an example of a horopter curve. Horopter curves and orthogonal collineations are closely related, as shown in the next theorem.

Theorem 2.29. *The space curve of degree three contained in the intersection of the critical surface pair* ψ, ϕ *is the locus of points* $\mathbf{x}$ *such that*

$$\omega\langle \mathbf{a}, \mathbf{x} \rangle = \langle \mathbf{b}, \mathbf{x} \rangle \tag{2.130}$$

Proof. It follows from Proposition 2.28 that $\psi \cap \phi = g \cup c$, where g is a line through $\mathbf{o}$ and c is a space curve of degree three. The surface ψ is the locus of points $\mathbf{x}$ such that $\langle \mathbf{o}, \mathbf{x} \rangle$ corresponds to $\langle \mathbf{a}, \mathbf{x} \rangle$. The surface ϕ is the locus of points $\mathbf{y}$ such that $\langle \mathbf{o}, \mathbf{y} \rangle$ corresponds to $\langle \mathbf{b}, \mathbf{y} \rangle$. If $\mathbf{x}$ is in $\psi \cap \phi$ but not on the common generator then $\langle \mathbf{o}, \mathbf{x} \rangle$ corresponds to $\langle \mathbf{a}, \mathbf{x} \rangle$ and to $\langle \mathbf{b}, \mathbf{x} \rangle$. It follows that $\omega\langle \mathbf{a}, \mathbf{x} \rangle = \langle \mathbf{b}, \mathbf{x} \rangle$. Conversely if $\omega\langle \mathbf{a}, \mathbf{x} \rangle = \langle \mathbf{b}, \mathbf{x} \rangle$ then $\langle \mathbf{o}, \mathbf{x} \rangle$ corresponds to both $\langle \mathbf{a}, \mathbf{x} \rangle$ and $\langle \mathbf{b}, \mathbf{x} \rangle$, thus $\mathbf{x}$ is in $\psi \cap \phi$. □

The common generator g obtained in Proposition 2.28 can be parameterised as follows.

Proposition 2.30. *Let ψ, ϕ be a critical surface pair with equations*

$$\begin{aligned}(U\mathbf{x} \times \mathbf{x}).\mathbf{b} + (\mathbf{x} \times U\mathbf{a}).\mathbf{b} &= 0\\ (U^\top\mathbf{x} \times \mathbf{x}).\mathbf{a} + (\mathbf{x} \times U^\top\mathbf{b}).\mathbf{a} &= 0\end{aligned}$$

where U is an orthogonal matrix. The line g contained in $\psi \cap \phi$ is given in parameterised form by

$$t \mapsto t(\mathbf{a} \times U^\top\mathbf{b}) \times (U\mathbf{a} \times \mathbf{b}) \tag{2.131}$$

Proof. Let k be the line defined by (2.131). The vector $\mathbf{a} \times U^\top\mathbf{b}$ is normal to the tangent plane of ϕ at $\mathbf{o}$ and $\mathbf{b} \times U\mathbf{a}$ is normal to the tangent plane of ψ at $\mathbf{o}$. Thus k is the intersection of the tangent planes of ψ and ϕ at $\mathbf{o}$. It follows from Proposition 2.28 that the tangent planes to ψ and ϕ at $\mathbf{o}$ intersect in the common generator g. The result $g = k$ then follows. □

In proving Proposition 2.30 the result of Proposition 2.28 is assumed, namely that $\psi \cap \phi$ contains a line passing through $\mathbf{o}$. This assumption can be dispensed with at the cost of some algebra. To prove directly that the line (2.131) is contained in ψ, it suffices to show that for all values of t the point

$$\mathbf{x} = t(\mathbf{a} \times U^\top\mathbf{b}) \times (U\mathbf{a} \times \mathbf{b})$$

satisfies the two equations

$$\begin{aligned}(U\mathbf{a} \times \mathbf{x}).\mathbf{b} &= 0\\ (U\mathbf{x} \times \mathbf{x}).\mathbf{b} &= 0\end{aligned}$$

and similarly, to show that $\mathbf{x}$ is contained in ϕ it suffices to show that $\mathbf{x}$ satisfies the two equations

$$\begin{aligned}(U^\top\mathbf{x} \times \mathbf{x}).\mathbf{a} &= 0\\ (U^\top\mathbf{b} \times \mathbf{x}).\mathbf{a} &= 0\end{aligned}$$

The details are omitted.

2.4 Reconstruction up to a Collineation

In reconstruction up to a collineation it is assumed that the two images are taken by different cameras with unknown calibrations. It is known only that the mappings from space to each image are linear. The epipolar transformation can still be obtained from pairs of corresponding points because it depends only on the linear properties of the camera projection.

Reconstruction up to a collineation is formulated in terms of the epipoles. Let $\mathbf{q}_i \leftrightarrow \mathbf{q}'_i$ for $1 \leq i \leq n$ be a set of image correspondences. Then reconstruction up to a collineation is possible if and only if there exist points $\mathbf{p}$ and $\mathbf{p}'$ in the first and second images respectively such that

$$\langle \mathbf{p}, \mathbf{q}_i \rangle \,\overline{\wedge}\, \langle \mathbf{p}', \mathbf{q}_i' \rangle \qquad (1 \le i \le n) \tag{2.132}$$

The homography (2.132) is not required to preserve the angles between lines. Hesse (1863) showed that if $n = 6$ then the points $\mathbf{p}$ such that (2.132) holds lie on a cubic plane curve in the first image. The points $\mathbf{p}'$ lie on a cubic plane curve in the second image.

The equations underlying reconstruction up to a collineation are simpler than those underlying reconstruction with known camera calibration, but more pairs of corresponding points are required and the amount of unavoidable ambiguity is greater. These aspects of the reconstruction are discussed in Sect. 2.4.1. The ambiguous case is discussed in Sect. 2.4.2 and the associated critical surfaces are described in Sect. 2.4.3.

2.4.1 Reconstruction Based on the Epipolar Transformation

It is shown first that (2.132) is an acceptable formulation of the reconstruction problem, in that if the epipoles $\mathbf{p}$, $\mathbf{p}'$ are given such that (2.132) holds, then points $\mathbf{x}_i$ can be found in space that project to $\mathbf{q}_i$ in the first image and to $\mathbf{q}_i'$ in the second image. To do this let the origin $\mathbf{o}$ of $\mathbb{R}^3$ be the optical centre of the camera before the displacement and let $\mathbf{a}$ be any point of $\mathbb{R}^3$ distinct from $\mathbf{o}$. The point $\mathbf{a}$ is the optical centre of the camera after the displacement. The projection lines for each image are embedded linearly as stars of lines in $\mathbb{R}^3$. The star of lines for the first image has centre $\mathbf{o}$ and the star of lines for the second image has centre $\mathbf{a}$. The epipolar transformation induces a self-homography of the pencil of planes with axis $\langle \mathbf{o}, \mathbf{a} \rangle$. Because the calibration of each camera is unknown the embedding of the sets of projection lines can be freely adjusted to ensure that the homography induced by the epipolar transformation is the identity. Once the embedding is carried out, each pair of corresponding image points $\mathbf{q}_i$, $\mathbf{q}_i'$ yields lines $\langle \mathbf{o}, \mathbf{q}_i \rangle$, $\langle \mathbf{a}, \mathbf{q}_i' \rangle$ which are both contained in the same plane Π_i in $\mathbb{P}^3$. The two lines thus meet at a point $\mathbf{x}_i$. In general, the intersection $\mathbf{x}_i$ is a point of $\mathbb{R}^3$, but if necessary $\mathbb{R}^3$ can be extended to $\mathbb{P}^3$ to cover the case in which $\langle \mathbf{o}, \mathbf{q}_i \rangle$ and $\langle \mathbf{a}, \mathbf{q}_i' \rangle$ are parallel. The point $\mathbf{x}_i$ projects to $\mathbf{q}_i$ in the first image and to $\mathbf{q}_i'$ in the second image, as required.

The next task is to describe the unavoidable ambiguity associated with reconstruction up to a collineation. Here unavoidable ambiguity means the range of different reconstructions that are possible for a general set of image correspondences when at least one reconstruction is known to exist. It is assumed that the epipoles are known, since they are uniquely determined by eight or more of the image correspondences. To describe the unavoidable ambiguities a single reconstruction is chosen and the reconstructions related to it by a collineation are described.

Theorem 2.31. *Let* $\mathbf{q}_i \leftrightarrow \mathbf{q}_i'$, $1 \le i \le n$, *be a general set of* $n \ge 3$ *image correspondences and let* $\mathbf{p}$, $\mathbf{p}'$ *be epipoles such that (2.132) holds. Let the optical centre of the camera before the displacement be at* $\mathbf{o}$ *and let the optical centre after the displacement be at* $\mathbf{a}$. *Then there is a six parameter family of possible reconstructions with the same epipoles and the same optical centres.*

Proof. It follows from the construction described at the beginning of this subsection that there is at least one reconstruction yielding points $\mathbf{x}_i$, $1 \le i \le n$, in $\mathbb{P}^3$ such that $\mathbf{x}_i$ projects to $\mathbf{q}_i$ in the first image and to $\mathbf{q}_i'$ in the second image. Let ω be any collineation of $\mathbb{P}^3$ such that $\omega(\mathbf{o}) = \mathbf{o}$, $\omega(\mathbf{a}) = \mathbf{a}$, and such that if Π is any plane containing $\langle \mathbf{o}, \mathbf{a} \rangle$ then $\omega(\Pi) = \Pi$. The points $\omega(\mathbf{x}_i)$ are a second reconstruction compatible with the $\mathbf{q}_i \leftrightarrow \mathbf{q}_i'$. In effect, ω changes the calibration of each camera. The resulting reconstruction is acceptable because the camera calibration is unknown prior to reconstruction. It is shown that the collineations ω form a linear space of dimension six.

Let coordinates be chosen such that $\mathbf{o} = (0,0,0,1)^\top$ and $\mathbf{a} = (1,0,0,1)^\top$, and let $A = (a_{ij})$ be the 4×4 matrix describing ω. On applying the conditions $\omega(\mathbf{o}) = \mathbf{o}$, $\omega(\mathbf{a}) = \mathbf{a}$, it follows that

$$A = \begin{pmatrix} a_{41} + a_{44} & a_{12} & a_{13} & 0 \\ 0 & a_{22} & a_{23} & 0 \\ 0 & a_{32} & a_{33} & 0 \\ a_{41} & a_{42} & a_{43} & a_{44} \end{pmatrix} \tag{2.133}$$

The pencil of planes with axis $\langle \mathbf{o}, \mathbf{a} \rangle$ is parameterised by points $\mathbf{x}$ of the form $\mathbf{x} = (0, x_2, x_3, 0)^\top$. Let Π_x be the plane defined by $\Pi_x = \langle \mathbf{o}, \mathbf{a}, \mathbf{x} \rangle$. Then $\omega(\Pi_x) = \Pi_x$ if and only if $\omega(\mathbf{x})$ is a point of Π_x. It follows that $\omega(\Pi_x) = \Pi_x$ if and only if the points at which the lines $\langle \mathbf{o}, \mathbf{x} \rangle$, $\langle \mathbf{o}, \mathbf{a} \rangle$ and $\langle \mathbf{o}, \omega(\mathbf{x}) \rangle$ meet the plane at infinity are collinear. Thus $\omega(\Pi_x) = \Pi_x$ if and only if

$$-x_3(a_{22}x_2 + a_{23}x_3) + x_2(a_{32}x_2 + a_{33}x_3) = 0 \tag{2.134}$$

It is required that (2.134) hold for all x_1, x_2. It follows that $a_{23} = a_{32} = 0$, and $a_{22} = a_{33}$. Thus (2.133) yields

$$A = \begin{pmatrix} a_{41} + a_{44} & a_{12} & a_{13} & 0 \\ 0 & a_{22} & 0 & 0 \\ 0 & 0 & a_{22} & 0 \\ a_{41} & a_{42} & a_{43} & a_{44} \end{pmatrix}$$

The matrix A is determined up to a scale factor by the following point of $\mathbb{P}^6$,

$$(a_{12}, a_{13}, a_{22}, a_{41}, a_{42}, a_{43}, a_{44})^\top$$ □

The ambiguity described in Theorem 2.31 does not apply if the images are taken by a single camera with an unknown calibration. Reconstruction can be achieved up to a single unknown scale factor using the epipolar transformations associated with three different displacements of a single camera (Maybank & Faugeras 1991).

In general, the six parameter family described in the proof of Theorem 2.31 includes all possible reconstructions based on the optical centres $\mathbf{o}$, $\mathbf{a}$. To see this let $\mathbf{x}_i$ for $1 \le i \le n$, and $\mathbf{y}_i$, $1 \le i \le n$ be two sets of reconstructed points that are compatible with the same set of image correspondences $\mathbf{q}_i \leftrightarrow \mathbf{q}_i'$ for $1 \le i \le n$. It is assumed that $n \ge 8$, and that the image points $\mathbf{q}_i$, $\mathbf{q}_i'$ are in general position subject to the condition that there is at least one reconstruction. It follows from (2.132) and the remarks following (2.132) that both reconstructions have the same epipoles. There exists a unique collineation ω of $\mathbb{P}^3$ such that $\omega(\mathbf{o}) = \mathbf{o}$, $\omega(\mathbf{a}) = \mathbf{a}$ and such that $\omega(\mathbf{x}_i) = \omega(\mathbf{y}_i)$ for $i = 1, 2, 3$. It is thus supposed without loss of generality that $\mathbf{x}_i = \mathbf{y}_i$ for $i = 1, 2, 3$. Let ω_x, ω_y be the collineations from the star of lines with centre $\mathbf{o}$ to the image plane defined respectively by the reconstructions $\mathbf{x}_i$ for $1 \le i \le n$, and $\mathbf{y}_i$ for $1 \le i \le n$. The collineations ω_x and ω_y take the same values on the four lines $\langle \mathbf{o}, \mathbf{a} \rangle$, $\langle \mathbf{o}, \mathbf{x}_i \rangle$, $i = 1, 2, 3$. It follows that $\omega_x = \omega_y$. Similarly, the collineations ω_x', ω_y' from the star of lines with centre $\mathbf{a}$ to the second image are identical. It follows that $\mathbf{x}_i = \mathbf{y}_i$ for $1 \le i \le n$, because

$$\begin{aligned} \mathbf{x}_i &= \omega_x(\mathbf{q}_i) \cap \omega_x'(\mathbf{q}_i') \\ &= \omega_y(\mathbf{q}_i) \cap \omega_y'(\mathbf{q}_i') \\ &= \mathbf{y}_i \end{aligned}$$

The rank two matrices play a role in reconstruction up to a collineation analogous to the role played by the essential matrices in reconstruction with known camera calibration. The details are as follows.

Theorem 2.32. *Let $\mathbf{q}_i \leftrightarrow \mathbf{q}_i'$, $1 \le i \le n$, be a set of image correspondences. Then the homography (2.132) holds for two points $\mathbf{p}$, $\mathbf{p}'$ if and only if there exists a rank two matrix E such that*

$$\mathbf{q}_i'^{\top} E \mathbf{q}_i = 0 \qquad (1 \le i \le n) \qquad (2.135)$$

Proof. Let $\mathbf{p}$, $\mathbf{p}'$ be points such that (2.132) holds, and let A be a collineation from the first image to the second image such that $A\mathbf{p} = \mathbf{p}'$, and $A\mathbf{q}_i = \mathbf{q}_i'$ for $i = 1, 2, 3$. It follows from (2.132) and the definition of A that

$$A\langle \mathbf{p}, \mathbf{q}_i \rangle = \langle \mathbf{p}', \mathbf{q}_i' \rangle \qquad (1 \le i \le 3) \qquad (2.136)$$

A homography is uniquely determined by three correspondences, thus the homography induced by A between the epipolar lines agrees with the epipolar transformation. Equation (2.136) thus holds for $1 \le i \le n$. Let $E = T_{p'}A$. Let r_i, s_i for $1 \le i \le n$ be scalars such that $A\mathbf{q}_i = r_i\mathbf{p}' + s_i\mathbf{q}_i'$. It follows that

$$\begin{aligned} E\mathbf{q}_i &= T_{p'}A\mathbf{q}_i \\ &= T_{p'}(r_i\mathbf{p}' + s_i\mathbf{q}_i') \\ &= s_i\mathbf{q}_i' \times \mathbf{p}' \qquad (1 \le i \le n) \qquad (2.137) \end{aligned}$$

Equation (2.135) follows from (2.137). The matrix E has rank two because $T_{p'}$ has rank two and A is invertible.

Conversely, let E be a matrix of rank two such that (2.135) holds. Let $\mathbf{p}$, $\mathbf{p}'$ be points such that $E\mathbf{p} = 0$, $\mathbf{p}'^{\top} E = 0$. Let A be an invertible matrix such that $E = T_{p'}A$. Then A defines a collineation from the first image to the second image. It follows from the definition of A that $A\mathbf{p} = \mathbf{p}'$. Let $r_i\mathbf{p} + s_i\mathbf{q}_i$ be any point on the line $\langle \mathbf{p}, \mathbf{q}_i \rangle$. It follows that

$$\begin{aligned}(\mathbf{p}' \times \mathbf{q}_i')^{\top} A(r_i\mathbf{p} + s_i\mathbf{q}_i) &= (\mathbf{p}' \times \mathbf{q}_i').(r_i\mathbf{p}' + s_iA\mathbf{q}_i)\\ &= (\mathbf{p}' \times \mathbf{q}_i')^{\top} A\mathbf{q}_i\\ &= (A\mathbf{q}_i \times \mathbf{p}').\mathbf{q}_i'\\ &= (T_{p'}A\mathbf{q}_i).\mathbf{q}_i'\\ &= \mathbf{q}_i'^{\top} E\mathbf{q}_i\\ &= 0\end{aligned}$$

The image of the line $\langle \mathbf{p}, \mathbf{q}_i \rangle$ under A is thus the line $\langle \mathbf{p}', \mathbf{q}_i' \rangle$. It follows that $\langle \mathbf{p}, \mathbf{q}_i \rangle \barwedge \langle \mathbf{p}', \mathbf{q}_i' \rangle$. □

2.4.2 Ambiguity

It is shown in Theorem 2.30 that each reconstruction up to a collineation is a member of a six parameter family of reconstructions, all with the same epipolar transformation. If the image correspondences are sufficiently general then no further ambiguity is possible. However, certain special sets of image correspondences yield reconstructions with different epipoles. The term 'ambiguity' is reserved for this case only. Formally, a set of image correspondences $\mathbf{q}_i \leftrightarrow \mathbf{q}_i'$ for $1 \le i \le n$, is ambiguous for reconstruction up to a collineation if and only if there exist pairs of points $\mathbf{p}_1$, $\mathbf{p}_1'$ and $\mathbf{p}_2$, $\mathbf{p}_2'$ such that $\mathbf{p}_1 \ne \mathbf{p}_2$ or $\mathbf{p}_1 \ne \mathbf{p}_2'$, and such that

$$\begin{aligned}\langle \mathbf{p}_1, \mathbf{q}_i \rangle &\barwedge \langle \mathbf{p}_1', \mathbf{q}_i' \rangle \qquad (1 \le i \le n)\\ \langle \mathbf{p}_2, \mathbf{q}_i \rangle &\barwedge \langle \mathbf{p}_2', \mathbf{q}_i' \rangle \qquad (1 \le i \le n)\end{aligned}$$

The case in which the image correspondences $\mathbf{q} \leftrightarrow \mathbf{q}'$ define a collineation from the first image to the second image is degenerate. A general point from the first image and a general point from the second image can be chosen as epipoles. A set of image correspondences $\mathbf{q}_i \leftrightarrow \mathbf{q}_i'$ for $1 \le i \le n$, such that $\mathbf{q}_i' = A\mathbf{q}_i$, $1 \le i \le n$, for some matrix A is said to be degenerate (for reconstruction up to a collineation). If no such A exists then the image correspondences are said to be non-degenerate. In the next theorem a necessary and sufficient condition for ambiguity is obtained for the non-degenerate case.

Theorem 2.33. *A non-degenerate set of image correspondences* $\mathbf{q}_i \leftrightarrow \mathbf{q}_i'$, $1 \le i \le n$, *is compatible with two distinct pairs of epipoles if and only if the mapping*

$\mathbf{q}_i \mapsto \mathbf{q}_i'$ between the two image planes is part of a quadratic transformation. The fundamental points of the quadratic transformation are the possible epipoles.

Proof. Let the image correspondences be compatible with two distinct epipoles $\mathbf{p}_1$, $\mathbf{p}_2$ in the first image plane. It follows from Theorem 2.32 that there exist rank two matrices E_1, E_2 such that $E_1\mathbf{p}_1 = E_2\mathbf{p}_2 = 0$ and such that

$$\mathbf{q}_i'^{\top} E_1 \mathbf{q}_i = \mathbf{q}_i'^{\top} E_2 \mathbf{q}_i = 0 \qquad (1 \leq i \leq n)$$

and such that $E_1\mathbf{p}_1 = 0$, $E_2\mathbf{p}_2 = 0$. The epipoles $\mathbf{p}_1$, $\mathbf{p}_2$ are distinct, thus E_1 and E_2 are linearly independent. It follows that for a general pair of corresponding points $\mathbf{q}_i \leftrightarrow \mathbf{q}_i'$

$$\mathbf{q}_i' = E_1\mathbf{q}_i \times E_2\mathbf{q}_i \tag{2.138}$$

Let $\mathbf{q} \mapsto \Phi(\mathbf{q})$ be the map defined by

$$\Phi(\mathbf{q}) = E_1\mathbf{q} \times E_2\mathbf{q} \tag{2.139}$$

It follows from (2.138) and (2.139) that $\mathbf{q}_i' = \Phi(\mathbf{q}_i)$ for $1 \leq i \leq n$. The transformation Φ is quadratic in the components of $\mathbf{q}$. The image correspondents are by hypothesis non-degenerate, thus Φ does not reduce to a collineation. The transformation Φ is invertible because of the symmetry between the two images. It follows that Φ is a quadratic transformation.

Conversely, let the correspondences $\mathbf{q}_i \mapsto \mathbf{q}_i'$ form part of a quadratic transformation $\mathbf{q} \mapsto \Phi(\mathbf{q})$. Let $\mathbf{p}_1$, $\mathbf{p}_2$, $\mathbf{p}_3$ be the fundamental points of Φ and let $\mathbf{p}_1'$, $\mathbf{p}_2'$, $\mathbf{p}_3'$ be the fundamental points of Φ^{-1}. Let coordinates be chosen in the two images such that

$$\Phi(\mathbf{x}) = (y\,z, x\,z, x\,y)^{\top}$$

A general line l through the point $(1,0,0)^{\top}$ has the parameterised form $t \mapsto (1, t, t\,a)^{\top}$, where a is a coordinate specifying l. The image of l under Φ is parameterised by

$$t \mapsto \Phi(l) = (t^2 a, t\,a, t)^{\top} = (t\,a, a, 1)^{\top}$$

It follows that $\Phi(l)$ is a line through the point $(1,0,0)^{\top}$ in the range of Φ. The correspondence $l \leftrightarrow \Phi(l)$ is linear. It therefore defines a homography $l \barwedge \Phi(l)$. It follows that the points $(1,0,0)^{\top}$ in the first image and $(1,0,0)^{\top}$ in the second image are epipoles, and that Φ induces the associated epipolar transformation. A similar argument shows that each of the two remaining fundamental points of Φ is a possible epipole. □

Corollary. An ambiguous set of image correspondences $\mathbf{q} \leftrightarrow \mathbf{q}' = \Phi(\mathbf{q})$ is compatible with three sets of epipoles $\{\mathbf{p}_1, \mathbf{p}_1'\}$, $\{\mathbf{p}_2, \mathbf{p}_2'\}$, $\{\mathbf{p}_3, \mathbf{p}_3'\}$, where $\mathbf{p}_1$, $\mathbf{p}_2$, $\mathbf{p}_3$ are the fundamental points of Φ and $\mathbf{p}_1'$, $\mathbf{p}_2'$, $\mathbf{p}_3'$ are the fundamental points of Φ^{-1}. If the image correspondences are dense then no further pairs of epipoles are possible.

A direct proof that the rational map Φ defined by (2.139) is invertible is given. Let ϕ_i for $i = 1, 2, 3$ be homogeneous quadratic polynomials in $\mathbf{q}$ such that

$$\Phi(\mathbf{q}) = (\phi_1(\mathbf{q}), \phi_2(\mathbf{q}), \phi_3(\mathbf{q}))^\top$$

There are three points $\mathbf{q}$ at which $\Phi(\mathbf{q})$ is not defined, namely $\mathbf{p}_1$, $\mathbf{p}_2$ and a third point $\mathbf{p}_3$, obtained as a solution to the generalised eigenvalue equation

$$(E_1 - \lambda E_2)\mathbf{q} = 0$$

It follows that the conics ϕ_i intersect at $\mathbf{p}_1$, $\mathbf{p}_2$, $\mathbf{p}_3$. The ϕ_i are not linearly dependent and the points $\mathbf{p}_i$ are not collinear because Φ does not define a collineation. It follows that Φ is a quadratic transformation.

2.4.3 Critical Surfaces

There are many similarities between the critical surfaces arising in reconstruction up to a collineation and the critical surfaces arising in reconstruction with known camera calibration. For example, a pair of critical surfaces intersects in a space curve which splits into a cubic space curve and a straight line. The term 'twisted cubic' is used for a cubic space curve.

Theorem 2.34. *Let ψ, ϕ be a critical surface pair arising in an ambiguous case of reconstruction up to a collineation. Let $\mathbf{o}$ be the optical centre of the first camera and let $\mathbf{a}$, $\mathbf{b}$ be two possible positions for the optical centre of the second camera. Then ψ, ϕ are hyperboloids of one sheet. The intersection $\psi \cap \phi$ is a space curve of degree four which splits into a line g through $\mathbf{o}$ and a twisted cubic which contains $\mathbf{a}$ and $\mathbf{b}$.*

Proof. The notation in Fig. 2.5 is used. The collineation ω from the lines through $\mathbf{a}$ to the lines through $\mathbf{b}$ is no longer required to be orthogonal. Cartesian coordinates are chosen with the origin at $\mathbf{o}$. Because $\mathbf{a}$, $\mathbf{b}$ are both possible optical centres for the second camera there is a collineation ω from the lines through $\mathbf{a}$ to the lines through $\mathbf{b}$ such that if q is an arbitrary line through $\mathbf{o}$ and r is the line through $\mathbf{a}$ corresponding to q, then $s = \omega(r)$ is the line through $\mathbf{b}$ corresponding to q. The line q is the unique common transversal of r and $\omega(r)$ which passes through $\mathbf{o}$. As q varies, the point $\mathbf{x} = q \cap r$ traces out ψ and the point $\mu\mathbf{x} = q \cap s$ traces out ϕ. Let $\mathbf{r}$ be a unit vector through $\mathbf{a}$ in the direction of r and let $\mathbf{x}$ be the point of ψ on r distinct from $\mathbf{a}$. Then there exist scalars λ_1, λ_2, μ such that

$$\begin{aligned} \mathbf{x} &= \mathbf{a} + \lambda_1 \mathbf{r} \\ \mu\mathbf{x} &= \mathbf{b} + \lambda_2 \omega(\mathbf{r}) \end{aligned} \tag{2.140}$$

The elimination of $\mathbf{r}$ from (2.140) yields

$$\mu\mathbf{x} = \mathbf{b} + \lambda_2 \lambda_1^{-1} \omega(\mathbf{x} - \mathbf{a}) \tag{2.141}$$

It follows from (2.141) that the equation for ψ is

$$(\mathbf{x} \times \mathbf{b}).\omega(\mathbf{x} - \mathbf{a}) = 0 \tag{2.142}$$

The surface ϕ is traced out by the point $s \cap q$ as q varies through the star of lines with centre $\mathbf{o}$. A calculation similar to that used to obtain (2.142) yields the following equation for ϕ:

$$(\mathbf{x} \times \mathbf{a}).\omega^{-1}(\mathbf{x} - \mathbf{b}) = 0 \tag{2.143}$$

The surfaces ψ and ϕ are each of degree two. They are hyperboloids of one sheet because each surface contains a real line and each surface intersects the plane at infinity in a non-singular conic with an infinite number of of real points. The surface ψ contains the line $t \mapsto t\mathbf{b}$ and ϕ contains the line $t \mapsto t\mathbf{a}$. It follows from (2.142) and (2.143) that ψ and ϕ each contain $\mathbf{o}$, $\mathbf{a}$ and $\mathbf{b}$. The intersection $\psi \cap \phi$ is a space curve of degree four because ψ and ϕ are each of degree two.

Let Π be the plane containing $\langle \mathbf{o}, \mathbf{a} \rangle$ such that $\omega(\Pi)$ contains $\mathbf{o}$. Let g be the line through $\mathbf{o}$ defined by $g = \Pi \cap \omega(\Pi)$. Let $\mathbf{x}$ be any point of g. Then $\mathbf{x} - \mathbf{a}$ defines a line through $\mathbf{a}$ contained in Π. It follows that $\omega(\mathbf{x} - \mathbf{a})$ is a line through $\mathbf{b}$ contained in $\omega(\Pi)$. The points $\mathbf{x}$, $\mathbf{a}$ are contained in Π, thus

$$(\mathbf{x} \times \mathbf{a}).\omega(\mathbf{x} - \mathbf{a}) = 0$$

It follows that g is contained in ψ. A similar argument establishes that g is contained in ϕ. Thus $\psi \cap \phi$ splits into g and a space curve of degree three. □

Let c be the space curve of degree three contained in $\psi \cap \phi$. The curve c is a twisted cubic but it is not in general a horopter curve. A point $\mathbf{x}$ is on c if and only if

$$\omega\langle \mathbf{a}, \mathbf{x} \rangle = \langle \mathbf{b}, \mathbf{x} \rangle \tag{2.144}$$

The proof of (2.144) is similar to the proof in Theorem 2.29 of the analogous result for reconstruction with known camera calibration.

The critical surface provides a geometrical interpretation of Theorem 2.33. If the mapping $\mathbf{q}_i \mapsto \mathbf{q}'_i$, $1 \leq i \leq n$, between the two image planes is part of a quadratic transformation Φ, then Φ is defined by the critical surface as follows. Let the point $\mathbf{q}$ in the first image be the projection of a line l through $\mathbf{o}$. Then l intersects ψ at a unique point $\mathbf{p}$ distinct from $\mathbf{o}$. The line $\langle \mathbf{a}, \mathbf{p} \rangle$ yields a point $\mathbf{q}'$ in the second image such that $\mathbf{q}' = \Phi(\mathbf{q})$. Let g_1, g_2 be the two generators of ψ through $\mathbf{o}$. Then the three possible epipoles in the first image are the projections of g_1, g_2 and $\langle \mathbf{o}, \mathbf{a} \rangle$. If the projection of $\langle \mathbf{o}, \mathbf{a} \rangle$ is selected as the epipole then the usual epipolar transformation associated with the camera displacement from $\mathbf{o}$ to $\mathbf{a}$ is obtained. If the projection of g_1 is chosen as the epipole then an epipolar transformation is constructed as follows. Let h_1 be the generator of ψ through $\mathbf{a}$ skew to g_1. Let $\mathbf{p}$ be a general point of ψ. There is a unique generator k of ψ through $\mathbf{p}$ that intersects both g_1 and h_1. Let $\Pi = \langle g_1, k \rangle$ and let $\Pi' = \langle h_1, k \rangle$. It is a property of the quadric surface ψ that as k varies, $\Pi \barwedge \Pi'$. The plane

Π projects to an epipolar line through the projection $\mathbf{q}$ of $\mathbf{p}$ in the first image and the plane Π' projects to an epipolar line through the projection $\mathbf{q}'$ of $\mathbf{p}$ in the second image. Let the point $\mathbf{g}_1$ be the projection of g_1 into the first image and let $\mathbf{h}_1$ be the projection of h_1 into the second image. Then it follows that $\langle \mathbf{g}_1, \mathbf{q} \rangle \barwedge \langle \mathbf{h}_1, \mathbf{q}' \rangle$.

References

Faugeras O.D. & Maybank S.J. 1990 Motion from point matches: multiplicity of solutions. *International J. Computer Vision* **4**, 225-246.

Hay J.C. 1966 Optical motions and space perception: an extension of Gibson's analysis. *Psychological Review* **73**, 550-565.

Hesse O. 1863 Die cubische Gleichung, von welcher die Lösung des Problems der Homographie von M. Chasles abhängt. *J. Reine Angew. Math.* **62**, 188-192.

Hofmann W. 1950 Das Problem der "Gefährlichen Flächen in Theorie und Praxis". *Dissertation, Fakultät für Bauwesen der Technischen Hochschule München, München, FR Germany. Published in Reihe C, No. 3 der Deutschen Geodatischen Kommission bei der Bayerischen Akademie der Wissenschaften, München 1953.*

Horn B.K.P. 1990 Relative orientation. *International J. Computer Vision* **4**, 59-78.

Krames J. 1940 Zur Ermittlung eines Objektes aus zwei Perspektiven. (Ein Beitrag zur Theorie der "gerfählichen Örter". *Monatsch. Math. Physik* **49**, 327-354.

Krames J. 1941 Über bemerkenswerte Sonderfälle des "Gefährlichen Ortes" der photogrammetrischen Hauptaufgabe. *Monatsch. Math. Physik* **50**.

Longuet-Higgins H.C. 1981 A computer algorithm for reconstructing a scene from two projections. *Nature* **293**, 133-135.

Longuet-Higgins H.C. 1986 The reconstruction of a plane surface from two perspective projections. *Proc. Royal Soc. London, Series B* **227**, 399-410.

Longuet-Higgins H.C. 1988 Multiple interpretations of a pair of images of a surface. *Proc. Royal Soc. London, Series A* **418**, 1-15.

Maybank S.J. 1990 The projective geometry of ambiguous surfaces. *Phil. Trans. Royal Soc. London, Series A* **332**, 1-47.

Mirsky L. 1961 *An Introduction to Linear Algebra.* Oxford: Clarendon Press.

Negahdaripour S. 1990 Multiple interpretations of the shape and motion of objects from two perspective images. *IEEE Trans. Pattern Analysis and Machine Intelligence* **12**, 1025-1039.

Sturm R. 1869 Das Problem der Projektivität und seine Anwendung auf die Flächen zweiten Grades. *Math. Annalen* **1**, 533-573.

Tsai R.Y. & Huang T.S. 1981 Estimating three-dimensional motion parameters of a rigid planar patch. *IEEE Trans. Acoustics, Speech, and Signal Processing* **29**, 1147-1152.

Tsai R.Y. & Huang T.S. 1984 Uniqueness and estimation of three-dimensional motion parameters of rigid objects with curved surfaces. *IEEE Trans. Pattern Analysis and Machine Intelligence* **6**, 13-27.

Wunderlich W. 1942 Zur Eindeutigkeitsfrage der Hauptaufgabe der Photogrammetrie. *Monatsch. Math. Physik* **50**, 151-164.

3 Critical Surfaces and Horopter Curves

In the ambiguous case of reconstruction the points giving rise to the image correspondences lie on certain surfaces of degree two known as critical surfaces. The critical surfaces compatible with the same ambiguous set of image correspondences are closely related to each other. In this chapter the geometry underlying the ambiguous case is explored in detail. If the camera calibration is known then the intersection of a critical surface pair contains a space curve of degree three known as a horopter curve. The horopter curve is of central importance, in that many of the properties of critical surfaces arise from the properties of horopter curves.

In the Euclidean framework for reconstruction the camera calibration is fixed by choosing the sphere as a projection surface. This is convenient for numerical calculations, but it is not the best choice when it comes to describing the geometry of reconstruction. In the projective geometric framework the camera calibration is described using the absolute conic. The camera calibration determines the image of the absolute conic and conversely the projected image of the absolute conic determines a unique camera calibration. The advantage of using the absolute conic is that it is not necessary to choose a special image coordinate system. The properties of the absolute conic are described in Sect. 3.1.

The critical surfaces are hyperboloids of one sheet. They belong to a particular class of quadrics known as rectangular quadrics. The rectangular quadrics are discussed in Sect. 3.2. The properties of horopter curves are described in Sect. 3.3, with particular emphasis on the rigid involutions of horopter curves. The role played by the horopter curves in reconstruction is described in Sect. 3.4. In the last section, Sect. 3.5, the ambiguous case of reconstruction up to a collineation is explored. The critical surfaces for reconstruction up to a collineation are still hyperboloids of one sheet, but they need not be rectangular. The intersection of a critical surface pair contains a space curve of degree three, but the space curve need not be a horopter curve.

3.1 The Absolute Conic

The absolute conic is a particular non-singular conic chosen once and for all in the plane at infinity, Π_∞. The equations defining the absolute conic are required to have real coefficients and the conic is required to have no real points. In the Cartesian coordinate system for $\mathbb{R}^3$ the absolute conic has a standard form defined by the equations

$$x_4 = 0 \qquad x_1^2 + x_2^2 + x_3^2 = 0 \tag{3.1}$$

The conic defined by (3.1) is contained in every sphere and conversely any non-singular quadric surface containing the conic (3.1) is a sphere.

If the camera calibration is known then the image w of the absolute conic Ω can be calculated. The conic w is defined in the image plane by an equation with real coefficients, but it does not have any real points. Although there is no absolute conic 'out there' in space such that a photograph of it can be taken, the task of calculating the image of the absolute conic is well defined mathematically. The crucial property of the absolute conic is that the image w determines the camera calibration. The problem of describing the camera calibration is thus equivalent to the problem of describing the image of the absolute conic.

The absolute conic is closely related to the Euclidean norm and to rigid displacements of $\mathbb{R}^3$. Each rigid displacement ρ of $\mathbb{R}^3$ extends to a unique collineation ρ' of $\mathbb{P}^3$. It is shown below that ρ' leaves Ω invariant, i.e. $\rho'(\Omega) = \Omega$. The converse is also proved: each collineation ρ' of $\mathbb{P}^3$ that leaves Ω invariant restricts to a linear transformation ρ of $\mathbb{R}^3$ that is the composition of a rigid displacement and a uniform change of scale. Collineations of $\mathbb{P}^3$ that leave Ω invariant are known as Euclidean transformations. As the name implies, they are transformations of space that preserve Euclidean properties such as angles and ratios of lengths. A transformation of space that preserves Euclidean length is said to be rigid. The terms rigid displacement or rigid motion are also used.

The Euclidean transformations form a group in which the group operation is the composition of maps. It is recalled that a set G equipped with a binary operation $\circ$ is a group if and only if $\circ$ has the following properties.

- **Closure**. The product $g_1 \circ g_2$ of any two members g_1, g_2 of G is in G.
- **Identity**. There is an identity element e in G such that $e \circ g = g \circ e = g$ for all g in G.
- **Inverse**. If g is any member of G then there exists a unique inverse g^{-1} of G such that $g^{-1} \circ g = g \circ g^{-1} = e$.
- **Associativity**. The product is associative, $g_1 \circ (g_2 \circ g_3) = (g_1 \circ g_2) \circ g_3$, for all g_1, g_2, g_3 in G.

The operation $\circ$ is called the group law. The product of two members g_1, g_2 of G is often written as $g_1 g_2$ rather that $g_1 \circ g_2$.

It is straightforward to check that the Euclidean transformations form a group. If ω_1, ω_2 are Euclidean transformations, then $\omega_1 \omega_2$ is a Euclidean transformation because

$$\omega_1\omega_2(\Omega) = \omega_1(\omega_2(\Omega)) = \omega_1(\Omega) = \Omega$$

The identity collineation is the identity of the group. If ω is a Euclidean transformation then the usual inverse collineation ω^{-1} is a Euclidean transformation because

$$\omega^{-1}(\Omega) = \omega^{-1}(\omega(\Omega)) = \Omega$$

Finally, associativity holds because the composition of maps is always associative.

The connection between Ω and the rigid displacements is seen most clearly in the Cartesian coordinate system, in which Ω is defined by (3.1). The point $\mathbf{o} = (0,0,0,1)^\top$ is chosen as the origin. The plane at infinity is the plane $x_4 = 0$ containing the absolute conic. Each rigid displacement ρ of $\mathbb{R}^3$ has the form

$$(x_1, x_2, x_3, 1) \mapsto ((x_1 + a_1, x_2 + a_2, x_3 + a_3)R^\top, 1)$$

where R is an orthogonal matrix and $\mathbf{a}$ is a translation vector. The extension of ρ to $\mathbb{P}^3$ is given by

$$(x_1, x_2, x_3, x_4) \mapsto ([x_1 + x_4a_1, x_2 + x_4a_2, x_3 + x_4a_3]R^\top, x_4) \tag{3.2}$$

On setting $x_4 = 0$ in (3.2) the restriction of ρ to Π_∞ is obtained,

$$(x_1, x_2, x_3, 0) \mapsto ([x_1, x_2, x_3]R^\top, 0) \tag{3.3}$$

It follows from (3.1) and (3.3) that $\rho(\Omega) = \Omega$, thus ρ extends to a Euclidean transformation of $\mathbb{P}^3$. The next theorem elaborates this result.

Theorem 3.1. *Let a Cartesian coordinate system be chosen in $\mathbb{P}^3$. Then a collineation ω of $\mathbb{P}^3$ is a Euclidean transformation if and only if ω is represented by the matrix*

$$\omega = \begin{pmatrix} r_{11} & r_{12} & r_{13} & a_1 \\ r_{21} & r_{22} & r_{23} & a_2 \\ r_{31} & r_{32} & r_{33} & a_3 \\ 0 & 0 & 0 & a_4 \end{pmatrix} \tag{3.4}$$

where $a_4 \neq 0$, and where the 3×3 matrix R formed by the r_{ij} satisfies $R^\top R = \mu I$ for some non-zero scalar μ.

Proof. It follows from the remarks preceding the statement of the theorem that if ω is given by (3.4) then $\omega(\Omega) = \Omega$. The collineation ω is thus a Euclidean transformation. Conversely, let ω be a Euclidean transformation of $\mathbb{P}^3$ and let the matrix representation of ω be

$$\omega = \begin{pmatrix} r_{11} & r_{12} & r_{13} & a_1 \\ r_{21} & r_{22} & r_{23} & a_2 \\ r_{31} & r_{32} & r_{33} & a_3 \\ r_{41} & r_{42} & r_{43} & a_4 \end{pmatrix} \tag{3.5}$$

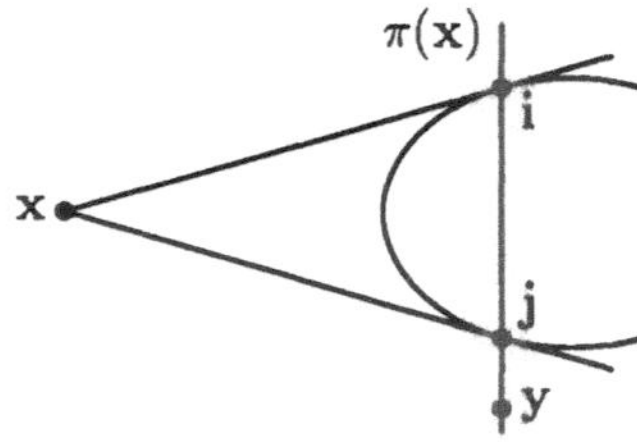

Fig. 3.1. The polar line of a point with respect to a conic

Let $\mathbf{r}_1$, $\mathbf{r}_2$, $\mathbf{r}_3$ be the vectors defined by $\mathbf{r}_1 = (r_{11}, r_{12}, r_{13})^\top$, $\mathbf{r}_2 = (r_{21}, r_{22}, r_{23})^\top$, $\mathbf{r}_3 = (r_{31}, r_{32}, r_{33})^\top$. The absolute conic Ω contains the points $\mathbf{p} = (1, i, 0, 0)^\top$, $\mathbf{q} = (1, -i, 0, 0)^\top$. By hypothesis, $\omega(\Omega) = \Omega$, thus

$$\begin{aligned} \omega(\mathbf{p}).\omega(\mathbf{p}) &= \omega(\mathbf{q}).\omega(\mathbf{q}) = 0 \\ \omega(\mathbf{p})_4 &= \omega(\mathbf{q})_4 = 0 \end{aligned}$$

It follows that

$$\begin{aligned} (\mathbf{r}_1 + i\mathbf{r}_2).(\mathbf{r}_1 + i\mathbf{r}_2) &= (\mathbf{r}_1 - i\mathbf{r}_2).(\mathbf{r}_1 - i\mathbf{r}_2) = 0 \\ r_{41} + ir_{42} &= r_{41} - ir_{42} = 0 \end{aligned}$$

hence $\mathbf{r}_1.\mathbf{r}_1 = \mathbf{r}_2.\mathbf{r}_2$, and $r_{41} = r_{42} = 0$. Similarly, Ω contains the points $(0, 1, i, 0)^\top$, $(0, 1, -i, 0)^\top$, thus $\mathbf{r}_2.\mathbf{r}_2 = \mathbf{r}_3.\mathbf{r}_3$, $r_{43} = 0$. Let the scalar μ be defined by

$$\mu = \mathbf{r}_1.\mathbf{r}_1 = \mathbf{r}_2.\mathbf{r}_2 = \mathbf{r}_3.\mathbf{r}_3$$

By hypothesis, ω is a collineation of $\mathbb{P}^3$, thus the matrix of (3.5) which represents ω is non-singular. It follows that $\mu \neq 0$, $a_4 \neq 0$ as required. □

Corollary. The restriction of ω to $\mathbb{R}^3$ is a rigid displacement followed by a uniform change of scale. The application of ω to an arbitrary point $(\mathbf{x}, 1)^\top = (x_1, x_2, x_3, 1)^\top$ of $\mathbb{R}^3 \subset \mathbb{P}^3$ yields

$$\begin{pmatrix} r_{11} & r_{12} & r_{13} & a_1 \\ r_{21} & r_{22} & r_{23} & a_2 \\ r_{31} & r_{32} & r_{33} & a_3 \\ 0 & 0 & 0 & a_4 \end{pmatrix} \begin{pmatrix} x_1 \\ x_2 \\ x_3 \\ 1 \end{pmatrix} = \begin{pmatrix} \mathbf{r}_1.\mathbf{x} + a_1 \\ \mathbf{r}_2.\mathbf{x} + a_2 \\ \mathbf{r}_3.\mathbf{x} + a_3 \\ a_4 \end{pmatrix} = (a_4^{-1} R(\mathbf{x} + R^\top \mathbf{a}), 1)^\top$$

It follows that ω is the product of the rigid displacement $\{\mu^{-1/2} R, R^\top \mathbf{a}\}$ and a uniform change of scale by a factor $\sqrt{\mu}/a_4$.

The polar line of a point is a standard construction in the geometry of conics. If $\mathbf{x}$ is a point in the plane of a non-singular conic s then the polar line or polar of $\mathbf{x}$ with respect to s is the line joining the points of contact of the tangents drawn from $\mathbf{x}$ to s, as illustrated in Fig. 3.1. If $\mathbf{x}$ is on s then the polar of $\mathbf{x}$ is the tangent to s at $\mathbf{x}$. If l is any line in the plane of s then the pole of l with respect to s is the point at which the tangents to s at the two points $l \cap s$ intersect. The polar of $\mathbf{x}$ with respect to s is often denoted by $\pi(\mathbf{x})$. The correspondence $\mathbf{x} \mapsto \pi(\mathbf{x})$ is called a correlation. It is a projective linear transformation from the projective plane containing s to the dual projective plane formed by the lines of $\mathbb{P}^2$. To prove that $\mathbf{x} \mapsto \Pi(\mathbf{x})$ is linear, let $\mathbf{x}^\top M\mathbf{x} = 0$ be the equation of s, where M is a symmetric 3×3 matrix, and let $\mathbf{a}$, $\mathbf{b}$ be any two points in the plane of s. The line parameterised by $t \mapsto \mathbf{a} + t\,\mathbf{b}$ is tangent to s if and only if the quadratic equation in t

$$(\mathbf{a} + t\,\mathbf{b})^\top M(\mathbf{a} + t\,\mathbf{b}) = 0$$

has a repeated root. This yields the following condition on $\mathbf{b}$,

$$(\mathbf{a}^\top M\mathbf{b})^2 = (\mathbf{a}^\top M\mathbf{a})(\mathbf{b}^\top M\mathbf{b}) \tag{3.6}$$

Let $\mathbf{b}_1$, $\mathbf{b}_2$ be the two points at which a line through $\mathbf{a}$ touches s. The points $\mathbf{b}_1$, $\mathbf{b}_2$ are on s, thus

$$\mathbf{b}_1^\top M\mathbf{b}_1 = \mathbf{b}_2^\top M\mathbf{b}_2 = 0 \tag{3.7}$$

It follows from (3.6) and (3.7) that

$$\mathbf{a}^\top M\mathbf{b}_1 = \mathbf{a}^\top M\mathbf{b}_2 = 0 \tag{3.8}$$

The line $\langle \mathbf{b}_1, \mathbf{b}_2 \rangle$ is the polar $\pi(\mathbf{a})$. It follows from (3.8) that a point $\mathbf{x}$ is on $\pi(\mathbf{a})$ if and only if $\mathbf{a}^\top M\mathbf{x} = 0$. The coefficients of the line $\pi(\mathbf{a}) = \mathbf{a}^\top M$ depend linearly on $\mathbf{a}$ as required. The correspondence $\mathbf{a} \mapsto \mathbf{a}^\top M$ is invertible because M is a non-singular matrix.

Polarity with respect to the absolute conic is used to define a relation of orthogonality between pairs of vectors. The resulting definition appears to be different from the usual definition of orthogonality based on the scalar product of vectors but the differences are only superficial. The two definitions have the same formal properties. Orthogonality is first defined for two points $\mathbf{x}$, $\mathbf{y}$ in Π_∞. The point $\mathbf{x}$ is orthogonal to $\mathbf{y}$ if and only if $\mathbf{y}$ is on the polar line of $\mathbf{x}$ with respect to Ω, as illustrated in Fig. 3.1. Let $\mathbf{o}$ be the origin of $\mathbb{R}^3 \subset \mathbb{P}^3$. Two non-zero vectors $\mathbf{x}$, $\mathbf{y}$ in $\mathbb{P}^3$ are said to be orthogonal if and only if the points $\langle \mathbf{o}, \mathbf{x} \rangle \cap \Pi_\infty$, $\langle \mathbf{o}, \mathbf{y} \rangle \cap \Pi_\infty$ are orthogonal. If coordinates are chosen such that $\mathbf{o} = (0,0,0,1)^\top$ and if $\mathbf{x} = (x_1, x_2, x_3, x_4)^\top$ is a point distinct from $\mathbf{o}$ then the line $\langle \mathbf{o}, \mathbf{x} \rangle$ intersects Π_∞ at $(x_1, x_2, x_3, 0)^\top$.

This geometric definition of orthogonality agrees with the usual definition if a Cartesian coordinate system is chosen. If s is a conic (in Π_∞) with equation $\mathbf{x}^\top A\mathbf{x} = 0$ then a point $\mathbf{a}$ is on the the polar line $\pi(\mathbf{b})$ of a point $\mathbf{b}$ if and only if $\mathbf{b}^\top A\mathbf{a} = 0$. In the case of a Cartesian coordinate system, the matrix defining the absolute conic is the identity matrix, I. Thus $\mathbf{a}$ is on the polar line of $\mathbf{b}$ with respect to Ω if and only if $\mathbf{b}.\mathbf{a} = 0$. The polar line of $\mathbf{b}$ consists of those vectors orthogonal to $\mathbf{b}$ in the usual sense. It is easy to check, using the geometric definition of orthogonality, that $\mathbf{a}$ is orthogonal to $\mathbf{b}$ if and only if $\mathbf{b}$ is orthogonal to $\mathbf{a}$; also if $\mathbf{b}_1$, $\mathbf{b}_2$ are two distinct vectors, then there is a unique vector orthogonal to both $\mathbf{b}_1$ and $\mathbf{b}_2$, namely $\pi(\mathbf{b}_1) \cap \pi(\mathbf{b}_2)$.

In a Cartesian coordinate system a matrix is orthogonal up to a scale factor if and only if it defines a collineation of Π_∞ that leaves Ω invariant. Many of the properties of orthogonal matrices are just special cases of properties of collineations of the plane that leave a specified non-singular conic invariant. An example of the use of this geometric approach to orthogonal matrices is given in the following proposition.

Proposition 3.2. *Let ω be an orthogonal collineation of the plane at infinity such that ω^2 is not the identity, and let $\mathbf{n}$ be a fixed point of ω not contained in the absolute conic. Then the polar of $\mathbf{n}$ with respect to the absolute conic intersects the absolute conic at two fixed points of ω.*

Proof. Let $\pi(\mathbf{n})$ be the polar of $\mathbf{n}$ with respect to the absolute conic Ω, and let $\mathbf{i}_n$, $\mathbf{j}_n$ be the two points of $\pi(\mathbf{n}) \cap \Omega$. Both Ω and $\pi(\mathbf{n})$ are invariant under ω, thus either $\mathbf{i}_n$, $\mathbf{j}_n$ are fixed by ω, or $\mathbf{i}_n$, $\mathbf{j}_n$ are interchanged by ω.

It is shown that $\mathbf{i}_n$, $\mathbf{j}_n$ are not interchanged by ω. Suppose, if possible, that $\omega(\mathbf{i}_n) = \mathbf{j}_n$. The collineation ω induces a homography on the pencil of lines through $\mathbf{n}$. Each linear transformation of a projective space has at least one fixed point, thus there exists a line k through $\mathbf{n}$ which is fixed by ω. The line k is distinct from $\langle \mathbf{n}, \mathbf{i}_n \rangle$ and $\langle \mathbf{n}, \mathbf{j}_n \rangle$, thus k is not tangent to Ω. It follows that $k \cap \Omega$ contains two points. Both k and Ω are invariant under ω, thus the points of $k \cap \Omega$ are either fixed by ω or interchanged by ω. The collineation ω^2 thus has four fixed points on Ω, namely, $\mathbf{i}_n$, $\mathbf{j}_n$ and the two points of $k \cap \Omega$. The conic Ω is, by definition, non-singular, thus no three of these points are collinear. The collineation ω^2 is then the identity, because it has four fixed points, no three of which are collinear. It follows from this contradiction with the hypotheses of the theorem that $\mathbf{i}_n$, $\mathbf{j}_n$ are fixed by ω. □

Corollary 1. If ω is a non-trivial orthogonal collineation of Π_∞ with a fixed point $\mathbf{n}$ not on Ω such that ω interchanges the two points of $\pi(\mathbf{n}) \cap \Omega$ then $\omega = \omega^2$, i.e. ω is an involution.

Corollary 2. The coordinates of $\mathbf{n}$, $\mathbf{i}_n$, $\mathbf{j}_n$ yield eigenvectors of the matrix representing ω.

In Corollary 2 of Proposition 3.2 the fixed points of ω are referred to as eigenvectors of the matrix W representing ω. This is possible because in projective geometry the eigenvalues of W appear as scale factors which do not affect the eigenvector, considered as a point of projective space. In detail, it follows from Proposition 3.2 that there exist non-zero scalars λ_i, λ_j such that

$$\begin{aligned} W\mathbf{i}_n &= \lambda_i \mathbf{i}_n \\ W\mathbf{j}_n &= \lambda_j \mathbf{j}_n \end{aligned}$$

In projective geometry $\lambda_i \mathbf{i}_n$ is identified with $\mathbf{i}_n$, and $\lambda_j \mathbf{j}_n$ is identified with $\mathbf{j}_n$. Thus in $\mathbf{P}^2$ the following equations are obtained:

$$W\mathbf{i}_n = \mathbf{i}_n$$

$$W\mathbf{j}_n = \mathbf{j}_n$$

3.1.1 The Absolute Conic and Camera Calibration

In camera calibration a correspondence is established between the lines in $\mathbb{P}^3$ through the optical centre of the camera and the points of the image. The image point $\mathbf{q}$ corresponding to a line l through the optical centre is the projection of l to the image. Once the calibration is known the angle between lines l_1, l_2 corresponding to any two given image points $\mathbf{q}_1$, $\mathbf{q}_2$ can be calculated. Conversely, if the angle between the lines projecting to any two given image points $\mathbf{q}_1$, $\mathbf{q}_2$ is known then the camera calibration can be deduced. It is shown that if the image is formed by a linear projection from $\mathbb{P}^3$ then the task of establishing the camera calibration is equivalent to the task of finding the image of the absolute conic.

The link between camera calibration and the absolute conic depends on the properties of the cross ratio. The cross ratio is of fundamental importance in projective geometry. Let $\mathbf{x}_1$, $\mathbf{x}_2$, $\mathbf{x}_3$, $\mathbf{x}_4$ be an ordered set of four distinct points in $\mathbb{P}^1$. The cross ratio is a single number $\{\mathbf{x}_1, \mathbf{x}_2;\ \mathbf{x}_3, \mathbf{x}_4\}$ computed from the coordinates of the four points $\mathbf{x}_i$. If the coordinates of $\mathbf{x}_i$ are $(\theta_i, 1)^\top$ for $1 \le i \le 4$ then

$$\{\mathbf{x}_1, \mathbf{x}_2; \mathbf{x}_3, \mathbf{x}_4\} = \left(\frac{\theta_1 - \theta_3}{\theta_1 - \theta_4}\right) \Big/ \left(\frac{\theta_2 - \theta_3}{\theta_2 - \theta_4}\right)$$

The cross ratio is invariant under any linear transformation ω of $\mathbb{P}^1$,

$$\{\omega(\mathbf{x}_1), \omega(\mathbf{x}_2); \omega(\mathbf{x}_3), \omega(\mathbf{x}_4)\} = \{\mathbf{x}_1, \mathbf{x}_2; \mathbf{x}_3, \mathbf{x}_4\} \tag{3.9}$$

One way of proving (3.9) is to write ω as a product of simpler linear transformations such as

$$\begin{aligned}(x_1, x_2)^\top &\mapsto (ax_1 + bx_2, x_2)^\top \\ (x_1, x_2)^\top &\mapsto (x_2, x_1)^\top\end{aligned}$$

It can also be shown that if $\mathbf{y}_1$, $\mathbf{y}_2$, $\mathbf{y}_3$, $\mathbf{y}_4$ are four distinct points of $\mathbb{P}^1$ such that

$$\{\mathbf{x}_1, \mathbf{x}_2; \mathbf{x}_3, \mathbf{x}_4\} = \{\mathbf{y}_1, \mathbf{y}_2; \mathbf{y}_3, \mathbf{y}_4\}$$

then there exists a unique linear transformation ω such that $\omega(\mathbf{x}_i) = \mathbf{y}_i$ for $1 \le i \le 4$.

Theorem 3.3. *Let a camera take an image of $\mathbb{P}^3$ by a general linear projection with optical centre $\mathbf{o}$ and let w be the image of the absolute conic. Then w is independent of $\mathbf{o}$ and of the orientation of the camera in space. Let $\mathbf{q}_1$, $\mathbf{q}_2$ be two points in the image and let h_1, h_2 be the two lines of $\mathbb{P}^3$ through $\mathbf{o}$ which project to $\mathbf{q}_1$ and $\mathbf{q}_2$ respectively. Then $\mathbf{q}_1$, $\mathbf{q}_2$ and w together determine the angle between h_1 and h_2. Conversely, if for each pair of image points, $\mathbf{q}_1$, $\mathbf{q}_2$, the angle between the lines h_1 and h_2 projecting to $\mathbf{q}_1$ and $\mathbf{q}_2$ is known then w can be found.*

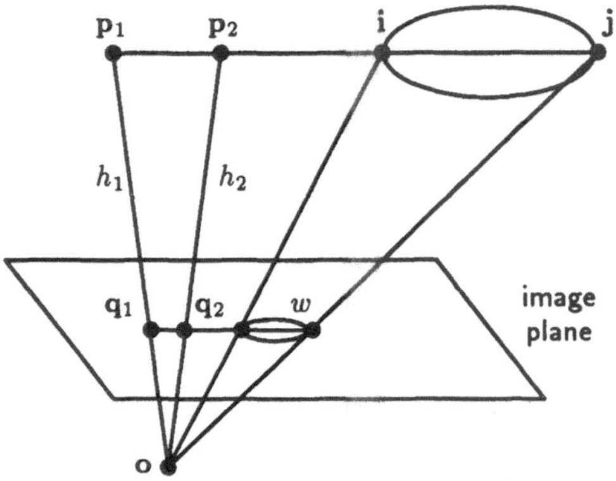

Fig. 3.2. The projection of a line $\langle \mathbf{p}_1, \mathbf{p}_2 \rangle$ contained in Π_∞

Proof. Let P be the camera projection from $\mathbb{P}^3$ to the image $\mathbb{P}^2$ when the camera is placed at some convenient position in $\mathbb{R}^3$. Let Ω be the absolute conic and let ω be a Euclidean transformation of $\mathbb{P}^3$. To prove that s is independent of the position and orientation of the camera it is sufficient to show that $s = P\omega(s)$. The result is immediate because $s = P(\Omega)$ and $\omega(\Omega) = \Omega$.

To prove the second part of the theorem, let $\mathbf{p}_1 = h_1 \cap \Pi_\infty$, $\mathbf{p}_2 = h_2 \cap \Pi_\infty$, as illustrated in Fig. 3.2. Let $\mathbf{i}$, $\mathbf{j}$ be the two points at which $\langle \mathbf{p}_1, \mathbf{p}_2 \rangle$ intersects Ω. The angle θ between h_1 and h_2 is given in terms of $\mathbf{p}_1$, $\mathbf{p}_2$, $\mathbf{i}$, $\mathbf{j}$ by the Laguerre formula,

$$\theta = \frac{1}{2i} \log(\{\mathbf{p}_1, \mathbf{p}_2; \mathbf{i}, \mathbf{j}\}) \tag{3.10}$$

The cross ratio is invariant under the collineation from Π_∞ to the image plane induced by P. It follows that

$$\begin{aligned} \{\mathbf{p}_1, \mathbf{p}_2; \mathbf{i}, \mathbf{j}\} &= \{P(\mathbf{p}_1), P(\mathbf{p}_2); P(\mathbf{i}), P(\mathbf{j})\} \\ &= \{\mathbf{q}_1, \mathbf{q}_2; P(\mathbf{i}), P(\mathbf{j})\} \end{aligned} \tag{3.11}$$

It follows from the definition of $\mathbf{i}$ and $\mathbf{j}$ that $P(\mathbf{i})$, $P(\mathbf{j})$ are the intersections of $\langle \mathbf{q}_1, \mathbf{q}_2 \rangle$ with s. Hence (3.10) and (3.11) determine θ as a function of $\mathbf{q}_1$, $\mathbf{q}_2$ and s.

Conversely, if the angle θ between the lines h_1, h_2 projecting to $\mathbf{q}_1$ and $\mathbf{q}_2$ respectively is known for all pairs $\mathbf{q}_1$, $\mathbf{q}_2$, then the projection P can be found after fixing the camera in any convenient position. The conic w is then given by $s = P(\Omega)$. □

It is instructive to look at the Laguerre formula in detail in the Cartesian case. The notation introduced in the proof of Theorem 3.3 is used. The line $\langle \mathbf{p}_1, \mathbf{p}_2 \rangle$ in Π_∞ is parameterised by the points $(r, s)^\top$ of $\mathbb{P}^1$, such that a general point of $\langle \mathbf{p}_1, \mathbf{p}_2 \rangle$ has the form $r\mathbf{p}_1 + s\mathbf{p}_2$. The intersections $\mathbf{i}$, $\mathbf{j}$ of $\langle \mathbf{p}_1, \mathbf{p}_2 \rangle$ with

Ω are obtained by solving the equation

$$(r\mathbf{p}_1 + s\mathbf{p}_2).(r\mathbf{p}_1 + s\mathbf{p}_2) = 0 \tag{3.12}$$

The expansion of (3.12) yields

$$r^2\mathbf{p}_1.\mathbf{p}_1 + 2rs\mathbf{p}_1.\mathbf{p}_2 + s^2\mathbf{p}_2.\mathbf{p}_2 = 0 \tag{3.13}$$

The coordinates of $\mathbf{p}_1$ and $\mathbf{p}_2$ are scaled such that $\mathbf{p}_1.\mathbf{p}_1 = \mathbf{p}_2.\mathbf{p}_2 = 1$. Let ϕ be the angle such that $\cos(\theta) = \mathbf{p}_1.\mathbf{p}_2$. Equation (3.13) reduces to

$$r^2 + 2rs\cos(\phi) + s^2 = 0$$

from which it follows that $s^{-1}r = -\cos(\phi) \pm i\sin(\phi)$. The inhomogeneous coordinate θ of the point $\mathbf{p} = r\mathbf{p}_1 + s\mathbf{p}_2$ is defined by $\theta = s^{-1}r$. Let θ_1, θ_2, θ_3, θ_4 be the inhomogeneous coordinates of the points $\mathbf{p}_1$, $\mathbf{p}_2$, $\mathbf{i}$, $\mathbf{j}$, taken in order. It follows from the definition of the θ_i that

$$\begin{aligned}
\theta_1 &= \infty \\
\theta_2 &= 0 \\
\theta_3 &= -\cos(\theta) - i\sin(\theta) = -e^{i\theta} \\
\theta_4 &= -\cos(\theta) + i\sin(\theta) = -e^{-i\theta}
\end{aligned} \tag{3.14}$$

It follows from (3.14) that $\{\mathbf{p}_1, \mathbf{p}_2; \mathbf{i}, \mathbf{j}\}$ is given by

$$\begin{aligned}
\{\mathbf{p}_1, \mathbf{p}_2; \mathbf{i}, \mathbf{j}\} &= \left(\frac{\theta_1 - \theta_3}{\theta_1 - \theta_4}\right) \Big/ \left(\frac{\theta_2 - \theta_3}{\theta_2 - \theta_4}\right) \\
&= \theta_3\theta_4^{-1} \\
&= e^{2i\theta}
\end{aligned} \tag{3.15}$$

The Laguerre formula for θ follows from (3.15),

$$\theta = \frac{1}{2i}\log(\{\mathbf{p}_1, \mathbf{p}_2; \mathbf{i}, \mathbf{j}\})$$

3.1.2 Involutions and the Absolute Conic

Involutions are collineations of a particularly simple type. They play an important role in the ambiguous case of reconstruction. An involution is a linear transformation of projective space that is equal to its own inverse. An example of an involution of $\mathbb{P}^3$ is a rotation through 180°. The aim in this subsection is to describe the involutions of the absolute conic. Some brief remarks are made concerning the involutions of $\mathbb{P}^1$, $\mathbb{P}^2$, $\mathbb{P}^3$ and the involutions of a general conic. Further information about involutions can be found in Maybank (1990) and Semple & Kneebone (1952).

A trivial involution is an involution that is equal to the identity. A non-trivial involution of $\mathbb{P}^1$ has exactly two fixed points and it is uniquely determined by these two points. If τ is a non-trivial involution of $\mathbb{P}^1$ and if coordinates x_1, x_2 are chosen in $\mathbb{P}^1$ such that the fixed points of τ are $(1,0)^\top$ and $(0,1)^\top$ then τ is given by

$$\tau(x_1, x_2)^\top = (x_1, -x_2)^\top$$

Let τ be an involution of $\mathbb{P}^n$, and let $\mathbf{a}$ be any point of $\mathbb{P}^n$ such that $\tau(\mathbf{a}) \neq \mathbf{a}$. The line $\langle \mathbf{a}, \tau(\mathbf{a})\rangle$ is invariant under τ. The action of τ on $\langle \mathbf{a}, \tau(\mathbf{a})\rangle$ is determined by the fixed points of τ on $\langle \mathbf{a}, \tau(\mathbf{a})\rangle$. It follows that τ is completely determined by its fixed points in $\mathbb{P}^n$.

The fixed points of a non-trivial involution of $\mathbb{P}^2$ consist of a line and a single point not contained in the line. The isolated fixed point is called the vertex of the involution. If the involution is a rotation through 180° then the vertex may also be referred to as the centre of rotation. In $\mathbb{P}^3$ there are two different types of involution. A non-trivial involution of $\mathbb{P}^3$ is either a skew involution, in which case it has two skew lines of fixed points, or it is a harmonic homology, in which case it has a plane of fixed points and a single additional fixed point not on the plane. A rotation through 180° is an example of a skew involution and a reflection is an example of a harmonic homology.

An involution of a conic is, by definition, an involution of the plane containing the conic that leaves the conic invariant. Let τ be a non-trivial involution of a non-singular conic s, let $\mathbf{v}$ be the vertex of τ and let g be the line of fixed points of τ. It is shown that $\mathbf{v}$ is not a point of s. Suppose if possible that $\mathbf{v}$ is a point of s. Each line l through $\mathbf{v}$ is invariant under τ, thus the point of $l \cap s$ distinct from $\mathbf{v}$ is fixed by τ. It follows that τ is the identity on all points of s, thus τ is the identity, contrary to hypothesis.

Let h, k be the two lines through $\mathbf{v}$ tangent to s. The lines h, k are each invariant under τ, thus each of the points $h \cap s$, $k \cap s$ is a fixed point of τ. It follows that the line g of fixed points of τ is given by $g = \langle h \cap s, k \cap s\rangle$. The line g is thus the polar of $\mathbf{v}$ with respect to s.

A collineation of Π_∞ that leaves the absolute conic Ω invariant is called an orthogonal collineation. In the remaining propositions of this subsection certain useful properties of involutions of Ω are obtained.

Proposition 3.4. *Let two distinct points be given in the plane at infinity such that neither point is on the absolute conic and such that the line joining the two points is not tangent to the absolute conic. Then there are exactly two orthogonal involutions of the plane at infinity that interchange the two points.*

Proof. Let $\mathbf{m}$, $\mathbf{n}$ be distinct points of Π_∞, not contained in Ω, such that $\langle \mathbf{m}, \mathbf{n}\rangle$ is not tangent to Ω, and let $\mathbf{i}, \mathbf{j}$ be the points at which $\langle \mathbf{m}, \mathbf{n}\rangle$ intersects Ω. There is a unique involution σ of $\langle \mathbf{m}, \mathbf{n}\rangle$ such that $\sigma(\mathbf{m}) = \mathbf{n}$, $\sigma(\mathbf{i}) = \mathbf{j}$. Let $\mathbf{p}_1$, $\mathbf{p}_2$ be the fixed points of σ and let τ_1, τ_2 be orthogonal involutions of Π_∞ with vertices at $\mathbf{p}_1$ and $\mathbf{p}_2$ respectively. The restriction of each involution τ_i to $\langle \mathbf{m}, \mathbf{n}\rangle$ is equal to σ because $\tau_i(\mathbf{i}) = \mathbf{j}$, and because each τ_i shares a fixed point on $\langle \mathbf{m}, \mathbf{n}\rangle$ with

σ. It follows that $\tau_1(\mathbf{m}) = \mathbf{n}$, $\tau_2(\mathbf{m}) = \mathbf{n}$.

The proof is completed by showing that τ_1 and τ_2 are the only orthogonal involutions of Π_∞ that interchange $\mathbf{m}$ and $\mathbf{n}$. Let τ be any orthogonal involution of Π_∞ such that $\tau(\mathbf{m}) = \mathbf{n}$. The line $\langle \mathbf{m}, \mathbf{n} \rangle$ is invariant under τ, thus it contains two fixed points of τ. One of these points is the vertex $\mathbf{p}_\tau$ of τ. The point $\mathbf{p}_\tau$ is not equal to $\mathbf{i}$ or $\mathbf{j}$. The involution τ has exactly two fixed points on $\langle \mathbf{m}, \mathbf{n} \rangle$, thus $\tau(\mathbf{i}) = \mathbf{j}$. It follows that the restriction of τ to $\langle \mathbf{m}, \mathbf{n} \rangle$ is equal to σ and that as a consequence $\mathbf{p}_\tau$ is one of the points $\mathbf{p}_1$, $\mathbf{p}_2$. If $\mathbf{p}_\tau = \mathbf{p}_1$ then $\tau = \tau_1$, and if $\mathbf{p}_\tau = \mathbf{p}_2$ then $\tau = \tau_2$. □

If a Cartesian coordinate system is chosen then the fixed points $\mathbf{p}_i$ of σ in the proof of Proposition 3.4 are the external and internal bisectors of the angles between $\mathbf{m}$ and $\mathbf{n}$.

The next proposition examines the orthogonal collineations that fix two given points.

Proposition 3.5. *Let two distinct points be given in the plane at infinity such that the points are not orthogonal and such that neither point is contained in the absolute conic. Then there is exactly one non-trivial orthogonal involution of the plane at infinity that fixes both points.*

Proof. Let $\mathbf{m}$, $\mathbf{n}$ be distinct non-orthogonal points of Π_∞ such that neither point is contained in Ω. The orthogonal involution τ of Π_∞ with a line $g_\tau = \langle \mathbf{m}, \mathbf{n} \rangle$ of fixed points necessarily fixes $\mathbf{m}$ and $\mathbf{n}$. It remains to show that no other non-trivial orthogonal involution of Π_∞ fixes $\mathbf{m}$ and $\mathbf{n}$.

Let σ be a non-trivial orthogonal involution of Π_∞ with vertex $\mathbf{p}_\sigma$ and a line of fixed points g_σ such that $\mathbf{m}$, $\mathbf{n}$ are fixed points of σ. If $\mathbf{m} = \mathbf{p}_\sigma$ then $\mathbf{n}$ is on the polar line g_σ of $\mathbf{m} = \mathbf{p}_\sigma$ with respect to Ω. In this case $\mathbf{n}$ is orthogonal to $\mathbf{m}$, contrary to hypothesis. It follows that $\mathbf{m} \neq \mathbf{p}_\sigma$. A similar argument establishes that $\mathbf{n} \neq \mathbf{p}_\sigma$. It follows that $\mathbf{m}$, $\mathbf{n}$ are contained in g_σ, thus $g_\sigma = g_\tau$. The vertex $\mathbf{p}_\sigma$ is the polar of g_σ with respect to Ω, thus $\mathbf{p}_\sigma = \mathbf{p}_\tau$. The involutions σ, τ have the same fixed points, thus $\sigma = \tau$. □

Proposition 3.6. *A non-trivial rigid skew involution of $\mathbb{P}^3$ is uniquely determined by the line of fixed points of the involution not contained in the plane at infinity.*

Proof. Let τ be a non-trivial rigid skew involution of $\mathbb{P}^3$ and let g_τ, h_τ be the two skew lines of fixed points of τ, chosen such that h_τ is not included in Π_∞. The involution τ induces a non-trivial orthogonal involution of Π_∞ with a line of fixed points g_τ, and vertex $h_\tau \cap \Pi_\infty$. It follows that g_τ is the polar of $h_\tau \cap \Pi_\infty$ with respect to Ω. Thus h_τ determines g_τ. The result follows because the fixed points $g_\tau \cup h_\tau$ of τ determine τ uniquely. □

Proposition 3.6 is just a restatement of the fact that a rotation through 180° is uniquely determined by its axis. The next result is useful for showing that certain involutions are orthogonal.

Proposition 3.7. *A non-trivial involution of the plane at infinity is orthogonal if it fixes a point not on the absolute conic and if it also interchanges the points of contact of the tangents drawn from the fixed point to the absolute conic.*

Proof. Let τ be a non-trivial involution of Π_∞ with a fixed point $\mathbf{n}$ not on Ω such that τ interchanges the two points of contact $\mathbf{i}_n$, $\mathbf{j}_n$ of the tangents drawn from $\mathbf{n}$ to Ω. Let g_τ be the line of fixed points of τ. The conic $\tau(\Omega)$ contains $g_\tau \cap \Omega$ and in addition, $\tau(\Omega)$ is tangent to $\langle \mathbf{n}, \mathbf{i}_n \rangle$ and $\langle \mathbf{n}, \mathbf{j}_n \rangle$ at $\mathbf{i}_n$ and $\mathbf{j}_n$ respectively. The two tangencies and the constraint $g_\tau \cap \tau(\Omega) = g_\tau \cap \Omega$ impose six constraints on $\tau(\Omega)$. A general plane conic has five degrees of freedom, thus $\tau(\Omega)$ is uniquely determined. The only possibility is $\tau(\Omega) = \Omega$. □

Involutions are important in the study of ambiguity because each orthogonal collineation is expressible as a product of two orthogonal involutions in an infinite number of ways. This result is proved by Semple & Kneebone (1952) as part of the general theory of conics (Theorem 17 of Chapter VI). An alternative proof is given in the following proposition.

Proposition 3.8. *An orthogonal collineation of the plane at infinity with exactly one fixed point not on the absolute conic is expressible as the product of two orthogonal involutions with vertices on the polar line of the fixed point with respect to the absolute conic. Any point on the polar line, but not on the absolute conic, can serve as the vertex of one of the involutions. The vertex of the other involution is then uniquely determined.*

Proof. Let ω be an orthogonal collineation of Π_∞ with exactly one fixed point $\mathbf{n}$ not on Ω as illustrated in Fig. 3.3, and let τ be an orthogonal involution of Π_∞ with vertex $\mathbf{p}_\tau$ on the polar line, $\pi(\mathbf{n})$, of $\mathbf{n}$ with respect to Ω. The line $\pi(\mathbf{n})$ is invariant under τ. The point $\mathbf{n}$ is on $\pi(\mathbf{p}_\tau)$, thus $\tau(\mathbf{n}) = \mathbf{n}$. It follows that $\omega\tau(\mathbf{n}) = \mathbf{n}$. Let $\mathbf{i}_n$, $\mathbf{j}_n$ be the points at which $\pi(\mathbf{n})$ intersects Ω. The involution τ does not restrict to the identity on the line $\pi(\mathbf{n})$, thus $\mathbf{i}_n$ and $\mathbf{j}_n$ are not fixed by τ. It follows that $\tau(\mathbf{i}_n) = \mathbf{j}_n$. It follows from Proposition 3.2 that $\mathbf{i}_n$, $\mathbf{j}_n$ are fixed by ω, thus $\omega\tau(\mathbf{i}_n) = \mathbf{j}_n$, $\omega\tau(\mathbf{j}_n) = \mathbf{i}_n$. Let σ be the collineation defined by $\sigma = \omega\tau$. It follows from Corollary 1 to Proposition 3.2 that σ is an involution, thus $\omega = \sigma\tau$ is a product of involutions. The involution σ is uniquely determined by ω and τ. Let $\mathbf{p}_\sigma$ be the vertex of σ. It follows from the definition of σ that $\sigma(\mathbf{i}_n) = \mathbf{j}_n$. Thus $\pi(\mathbf{n}) = \langle \mathbf{i}_n, \mathbf{j}_n \rangle$ is invariant under σ. The restriction of σ to $\pi(\mathbf{n})$ is not the identity, thus $\mathbf{p}_\sigma$ is a point of $\pi(\mathbf{n})$.

A similar argument applied to ω^{-1} establishes that in the product $\sigma\tau = \omega$ the vertex of σ can first be chosen to be any point on $\pi(\mathbf{n})$ and that ω, σ then uniquely determine τ. □

3.2 Rectangular Quadrics

The critical surfaces arising in the ambiguous case of reconstruction with known camera calibration are examples of a particular type of quadric known as a rect-

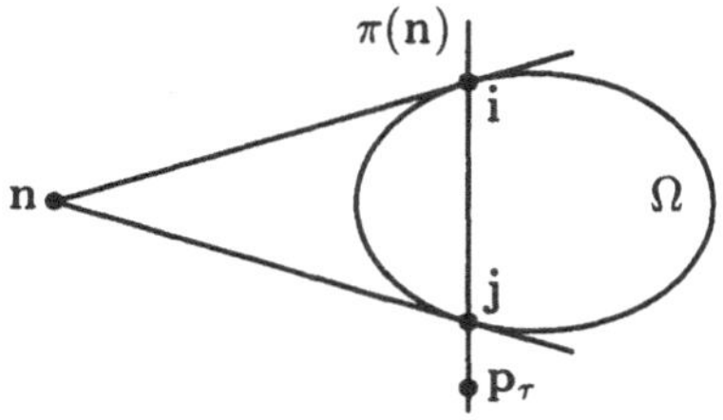

Fig. 3.3. Illustration to Proposition 3.8

angular quadric. In this section the basic properties of rectangular quadrics are described, beginning with the properties that they share with a general quadric.

Each quadric in $\mathbb{P}^3$ is described by an equation of the form $\mathbf{x}^\top M\mathbf{x} = 0$, where M is a symmetric 4×4 matrix. The quadric is singular if and only if M has determinant zero. It is recalled from Sect. 2.3.2 that a non-singular quadric contains two disjoint one-parameter families of lines. The lines are called the generators of the quadric. Each point of the quadric is contained in two generators, one from each family. Any two generators from the same family are skew and any two generators from different families intersect.

The two families of generators provide a way of classifying the curves contained in a non-singular quadric. A curve which meets a general member of the first family of generators m times and a general member of the second family n times is called an (m,n) curve. For example the generators in the first family are $(0,1)$ curves and any conic contained in the quadric is a $(1,1)$ curve . The twisted cubics (cubic space curve) contained in a quadric divide into two families, the $(2,1)$ curves and the $(1,2)$ curves.

There are several equivalent ways of defining rectangular quadrics. The first one given below as Definition 3.10 is a geometric definition, in which the rectangular quadrics are characterised by their position with respect to the absolute conic. As a consequence of this definition a general rectangular quadric contains two distinguished points referred to as the principal points of the quadric. Many of the properties of rectangular quadrics can be obtained directly by considering the principal points. For example, the principal points are used to show that each non-singular rectangular quadric is invariant under exactly three non-trivial rigid skew involutions.

The principal points are defined as follows.

Definition 3.9. *A principal point of a quadric is a real point contained in the intersection of the quadric with the plane at infinity such that the quadric contains the two points of contact of the tangents drawn from the principal point to the absolute conic.*

Definition 3.10. *A real quadric is rectangular if and only if it contains a principal point.*

Definitions 3.9 and 3.10 are illustrated in Fig. 3.4. The point $\mathbf{n}$ in the figure is a principal point.

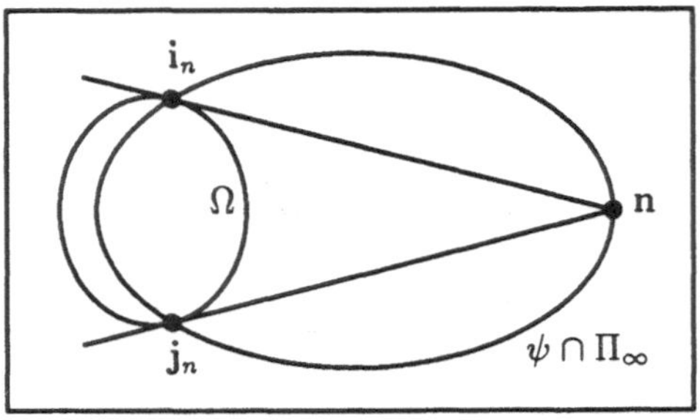

Fig. 3.4. Rectangular quadrics

It is specified in Definition 3.9 that the principal points are real because this is the case of most interest in reconstruction. However, many of the properties of rectangular quadrics obtained below apply even if the principal points are allowed to have complex coordinates.

Theorem 3.11. *A general rectangular quadric contains two principal points.*

Proof. Let ψ be a general rectangular quadric and let $\mathbf{i}_m$, $\mathbf{j}_m$, $\mathbf{i}_n$, $\mathbf{j}_n$ be the four points of intersection of ψ with the absolute conic, Ω. It follows from the Definition 3.2 of a rectangular quadric that the points of $\psi \cap \Omega$ can be labelled such that the tangents to Ω at $\mathbf{i}_m$, $\mathbf{j}_m$ intersect at a point $\mathbf{m}$ on ψ. Let the tangents to ψ at $\mathbf{i}_n$, $\mathbf{j}_n$ intersect at a point $\mathbf{n}$. To prove the theorem, it suffices to show that $\mathbf{n}$ is contained in the conic $s_\infty = \psi \cap \Pi_\infty$.

Let τ be an orthogonal involution of Π_∞ such that $\tau(\mathbf{m}) = \mathbf{n}$. The absolute conic Ω is invariant under τ, thus τ takes the pair $\{\mathbf{i}_m, \mathbf{j}_m\}$ to the pair $\{\mathbf{i}_n, \mathbf{j}_n\}$. The points $\mathbf{i}_n$ and $\mathbf{j}_n$ are labelled such that $\tau(\mathbf{i}_m) = \mathbf{i}_n$, $\tau(\mathbf{j}_m) = \mathbf{j}_n$. The conic $\tau(s_\infty)$ contains the points $\{\mathbf{i}_m, \mathbf{j}_m, \mathbf{i}_n, \mathbf{j}_n\}$. Let g_τ be the line of fixed points of τ. Then the points of $s_\infty \cap g_\tau$ are fixed by τ, thus they are contained in $\tau(s_\infty)$. It follows that s_∞ and $\tau(s_\infty)$ intersect in at least five points, thus $s_\infty = \tau(s_\infty)$. The conic s_∞ contains $\mathbf{n} = \tau(\mathbf{m})$ as required. □

In the next theorem it is shown that a general critical surface is a rectangular quadric.

Theorem 3.12. *A critical surface arising in the ambiguous case of reconstruction with known camera calibration contains a principal point.*

Proof. The projective geometric construction of a critical surface ψ is recalled from Sect. 2.3.2. The notation is summarised briefly. The points $\mathbf{a}$, $\mathbf{b}$ are the two possible optical centres of the camera taking the second image. The construction of ψ depends on a collineation ω from the star of lines through $\mathbf{a}$ to the star of lines through $\mathbf{b}$. The surface ψ is swept out by the lines $\Pi \cap \omega(\Pi)$ as Π runs through a pencil of planes with axis h containing $\mathbf{a}$. The collineation ω induces an orthogonal collineation of Π_∞, also denoted by ω. If $\mathbf{x}$ is any point of Π_∞

then $\omega(\mathbf{x})$ is by definition the point of Π_∞ such that $\omega\langle\mathbf{a},\mathbf{x}\rangle = \langle\mathbf{b},\omega(\mathbf{x})\rangle$. Let $\mathbf{x}$ be any one of the three fixed points of ω in Π_∞. The point $\mathbf{x}$ is contained in $\langle\mathbf{x},h\rangle \cap \omega\langle\mathbf{x},h\rangle$ because

$$\begin{aligned}\mathbf{x} &= \langle\mathbf{a},\mathbf{x}\rangle \cap \langle\mathbf{b},\mathbf{x}\rangle \\ &= \langle\mathbf{a},\mathbf{x}\rangle \cap \langle\mathbf{b},\omega(\mathbf{x})\rangle\end{aligned}$$

It follows that $\mathbf{x}$ is contained in ψ. Let the three fixed points of ω in Π_∞ be $\mathbf{n}$, $\mathbf{i}_n$, $\mathbf{j}_n$. It follows from Proposition 3.2 that the fixed points can be labelled such that $\mathbf{i}_n$, $\mathbf{j}_n$ are the points of contact of the tangents drawn from $\mathbf{n}$ to Ω. The collineation ω is real, thus $\mathbf{n}$ is real and $\mathbf{i}_n$, $\mathbf{j}_n$ are complex conjugate. It follows that $\mathbf{n}$ is a principal point of ψ. □

3.2.1 Algebraic Characterisations of Rectangular Quadrics

Rectangular quadrics are defined geometrically in terms of the position of the quadric relative to the absolute conic. In this subsection an algebraic definition of a rectangular quadric is given. It is recalled from Sect. 3.1 that a Cartesian coordinate system is one in which the absolute conic is defined by the equations

$$x_4 = 0 \qquad\qquad x_1^2 + x_2^2 + x_3^2 = 0 \tag{3.16}$$

Theorem 3.13. *Let Cartesian coordinates be chosen in $\mathbb{P}^3$. Let* m, n *be real points in the plane at infinity, Π_∞, and let M_{mn} be the matrix defined by*

$$M_{mn} = \frac{1}{2}(\mathrm{m}\otimes\mathrm{n} + \mathrm{n}\otimes\mathrm{m}) - \mathrm{m}.\mathrm{n}I \tag{3.17}$$

A quadric in $\mathbb{P}^3$ is rectangular if and only if it intersects Π_∞ in a conic with an equation $\mathbf{x}^\top M_{mn}\mathbf{x} = 0$.

Proof. Let ψ be a rectangular quadric and let the conic $\psi\cap\Pi_\infty$ have the equation $\mathbf{x}^\top M\mathbf{x} = 0$, where $\mathbf{x} = (x_1, x_2, x_3)^\top$ and where M is a symmetric 3×3 matrix. The entries of M are

$$M = \begin{pmatrix} a & d & e \\ d & b & f \\ e & f & c \end{pmatrix}$$

Let $\mathbf{n}$ be a principal point of ψ. It is assumed that $\mathbf{n} = (1,0,0)^\top$. There is no loss of generality in making this assumption. The point $\mathbf{n}$ can be moved to the point $(1,0,0)^\top$ by an orthogonal collineation ω. The quadric ψ is rectangular if and only if the transformed quadric $\omega(\psi)$ is rectangular. Let U be an orthogonal matrix representing ω. The set of matrices of the form (3.17) is invariant under the transformation $M_{mn} \mapsto U^\top M_{mn} U$. In effect, to prove the theorem it is shown that $\omega(\psi)\cap\Pi_\infty$ is a conic with an equation $\mathbf{x}^\top U^\top M_{mn} U\mathbf{x} = 0$.

The points of contact, $\mathbf{i}_n$, $\mathbf{j}_n$ of the tangents drawn from $\mathbf{n} = (1,0,0)^\top$ to Ω are given by

$$\mathbf{i}_n = (0,1,i)^\top \qquad \mathbf{j}_n = (0,1,-i)^\top$$

The quadric ψ contains $\mathbf{i}_n$ and $\mathbf{j}_n$, thus $\mathbf{i}_n^\top M \mathbf{i}_n = 0$, $\mathbf{j}_n^\top M \mathbf{j}_n = 0$. It follows that

$$M = \begin{pmatrix} 0 & d & e \\ d & b & 0 \\ e & 0 & b \end{pmatrix}$$

A short calculation yields $M = M_{mn}$, where $\mathbf{n} = (1,0,0)^\top$ and $\mathbf{m}$ is defined by $\mathbf{m} = (-b, 2d, 2e)^\top$. Conversely, any quadric intersecting Π_∞ in a conic of the form $\mathbf{x}^\top M_{mn}\mathbf{x} = 0$, where M_{mn} is given by (3.17), has $\mathbf{m}$ and $\mathbf{n}$ as principal points. □

It follows from Theorem 3.13 that there are degenerate rectangular quadrics which possess only one principal point. It suffices to set $\mathbf{m} = \mathbf{n}$ in (3.17). A rectangular quadric with a single principal point $\mathbf{n}$ intersects the plane at infinity in the line pair

$$\mathbf{x.x} - (\mathbf{x.n})^2 \tag{3.18}$$

If coordinates are chosen such that $\mathbf{n} = (1,0,0)^\top$, then the line pair (3.18) becomes

$$\mathbf{x.x} - (\mathbf{x.n})^2 = y^2 + z^2 = (y+iz)(y-iz)$$

If the rectangular quadric is also non-singular then it follows from the classification of non-singular quadrics under real affine collineations described in Sect. 2.1.1 that it is an elliptic paraboloid. A critical surface cannot be an elliptic paraboloid because an elliptic paraboloid does not contain real lines. It follows that a critical surface with only one principal point is singular. It is thus either a cone or a plane pair. An algebraic argument is given to show that a critical surface with exactly one principal point is in general a circular cylinder.

The equation of a critical surface ψ is recalled from (2.24):

$$(U\mathbf{x} \times \mathbf{x}).\mathbf{b} + (\mathbf{x} \times U\mathbf{a}).\mathbf{b} = 0 \tag{3.19}$$

where U is an orthogonal matrix. Let $\mathbf{n}$ be the axis of U and let θ be the angle of rotation of U about $\mathbf{n}$. It follows from Rodrigues' formula (2.74) that

$$U\mathbf{x} \times \mathbf{x} = (1-\cos(\theta))(\mathbf{x.n})\mathbf{n} \times \mathbf{x} - \sin(\theta)(\mathbf{x} \times \mathbf{n}) \times \mathbf{x} \tag{3.20}$$

It is assumed that $\theta \neq 0$. The scalar product of (3.20) with $\mathbf{b}$ yields

$$\begin{aligned}(U\mathbf{x} \times \mathbf{x}).\mathbf{b} &= (1-\cos(\theta))(\mathbf{x.n})(\mathbf{n} \times \mathbf{x}).\mathbf{b} - \sin(\theta)[(\mathbf{x.x})(\mathbf{n.b}) - (\mathbf{x.n})(\mathbf{x.b})] \\ &= (\mathbf{x.n})[(1-\cos(\theta))(\mathbf{b} \times \mathbf{n}).\mathbf{x} + \sin(\theta)(\mathbf{x.b})] - \sin(\theta)(\mathbf{n.b})(\mathbf{x.x})\end{aligned} \tag{3.21}$$

Let $\mathbf{m}$ be the vector defined by

$$\mathbf{m} = (1-\cos(\theta))\mathbf{b} \times \mathbf{n} + \sin(\theta)\mathbf{b}$$

Let M_{mn} be the matrix defined by (3.17). It follows from (3.21) that

$$(U\mathbf{x} \times \mathbf{x}).\mathbf{b} = \mathbf{x}^{\top} M_{mn}\mathbf{x}$$

The hypothesis is that ψ has exactly one principal point. The vector $\mathbf{m}$ is thus parallel to $\mathbf{n}$. It follows that there exists a non-zero scalar λ such that

$$(1 - \cos(\theta))\mathbf{b} \times \mathbf{n} + \sin(\theta)\mathbf{b} = \lambda\mathbf{n} \tag{3.22}$$

The scalar product of (3.22) with the unit vector $\mathbf{n}$ yields $\lambda = \sin(\theta)\mathbf{b}.\mathbf{n}$. Equation (3.22) thus reduces to

$$(1 - \cos(\theta))\mathbf{b} \times \mathbf{n} + \sin(\theta)(\mathbf{b} - (\mathbf{b}.\mathbf{n})\mathbf{n}) = 0$$

The vectors $\mathbf{b} \times \mathbf{n}$ and $\mathbf{b} - (\mathbf{b}.\mathbf{n})\mathbf{n}$ are orthogonal and the coefficients $1 - \cos(\theta)$ and $\sin(\theta)$ are not both zero because the case $\theta = 0$ is excluded. It follows that

$$\mathbf{b} \times \mathbf{n} = \mathbf{b} - (\mathbf{b}.\mathbf{n})\mathbf{n} = 0$$

The vector $\mathbf{b}$ is thus parallel to $\mathbf{n}$.

A point of ψ is singular if and only if the gradient of the right-hand side of (3.19) vanishes at that point. Let F be the quadratic form in $(\mathbf{x}, x_4)$ defined by

$$F = (U\mathbf{x} \times \mathbf{x}).\mathbf{b} + x_4(\mathbf{x} \times U\mathbf{a}).\mathbf{b}$$

The vector $\mathbf{b}$ is parallel to $\mathbf{n}$. The scale of $\mathbf{b}$ is arbitrary, thus it is assumed without loss of generality that $\mathbf{b} = \mathbf{n}$. The form F reduces to

$$F = \lambda[(\mathbf{x}.\mathbf{n})^2 - \mathbf{x}.\mathbf{x}] + x_4(\mathbf{x} \times U\mathbf{a}).\mathbf{n} \tag{3.23}$$

The gradient of F is

$$(\frac{\partial F}{\partial x_1}, \frac{\partial F}{\partial x_2}, \frac{\partial F}{\partial x_3}, \frac{\partial F}{\partial x_4})^{\top} = (2\lambda[(\mathbf{x}.\mathbf{n})\mathbf{n} - \mathbf{x}] + x_4(U\mathbf{a} \times \mathbf{n}), (\mathbf{x} \times U\mathbf{a}).\mathbf{n})^{\top} \tag{3.24}$$

It follows from (3.24) that the gradient of F vanishes at $(\mathbf{x}, x_4)^{\top} = (\mathbf{n}, 0)^{\top}$.

The singular point of ψ is in Π_∞, thus ψ is a cylinder. Let coordinates be chosen such that $\mathbf{n} = (1, 0, 0)^{\top}$. It follows from (3.23) that the equation $F = 0$ of ψ reduces to

$$-\lambda(y^2 + z^2) + (U\mathbf{a} \times \mathbf{n}).\mathbf{x} = 0$$

The surface ψ is a circular cylinder because the projection of ψ from the point $(\mathbf{n}, 0)^{\top}$ onto any plane normal to $\mathbf{n}$ is a circle. The axis of the cylinder has direction $\mathbf{n}$.

The eigenvalues associated with the conic formed by the intersection of a critical surface with the plane at infinity are examined. It is assumed that a Cartesian coordinate system has been chosen. A general critical surface intersects Π_∞ in a conic with the equation

$$(U\mathbf{x} \times \mathbf{x}).\mathbf{b} = 0 \tag{3.25}$$

where U is an orthogonal matrix and $\mathbf{b}$ is a vector. Let E be the essential matrix defined by $E = U^\top T_b$. Then (3.25) is identical to $\mathbf{x}^\top E\mathbf{x} = 0$. It follows from Theorem 3.13 that $\text{sym}(E) = M_{mn}$ for a matrix M_{mn} of the form (3.17). It follows from the corollary to Proposition 2.20 that one eigenvalue of M_{mn} is equal to the sum of the other two eigenvalues. This constraint on the eigenvalues of M_{mn} can also be obtained directly from (3.17) by solving for the eigenvectors and eigenvalues of M_{mn} in terms of $\mathbf{m}$ and $\mathbf{n}$, as shown in the following theorem.

Theorem 3.14. *Let Cartesian coordinates be chosen in $\mathbb{P}^3$. Then a quadric is rectangular if and only if it intersects the plane at infinity in a conic with an equation $\mathbf{x}^\top M\mathbf{x} = 0$, where M is a real symmetric matrix with eigenvalues λ_1, λ_2, λ_3 such that $\lambda_1\lambda_2 \leq 0$ and such that $\lambda_1 + \lambda_2 = \lambda_3$.*

Proof. Let ψ be a quadric and let s_∞ be the conic $\psi \cap \Pi_\infty$. It follows from Theorem 3.13 that ψ is rectangular if and only if there exist real points $\mathbf{m}$, $\mathbf{n}$ in Π_∞ such that s_∞ has an equation $\mathbf{x}^\top M_{mn}\mathbf{x} = 0$. Thus to prove the theorem, it suffices to show that a symmetric matrix M has the form M_{mn} for real points $\mathbf{m}$, $\mathbf{n}$ in Π_∞ if and only if the eigenvalues of M satisfy the stated conditions.

It is shown first that the eigenvalues of M_{mn} have the required properties. If $\mathbf{m} = \mathbf{n}$ then $\lambda_1 = \lambda_3$, $\lambda_2 = 0$. Thus, the theorem holds in this case. If $\mathbf{m} \neq \mathbf{n}$ it follows from (3.17) that the eigenvectors of M_{mn} are

$$\|\mathbf{m}\|^{-1}\mathbf{m} + \|\mathbf{n}\|^{-1}\mathbf{n} \qquad \|\mathbf{m}\|^{-1}\mathbf{m} - \|\mathbf{n}\|^{-1}\mathbf{n} \qquad \mathbf{m} \times \mathbf{n}$$

The corresponding eigenvalues are

$$\frac{1}{2}(\|\mathbf{m}\|\ \|\mathbf{n}\| - \mathbf{m}.\mathbf{n}) \qquad -\frac{1}{2}(\|\mathbf{m}\|\ \|\mathbf{n}\| + \mathbf{m}.\mathbf{n}) \qquad -\mathbf{m}.\mathbf{n} \tag{3.26}$$

The first eigenvalue in (3.26) is positive, the second eigenvalue is negative and the third eigenvalue is the sum of the first two eigenvalues.

Conversely, let M be a real symmetric matrix with eigenvalues λ_1, λ_2, λ_3 such that $\lambda_1\lambda_2 \leq 0$ and such that $\lambda_1 + \lambda_2 = \lambda_3$. Let $\mathbf{e}_1$, $\mathbf{e}_2$, $\mathbf{e}_3$ be the corresponding eigenvectors, scaled such that $\mathbf{e}_1.\mathbf{e}_1 = \mathbf{e}_2.\mathbf{e}_2 = \mathbf{e}_3.\mathbf{e}_3$. The λ_i are labelled such that $\lambda_1 \geq 0$, $\lambda_2 \leq 0$. The real vectors $\mathbf{m}$, $\mathbf{n}$ are defined by

$$\begin{aligned} \mathbf{m} &= \alpha\mathbf{e}_1 + \beta\mathbf{e}_2 \\ \mathbf{n} &= \alpha\mathbf{e}_1 - \beta\mathbf{e}_2 \end{aligned} \tag{3.27}$$

where $\alpha = \sqrt{-\lambda_2}$, $\beta = \sqrt{\lambda_1}$. It follows from (3.17) and (3.27) that

$$M_{mn} = -\lambda_2\mathbf{e}_1 \otimes \mathbf{e}_1 - \lambda_1\mathbf{e}_2 \otimes \mathbf{e}_2 + (\lambda_1 + \lambda_2)I$$

thus

$$\begin{aligned} M_{mn}\mathbf{e}_1 &= \lambda_1\mathbf{e}_1 \\ M_{mn}\mathbf{e}_2 &= \lambda_2\mathbf{e}_2 \\ M_{mn}\mathbf{e}_3 &= \lambda_3\mathbf{e}_3 \end{aligned}$$

The matrices M and M_{mn} have the same eigenvectors and eigenvalues, thus $M = M_{mn}$. □

The final result of this subsection is an observation concerning the tangent planes to a rectangular quadric at the principal points.

Proposition 3.15. *If a rectangular quadric has two principal points then the tangent planes at the two principal points intersect in a line with direction orthogonal to both principal points of the quadric.*

Proof. Let ψ be a rectangular quadric with principal points $\mathbf{m}$, $\mathbf{n}$, let s_∞ be the conic $\psi \cap \Pi_\infty$ and let g be the line $\langle \mathbf{m}, \mathbf{n} \rangle$. Let the line formed by the intersection of the tangent planes to ψ at $\mathbf{m}$, $\mathbf{n}$ meet Π_∞ at $\mathbf{p}$. Then it is required to show that $\mathbf{p}$ is orthogonal to $\mathbf{m}$, $\mathbf{n}$, or equivalently, that $\mathbf{p}$ is the pole of g with respect to Ω.

Let τ be the rigid involution of Π_∞ with line of fixed points g. Then $\tau(s_\infty) \cap s_\infty$ contains the six points $\{\mathbf{m}, \mathbf{i}_m, \mathbf{j}_m, \mathbf{n}, \mathbf{i}_n, \mathbf{j}_n\}$, thus $\tau(s_\infty) = s_\infty$. The points $\mathbf{m}$, $\mathbf{n}$ are fixed points of τ on s_∞. It follows from the definition of $\mathbf{p}$ that $\mathbf{p}$ is the pole of g with respect to s_∞. The conic s_∞ is invariant under τ, thus $\mathbf{p}$ is the vertex of τ. The absolute conic Ω is invariant under τ, thus $\mathbf{p}$ is also the pole of g with respect to Ω. □

3.2.2 Rigid Involutions of Rectangular Quadrics

There are exactly three non-trivial rigid skew involutions that leave invariant a non-singular rectangular quadric ψ with two principal points. The three involutions are denoted by τ_ψ, τ_1, τ_2. The principal points of ψ are fixed by τ_ψ and they are interchanged by τ_1 and τ_2. The involution τ_ψ is the one most relevant to reconstruction. It is used in Sect. 3.4.2 to show that the critical surfaces are subject to a cubic polynomial constraint.

The involutions τ_ψ, τ_1, τ_2 are defined. Let $\mathbf{m}$, $\mathbf{n}$ be the principal points of ψ and let l_ψ be the line formed by the intersection of the tangent planes to ψ at $\mathbf{m}$ and $\mathbf{n}$, as illustrated in Fig. 3.5. Then τ_ψ is the non-trivial involution of $\mathbb{P}^3$ with two lines of fixed points l_ψ and $\langle \mathbf{m}, \mathbf{n} \rangle$. The involution τ_ψ is rigid because by Proposition 3.15 the direction $l_\psi \cap \Pi_\infty$ of l_ψ is orthogonal to every point of $\langle \mathbf{m}, \mathbf{n} \rangle$. Let $\mathbf{p}_1$, $\mathbf{p}_2$ be the fixed points of the unique rigid involution of the line $\langle \mathbf{m}, \mathbf{n} \rangle$ that interchanges $\mathbf{m}$ and $\mathbf{n}$, and let $\mathbf{d}$ be the mid-point of the line segment defined by the two intersections of l_ψ with ψ. Then τ_1 is defined to be the rigid skew involution of $\mathbb{P}^3$ with a line of fixed points $\langle \mathbf{d}, \mathbf{p}_1 \rangle$ and τ_2 is defined to be the rigid skew involution of $\mathbb{P}^3$ with a line of fixed points $\langle \mathbf{d}, \mathbf{p}_2 \rangle$.

It follows from the definitions of τ_ψ, τ_1, τ_2, that $\tau_\psi(\mathbf{m}) = \mathbf{m}$, $\tau_\psi(\mathbf{n}) = \mathbf{n}$ and $\tau_i(\mathbf{m}) = \mathbf{n}$ for $i = 1, 2$. In this subsection it is shown that τ_ψ, τ_1, τ_2 are rigid skew involutions of ψ and that any non-trivial rigid skew involution of ψ is equal to one of τ_ψ, τ_1, τ_2.

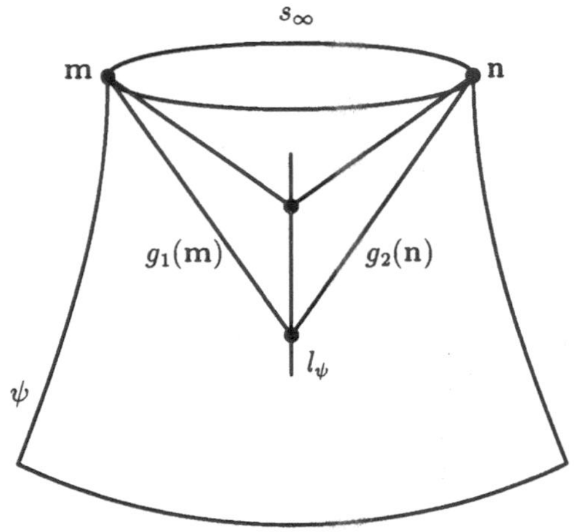

Fig. 3.5. The geometry of rectangular quadrics

Theorem 3.16. *A non-singular rectangular quadric ψ is invariant under the three non-trivial rigid skew involutions τ_ψ, τ_1, τ_2.*

Proof. The notation is illustrated in Fig. 3.5. Let $g_1(\mathbf{m})$, $g_2(\mathbf{m})$ be the generators of ψ through $\mathbf{m}$. The tangent plane to ψ at $\mathbf{m}$ contains $g_1(\mathbf{m})$, thus $g_1(\mathbf{m})$ intersects l_ψ. It follows that $g_1(\mathbf{m})$ contains two fixed points of τ_ψ, namely $\mathbf{m}$ and $g_1(\mathbf{m}) \cap l_\psi$. The line $g_1(\mathbf{m})$ is thus invariant under τ_ψ. It follows that $\tau_\psi(\psi)$ contains $g_1(\mathbf{m})$. Similarly, $\tau_\psi(\psi)$ contains the three lines $g_2(\mathbf{m})$, $g_1(\mathbf{n})$ and $g_2(\mathbf{n})$.

The generators $g_i(\mathbf{m})$, $g_i(\mathbf{n})$ meet l_ψ at the points of $l_\psi \cap \psi$. It follows from the definitions of τ_1 and τ_2 that the τ_i interchange the two points of $l_\psi \cap \psi$, thus the τ_i permute the $g_i(\mathbf{m})$, $g_i(\mathbf{n})$ amongst themselves. Hence the two quadrics $\tau_i(\psi)$ each contain the lines $g_1(\mathbf{m})$, $g_2(\mathbf{m})$, $g_1(\mathbf{n})$ and $g_2(\mathbf{n})$.

Let $s_\infty = \psi \cap \Pi_\infty$. Each of the three intersections $\tau_\psi(s_\infty) \cap s_\infty$, $\tau_1(s_\infty) \cap s_\infty$, $\tau_2(s_\infty) \cap s_\infty$ contains the six points $\{\mathbf{m}, \mathbf{i}_m, \mathbf{j}_m, \mathbf{n}, \mathbf{i}_n, \mathbf{j}_n\}$. Two distinct conics intersect at four points only, thus the conics in each of the three intersections are not distinct. In other words, $\tau_\psi(s_\infty) = s_\infty$, $\tau_1(s_\infty) = s_\infty$, $\tau_2(s_\infty) = s_\infty$.

It has been shown that the three intersections $\tau_\psi(\psi) \cap \psi$, $\tau_1(\psi) \cap \psi$, $\tau_2(\psi) \cap \psi$ each contain a (split) space curve of degree six, namely the union of s_∞ with the four generators $g_i(\mathbf{m})$, $g_i(\mathbf{n})$. The intersection of two distinct quadrics is a space curve of degree four rather than degree six, thus the quadrics in each of the pairs $\{\tau_\psi(\psi), \psi\}$, $\{\tau_1(\psi), \psi\}$, $\{\tau_2(\psi), \psi\}$ are not distinct. In other words $\tau_\psi(\psi) = \psi$, $\tau_1(\psi) = \psi$, $\tau_2(\psi) = \psi$. □

The generators $g_i(\mathbf{m})$, $g_i(\mathbf{n})$ in the proof of Theorem 3.16 are called the main generators (Haupterzeugende) of ψ (Hofmann 1950). Any two generators g, h of ψ such that $\tau_\psi(g) = h$ are adjoint generators (adjungierte Erzeugende) of ψ. The main generators of ψ are thus self-adjoint generators. Hofmann (1950) gives further information about τ_ψ, and references to the literature.

Proposition 3.17. *A non-trivial skew involution of $\mathbb{P}^3$ leaving invariant a non-singular quadric also leaves invariant the two families of generators on the quadric.*

Proof. Let ψ be a non-singular quadric invariant under a non-trivial skew involution τ and let $\mathcal{F}_1$, $\mathcal{F}_2$ be the two families of generators of ψ. Two generators, g and h, of ψ intersect if and only if the generators $\tau(g)$ and $\tau(h)$ intersect. It follows that g, h are in the same family of generators if and only if $\tau(g)$, $\tau(h)$ are in the same family of generators. Thus either τ interchanges $\mathcal{F}_1$ and $\mathcal{F}_2$, or τ leaves $\mathcal{F}_1$ and $\mathcal{F}_2$ invariant.

If ψ contains a line g_τ of fixed points of τ then τ leaves $\mathcal{F}_1$ and $\mathcal{F}_2$ invariant. If ψ does not contain a line of fixed points of τ then five generators g_i of ψ are chosen from $\mathcal{F}_1$. If τ interchanges $\mathcal{F}_1$ and $\mathcal{F}_2$ then the points $\tau(g_i) \cap g_i$ are fixed points of τ contained in ψ. The fixed points of τ form two lines in $\mathbb{P}^3$, because τ is, by hypothesis, a skew involution. Thus at least three of the $\tau(g_i) \cap g_i$ are on the same line of fixed points of τ. It follows that ψ contains a line of fixed points of τ, contrary to hypothesis. The families $\mathcal{F}_1$ and $\mathcal{F}_2$ are thus invariant under τ_ψ. □

Proposition 3.17 does not hold if the skew involution is replaced by a harmonic homology. A harmonic homology of a non-singular quadric interchanges the two families of generators of the quadric.

Theorem 3.18. *A non-singular rectangular quadric with distinct principal points is invariant under exactly three non-trivial rigid skew involutions.*

Proof. Let ψ be a non-singular rectangular quadric invariant under a rigid skew involution τ. In view of Theorem 3.16 it is sufficient to show that τ is equal to one of τ_ψ, τ_1, τ_2. Let $\mathbf{m}$, $\mathbf{n}$ be the principal points of ψ. The absolute conic Ω and the conic $s_\infty = \psi \cap \Pi_\infty$ are invariant under τ, thus either $\mathbf{m}$, $\mathbf{n}$ are fixed points of τ or $\mathbf{m}$, $\mathbf{n}$ are interchanged by τ. It follows from Propositions 3.4 and 3.5 that the restriction of τ to Π_∞ is equal to the restriction of one of τ_ψ, τ_1, τ_2 to Π_∞. Thus, by forming the appropriate product $\tau\tau_\psi$, $\tau\tau_1$ or $\tau\tau_2$, a Euclidean transformation ω is obtained which fixes every point of Π_∞ and which leaves ψ invariant.

It follows from Proposition 3.17 that the two skew involutions comprising ω each leave the two families $\mathcal{F}_1$ and $\mathcal{F}_2$ of generators of ψ invariant, thus ω also leaves $\mathcal{F}_1$ and $\mathcal{F}_2$ invariant. Each point of s_∞ lies on a single generator of $\mathcal{F}_1$ and the points of s_∞ are fixed points of ω, thus each individual generator of $\mathcal{F}_1$ is invariant under ω. Similarly, each generator of $\mathcal{F}_2$ is invariant under ω. Each point $\mathbf{p}$ of ψ is at the intersection of two generators $g_1(\mathbf{p})$, $g_2(\mathbf{p})$ of ψ belonging to $\mathcal{F}_1$ and $\mathcal{F}_2$ respectively, thus $\mathbf{p}$ is fixed by ω.

Let $\mathbf{a}$ be any point of $\mathbb{P}^3$ and let k be a line through $\mathbf{a}$. Then k contains three fixed points of ω, namely, $k \cap \Pi_\infty$ and the two points of $k \cap \psi$. It follows that k is invariant under ω and that the restriction of ω to k is the identity. Hence $\omega(\mathbf{a}) = \mathbf{a}$. It follows that ω is the identity, thus τ is equal to one of τ_ψ, τ_1, τ_2, as required. □

3.3 Horopter Curves

It is shown in Sect. 2.3.3 that the intersection of a critical surface pair contains a space curve of degree three known as a horopter curve. Horopter curves were first discovered by Helmholtz (1925). They appear in the photogrammetric literature (Wunderlich 1942), but they have not so far received much attention in computer vision. The properties of horopter curves underlie many of the results obtained in the ambiguous case of reconstruction. For example, a quadric which contains a horopter curve is necessarily rectangular.

In this section the properties of horopter curves are described in preparation for their application to the ambiguous case of reconstruction in Sect. 3.4. In Sect. 3.3.1 two different characterisations of horopter curves are given. In the first characterisation the horopter curve is generated by an orthogonal collineation. In the second characterisation a horopter curve is regarded as a twisted cubic which intersects the plane at infinity at three particular points. In Sect. 3.3.2 it is shown that each horopter curve is invariant under a unique non-trivial rigid skew involution. In the final two subsections, Sect. 3.3.6 and Sect. 3.3.7, it is shown that each non-singular rectangular quadric ψ with distinct principal points contains four one-parameter families of horopter curves, such that each horopter curve is invariant under the rigid skew involution τ_ψ.

3.3.1 Characterisations of Horopter Curves

The definition of a horopter curve is based on a method for generating the curve using an orthogonal collineation. The method is a special case of the star generation of a twisted cubic, as described by Semple & Kneebone (1952). It is recalled that a star of lines is the set of all lines in $\mathbb{P}^3$ containing a given point of $\mathbb{P}^3$. The given point is the centre of the star.

Definition 3.19. *Let two stars of lines with distinct real centres be given, together with a real orthogonal collineation between the two stars of lines that does not induce the identity collineation of* Π_∞*. Then the locus of the intersections of corresponding pairs of lines is a horopter curve.*

The identity collineation is excluded in Definition 3.19 because in this case the locus of intersections of corresponding lines consists of the whole plane Π_∞ together with the line joining the two centres.

Let $\mathbf{a}$, $\mathbf{b}$ be the two centres and let ω be the orthogonal collineation. Let each star of lines be parameterised by the points of Π_∞. A point $\mathbf{x}$ is on the locus of the intersections of corresponding pairs of lines if and only if there is a point $\mathbf{r}$ of Π_∞ depending on $\mathbf{x}$ such that

$$\mathbf{x} = \langle \mathbf{a}, \mathbf{r} \rangle \cap \langle \mathbf{b}, \omega(\mathbf{r}) \rangle \tag{3.28}$$

The notation used in Sect. 2.3 is recalled. Let ψ, ϕ be a critical surface pair and let $\mathbf{a}$, $\mathbf{b}$ be the two possible optical centres after the displacement of the camera. The optical centre of the camera prior to the displacement is at the origin $\mathbf{o}$. The construction of ψ and ϕ in Sect. 2.3.2 is based on an orthogonal collineation ω from the lines through $\mathbf{a}$ to the lines through $\mathbf{b}$. The points $\mathbf{x}$ such that (3.28) holds are contained in $\psi \cap \phi$ because $\langle \mathbf{o}, \mathbf{x} \rangle$ is the unique common transversal of $\langle \mathbf{a}, \mathbf{r} \rangle$ and $\langle \mathbf{b}, \omega(\mathbf{r}) \rangle$ passing through the optical centre $\mathbf{o}$ of the first camera. It follows that $\psi \cap \phi$ contains the horopter curve defined by (3.28). A horopter curve is in general a non-singular space curve of degree three. It is identical to the cubic factor of $\psi \cap \phi$ found in Proposition 2.28.

A horopter curve can be characterised as a twisted cubic which meets Π_∞ at three particular points.

Proposition 3.20. *The twisted cubic contained in the intersection of a critical surface pair meets the plane at infinity at three points* $\mathbf{n}$, $\mathbf{i}_n$, $\mathbf{j}_n$, *such that* $\mathbf{n}$ *is not on the absolute conic and such that* $\mathbf{i}_n$, $\mathbf{j}_n$ *are the points of contact of the tangents drawn from the* $\mathbf{n}$ *to the absolute conic. The point* $\mathbf{n}$ *is real and* $\mathbf{i}_n$, $\mathbf{j}_n$ *are complex conjugate.*

Proof. Let ψ, ϕ be a critical surface pair, and let ω be the orthogonal collineation of Π_∞ associated with ψ, as illustrated in Fig. 2.5. Let $\mathbf{n}$, $\mathbf{i}_n$, $\mathbf{j}_n$ be the fixed points of ω in Π_∞. It follows from (3.28) and the remarks preceding the statement of this proposition that $\mathbf{n}$, $\mathbf{i}_n$, $\mathbf{j}_n$ are contained in $\psi \cap \phi$. It is shown in Proposition 3.2 that $\mathbf{i}_n$, $\mathbf{j}_n$ are the points of contact of the tangents drawn from $\mathbf{n}$ to Ω. □

The equations defining a horopter curve in a Cartesian coordinate system are readily obtained from Definition 3.19. Let $\mathbf{a}$, $\mathbf{b}$ be the two centres of the stars of lines, and let ω be the orthogonal collineation from the lines through $\mathbf{a}$ to the lines through $\mathbf{b}$. A point $\mathbf{x}$ is on the horopter curve if and only if the vectors $\omega(\mathbf{x} - \mathbf{a})$ and $\mathbf{x} - \mathbf{b}$ are parallel. In Cartesian coordinates, $\mathbf{x}$ is on the horopter curve if and only if

$$\omega(\mathbf{x} - \mathbf{a}) \times (\mathbf{x} - \mathbf{b}) = 0 \tag{3.29}$$

Equation (3.29) yields three constraints of degree two on $\mathbf{x}$, one from each component of the equation. All three constraints are needed to define the horopter curve.

The converse to Proposition 3.20 is also true. Before proving it the properties of twisted cubics are recalled from Semple & Kneebone (1952). A twisted cubic is a curve in $\mathbb{P}^3$ of degree three, which is not contained in any plane in $\mathbb{P}^3$. Each twisted cubic is rational because it can be parameterised algebraically by the points of $\mathbb{P}^1$. The parameterisation of a twisted cubic has the form

$$t \mapsto A \begin{pmatrix} 1 \\ t \\ t^2 \\ t^3 \end{pmatrix}$$

where A is a non-singular 4×4 matrix. The twisted cubic inherits the projective geometric structure of $\mathbf{P}^1$ under this parameterisation. This inherited structure is compatible with the projective geometric structure of the space $\mathbf{P}^3$ containing the twisted cubic. In detail, let the following homographic correspondence be defined on a twisted cubic c,

$$t \mapsto (at + b)/(ct + d) \tag{3.30}$$

Then there exists a unique collineation ω of $\mathbf{P}^3$ such that $\omega(c) = c$, and such that ω induces the homography (3.30) on c. Further, every collineation of $\mathbf{P}^3$ which leaves c invariant arises from a unique homographic correspondence on c of the form (3.30). If a twisted cubic is projected onto a plane from a general point of $\mathbf{P}^3$ then a cubic plane curve with a node is obtained. If the twisted cubic is projected from a point on itself then a conic is obtained.

Proposition 3.21. *A non-singular real twisted cubic is a horopter curve if and only if it intersects the plane at infinity at three points such that one of these points is real and such that the other two points are the points of contact of the two tangents drawn from the first point to the absolute conic.*

Proof. An argument similar to that used in Proposition 3.20 establishes that a horopter curve satisfies the requirements of this proposition. Conversely, let c be a non-singular real twisted cubic intersecting Π_∞ at $\mathbf{n}$, $\mathbf{i}_n$, $\mathbf{j}_n$, where $\mathbf{n}$ is real and $\mathbf{i}_n$, $\mathbf{j}_n$ are the points of contact of the tangents drawn from $\mathbf{n}$ to the absolute conic Ω. It follows from the remarks preceding this proposition that there exists a unique involution τ of $\mathbf{P}^3$ which leaves c invariant and which has the properties $\tau(\mathbf{n}) = \mathbf{n}$, $\tau(\mathbf{i}_n) = \mathbf{j}_n$. The plane $\Pi_\infty = \langle \mathbf{n}, \mathbf{i}_n, \mathbf{j}_n \rangle$ is invariant under τ. It follows from Proposition 3.7 that τ is rigid. Let $\mathbf{a}$ be a real point of c such that $\tau(\mathbf{a}) \neq \mathbf{a}$, and let $\mathbf{b} = \tau(\mathbf{a})$. It follows from the star generation of twisted cubics that there exists a collineation ω of Π_∞ such that a point $\mathbf{x}$ is on c if and only if there is a point $\mathbf{r}$ of Π_∞, depending on $\mathbf{x}$, such that

$$\mathbf{x} = \langle \mathbf{a}, \mathbf{r} \rangle \cap \langle \mathbf{b}, \omega(\mathbf{r}) \rangle \tag{3.31}$$

As $\mathbf{x}$ varies on c, the points $\mathbf{r}$ of Π_∞ such that (3.31) holds vary on the conic s formed by projecting c from $\mathbf{a}$ to Π_∞.

The collineation ω exists for any pair of distinct points $\mathbf{a}$, $\mathbf{b}$ on c. It is shown that the condition $\tau(\mathbf{a}) = \mathbf{b}$ ensures that ω is real and orthogonal. Let $\overline{\mathbf{x}}$ be the complex conjugate of $\mathbf{x}$. Complex conjugation of (3.31) yields

$$\overline{\mathbf{x}} = \langle \mathbf{a}, \overline{\mathbf{r}} \rangle \cap \langle \mathbf{b}, \overline{\omega}(\overline{\mathbf{r}}) \rangle \tag{3.32}$$

The point $\overline{\mathbf{x}}$ is on c because $\mathbf{x}$ is on c and c is real. It follows from (3.32) that $\omega(\overline{\mathbf{r}}) = \overline{\omega}(\overline{\mathbf{r}})$ for all $\mathbf{r}$ in s. This yields $\omega = \overline{\omega}$ as required. To show that ω is orthogonal, note that the application of τ to (3.31) yields

$$\tau(\mathbf{x}) = \langle \mathbf{b}, \tau(\mathbf{r}) \rangle \cap \langle \mathbf{a}, \tau\omega(\mathbf{r}) \rangle \tag{3.33}$$

The point $\tau(\mathbf{x})$ is on c, because c is invariant under τ. A comparison of (3.31) and (3.33) yields $\omega\tau\omega(\mathbf{r}) = \tau(\mathbf{r})$ for all points $\mathbf{r}$ on s. It follows that $(\tau\omega)^2$ is the identity on s, thus $(\tau\omega)^2$ is the identity on the whole of Π_∞. Hence $\tau\omega$ is an involution of Π_∞. It follows from (3.31) that $\mathbf{n}$, $\mathbf{i}_n$, $\mathbf{j}_n$ are fixed points of ω, thus $\tau\omega(\mathbf{n}) = \mathbf{n}$ and $\tau\omega(\mathbf{i}_n) = \mathbf{j}_n$. Proposition 3.7 is applied to deduce that $\tau\omega$ is orthogonal. The involution τ is rigid, thus ω is orthogonal. The fact that c is a horopter curve now follows from Definition 3.19. □

Corollary 1. A horopter curve is invariant under a non-trivial rigid involution.

Corollary 2. The collineation from the star of lines through $\mathbf{a}$ to the star of lines through $\mathbf{b}$ defined by a horopter curve containing $\mathbf{a}$ and $\mathbf{b}$ is orthogonal if and only if $\tau(\mathbf{a}) = \mathbf{b}$.

3.3.2 Rigid Involutions of Horopter Curves

The non-trivial rigid involution of a horopter curve found in Corollary 1 to Proposition 3.21 is investigated further. An explicit construction of the involution is described. It is apparent from the construction that the involution is a rotation through 180°. The rigid involution of a horopter curve is thus a skew involution.

Theorem 3.22. *A horopter curve is invariant under a unique non-trivial rigid skew involution.*

Proof. Let c be a horopter curve generated by a real orthogonal collineation ω from the star of lines with centre $\mathbf{a}$ to the star of lines with centre $\mathbf{b}$, where $\mathbf{a}$, $\mathbf{b}$ are distinct points of $\mathbb{R}^3$. Let ω be described by its action on Π_∞. Let $\mathbf{n}$ be the real fixed point of ω in Π_∞, let $\mathbf{p}$ be the intersection of $\langle\mathbf{a},\mathbf{b}\rangle$ with Π_∞ and let $\mathbf{d}$ be the midpoint of the line segment $[\mathbf{a},\mathbf{b}]$.

The points of c are precisely those points $\mathbf{x}$ satisfying (3.31) for some point $\mathbf{r}$ in Π_∞ depending on $\mathbf{x}$. It follows from Proposition 3.8 that $\omega = \tau_1\tau_2$ where τ_1 and τ_2 are orthogonal involutions of Π_∞ with vertices $\mathbf{n}_1$ and $\mathbf{n}_2$, such that $\mathbf{n}_1$ and $\mathbf{n}_2$ are both orthogonal to $\mathbf{n}$. One of the vertices $\mathbf{n}_1$, $\mathbf{n}_2$ can be chosen at will, subject to the constraint that it is orthogonal to $\mathbf{n}$. Let $\mathbf{n}_1 = \mathbf{n}\times\mathbf{p}$ and let τ be the rigid skew involution of $\mathbb{P}^3$ with a line of fixed points passing through $\mathbf{d}$ with direction $\mathbf{n}_1$. It follows from the definition of τ that $\tau(\mathbf{a}) = \mathbf{b}$ and that τ_1 is the restriction of τ to Π_∞. Let $\mathbf{s} = \tau_2(\mathbf{r})$. The application of τ to (3.31) yields

$$
\begin{aligned}
\tau(\mathbf{x}) &= \langle\mathbf{b},\tau_1(\mathbf{r})\rangle \cap \langle\mathbf{a},\tau_1\omega(\mathbf{r})\rangle \\
&= \langle\mathbf{b},\tau_1(\mathbf{r})\rangle \cap \langle\mathbf{a},\tau_2(\mathbf{r})\rangle \\
&= \langle\mathbf{a},\mathbf{s}\rangle \cap \langle\mathbf{b},\omega(\mathbf{s})\rangle \qquad (3.34)
\end{aligned}
$$

On comparing (3.31) and (3.34) it is apparent that $\tau(\mathbf{x})$ is a point of c. The horopter curve c is thus invariant under the non-trivial rigid involution τ. It remains to show that τ is unique. The set $\{\mathbf{n},\mathbf{i}_n,\mathbf{j}_n\}$ is invariant under any rigid involution of c. It follows from the remarks preceding Proposition 3.21 that a

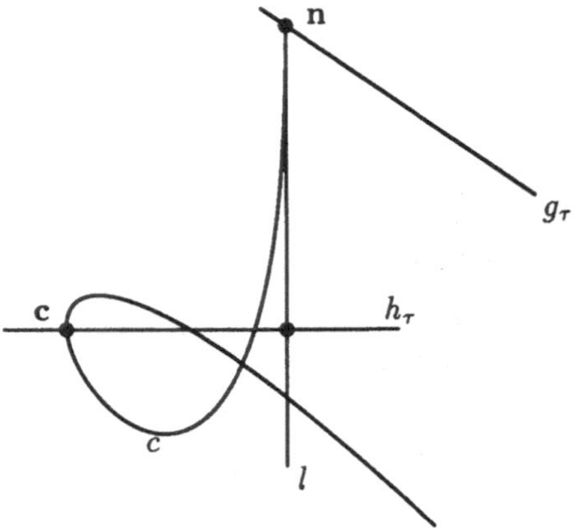

Fig. 3.6. The rigid involution of a horopter curve

rigid involution of c is uniquely determined by its action on $\{\mathbf{n}, \mathbf{i}_n, \mathbf{j}_n\}$. Any rigid involution of c fixes $\mathbf{n}$, because $\mathbf{n}$ is the only point of $c \cap \Pi_\infty$ not contained in the absolute conic. There remain only two possibilities. The involution fixes $\mathbf{i}_n$, $\mathbf{j}_n$, in which case it is the identity, or the involution interchanges $\mathbf{i}_n$, $\mathbf{j}_n$, in which case it is τ. □

The rigid involution τ constructed in the proof of Theorem 3.22 is illustrated in Fig. 3.6. The lines g_τ and h_τ are the two skew lines of fixed points of τ, labelled such that g_τ is contained in Π_∞. There are exactly two fixed points of τ on c. One of these points is the real point $\mathbf{n}$ of $c \cap \Pi_\infty$. The other point $\mathbf{c}$ is known as the centre of c. The point $\mathbf{n}$ lies on g_τ and $\mathbf{c}$ lies on h_τ. The tangent to c at $\mathbf{n}$ is the unique real asymptote of c. The tangents to c at $\mathbf{n}$ and $\mathbf{c}$ are both invariant under τ, thus each tangent is a common transversal of g_τ and h_τ.

3.3.3 The Centre of a Horopter Curve

The centre of a horopter curve is the unique fixed point of the non-trivial rigid involution of the curve that is not contained in the plane at infinity. The centre should not be confused with the optical centre of a camera.

Theorem 3.23. *The centre of a horopter curve is the unique point of the curve furthest from the real asymptote.*

Proof. Let c be a horopter curve invariant under a non-trivial rigid involution τ, and let $\mathbf{n}$, $\mathbf{i}_n$, $\mathbf{j}_n$ be the three intersections of c with Π_∞, such that $\mathbf{n}$ is not on the absolute conic Ω. Let l be the tangent to c at $\mathbf{n}$ (l is the real asymptote of c). Let ϕ_d be the circular cylinder of radius d with axis l. The surface ϕ_d is a quadric containing $\mathbf{n}$, $\mathbf{i}_n$, $\mathbf{j}_n$, with a singular point at $\mathbf{n}$. The line l is invariant under τ, thus ϕ_d is invariant under τ. It follows that $\phi_d \cap c$ is invariant under τ.

It is shown that for a general value of d, $\phi_d \cap c$ contains five distinct points.

Let $\mathbf{a}$, $\mathbf{b}$ be two real points on c such that c is the locus of the intersections of lines through $\mathbf{a}$ and $\mathbf{b}$, corresponding under an orthogonal collineation. Let r be the value of d for which ϕ_d contains $\mathbf{a}$. Then ϕ_r also contains $\mathbf{b}$ because ϕ_r is invariant under τ and because by Corollary 2 to Proposition 3.21, $\mathbf{b} = \tau(\mathbf{a})$. It follows that $c \cap \phi_r \supseteq \{\mathbf{n}, \mathbf{i}_n, \mathbf{j}_n, \mathbf{a}, \mathbf{b}\}$. The intersection $\phi_d \cap c$ for a general value of d contains at least five distinct points. The space curve c is of degree three and ϕ_d is of degree two, thus $\phi_d \cap c$ contains $6 = 3 \times 2$ points, counted with the correct multiplicities. The point $\mathbf{n}$ is counted at least twice in $\phi_d \cap c$ because it is a singular point of ϕ_d. Thus $\phi_d \cap c$ contains in general at most five distinct points. In consequence, $\phi_d \cap c$ contains in general exactly five distinct points.

Let $\phi_d \cap c = \{\mathbf{n}, \mathbf{i}_n, \mathbf{j}_n, \mathbf{p}_d, \mathbf{q}_d\}$. The points $\mathbf{p}_d = \mathbf{q}_d$ are not in general fixed by τ because τ has exactly two fixed points on c and neither of these fixed points is equal to $\mathbf{p}_d$ or $\mathbf{q}_d$. It follows that $\tau(\mathbf{p}_d) = \mathbf{q}_d$. If d is sufficently large then $\mathbf{p}_d$, $\mathbf{q}_d$ are complex conjugates, because c has only one real asymptotic line and this line is the axis of ϕ_d. There also exist values of d for which $\mathbf{p}_d$ and $\mathbf{q}_d$ are both real, for example $d = r$. Let d' be the largest value of d for which $\mathbf{p}_d$, $\mathbf{q}_d$ are real. The cone $\phi_{d'}$ is tangential to c at $\mathbf{p}_{d'}$ thus $\mathbf{p}_{d'}$ and $\mathbf{q}_{d'}$ coincide at a point $\mathbf{c}$, and $\tau(\mathbf{c}) = \mathbf{c}$. The point $\mathbf{c}$ is thus the centre of c. It follows from the definition of $\phi_{d'}$ that $\mathbf{c}$ is the unique point of c furthest from l. □

The centre of a horopter curve is used to obtain criteria for two horopter curves to be the same. These criteria are applied in Sect. 3.4 to the ambiguous case of reconstruction.

Theorem 3.24. *Let two non-singular horopter curves be given such that they are invariant under the same non-trivial rigid involution, they have the same centre and such that they have the same intersections with the plane at infinity. If either i) the two curves have at least one further point in common; or ii) the two curves have the same real asymptote and the same tangent line at the common centre, then the two curves are identical.*

Proof. Let c_1, c_2 be two horopter curves invariant under the same non-trivial rigid involution τ, such that $c_1 \cap \Pi_\infty = c_2 \cap \Pi_\infty = \{\mathbf{n}, \mathbf{i}_n, \mathbf{j}_n\}$, and such that c_1, c_2 have the same centre $\mathbf{c}$. To prove (i), let $\mathbf{a}$ be a point common to c_1 and c_2, but distinct from $\mathbf{n}$, $\mathbf{i}_n$, $\mathbf{j}_n$ and $\mathbf{c}$. Let ω_1, ω_2 be collineations of Π_∞ that define c_1 and c_2 respectively as the loci of the intersections of corresponding lines from stars of lines centred at $\mathbf{a}$ and $\tau(\mathbf{a})$. The horopter curves c_1 and c_2 intersect at $\mathbf{n}$, $\mathbf{i}_n$, $\mathbf{j}_n$ thus

$$\begin{aligned} \omega_1(\mathbf{n}) &= \omega_2(\mathbf{n}) = \mathbf{n} \\ \omega_1(\mathbf{i}_n) &= \omega_2(\mathbf{i}_n) = \mathbf{i}_n \\ \omega_1(\mathbf{j}_n) &= \omega_2(\mathbf{j}_n) = \mathbf{j}_n \end{aligned} \tag{3.35}$$

Let $\mathbf{r} = \langle \mathbf{a}, \mathbf{c} \rangle \cap \Pi_\infty$ and let $\mathbf{s} = \langle \tau(\mathbf{a}), \mathbf{c} \rangle \cap \Pi_\infty$. The centre $\mathbf{c}$ is on both c_1 and c_2 thus

$$\omega_1(\mathbf{r}) = \omega_2(\mathbf{r}) = \mathbf{s} \tag{3.36}$$

It follows from (3.35) and (3.36) that ω_1 and ω_2 take the same values at the four points $\mathbf{n}$, $\mathbf{i}_n$, $\mathbf{j}_n$, $\mathbf{r}$ of Π_∞. If $\mathbf{r}$, $\mathbf{n}$, $\mathbf{i}_n$ are collinear, then the plane $\Pi = \langle \mathbf{r}, \mathbf{n}, \mathbf{c} \rangle$ meets c at four points. This is impossible because c is by hypothesis non-singular. $\mathbf{r}$, $\mathbf{n}$, $\mathbf{j}_n$ are not collinear. Similar arguments establish that no three of the points $\mathbf{r}$, $\mathbf{n}$, $\mathbf{i}_n$, $\mathbf{j}_n$ are collinear. It follows that $\omega_1 = \omega_2$. Hence $c_1 = c_2$, as required.

To prove (ii), let c_1 and c_2 have the same real asymptotic line l and the same tangent line k at the common centre $\mathbf{c}$. Let ω_1, ω_2 be collineations from the star of lines centred at $\mathbf{c}$ to the star of lines centred at $\mathbf{n}$ that define c_1 and c_2 respectively as loci of the intersections of corresponding lines. It follows that

$$\begin{aligned} \omega_1\langle \mathbf{c}, \mathbf{i}_n \rangle &= \omega_2\langle \mathbf{c}, \mathbf{i}_n \rangle = \langle \mathbf{n}, \mathbf{i}_n \rangle \\ \omega_1\langle \mathbf{c}, \mathbf{j}_n \rangle &= \omega_2\langle \mathbf{c}, \mathbf{j}_n \rangle = \langle \mathbf{n}, \mathbf{j}_n \rangle \end{aligned} \tag{3.37}$$

It follows from the star generation of twisted cubics that

$$\begin{aligned} \omega_1(k) &= \omega_2(k) = \langle \mathbf{c}, \mathbf{n} \rangle \\ \omega_1\langle \mathbf{c}, \mathbf{n} \rangle &= \omega_2\langle \mathbf{c}, \mathbf{n} \rangle = l \end{aligned} \tag{3.38}$$

No three of the lines $\langle \mathbf{c}, \mathbf{i}_n \rangle$, $\langle \mathbf{c}, \mathbf{j}_n \rangle$, k, $\langle \mathbf{c}, \mathbf{n} \rangle$ are coplanar because otherwise c would have a component contained in a plane, leading to a contradiction of the hypothesis that c is non-singular. It follows from (3.37) and (3.38) that $\omega_1 = \omega_2$. Hence $c_1 = c_2$, as required. □

3.3.4 Examples

Examples are given of twisted cubics and horopter curves in which the parameterisation of the curves by cubic polynomials is made explicit. A Cartesian coordinate system is chosen. The curves are required to be invariant under the non-trivial rigid skew involution τ defined by

$$\tau = \begin{pmatrix} 1 & 0 & 0 & 0 \\ 0 & 0 & 1 & 0 \\ 0 & 1 & 0 & 0 \\ 0 & 0 & 0 & -1 \end{pmatrix}$$

It is also required that the curves pass through the fixed points $\mathbf{c}$, $\mathbf{n}$ of τ given by

$$\mathbf{c} = (0,0,0,1)^\top \qquad \mathbf{n} = (1,0,0,0)^\top$$

Let c be a twisted cubic with a parameterisation $t \mapsto (f(t), g(t), h(t), k(t))^\top$ where

$$\begin{aligned} f(t) &= f_0 + f_1 t + f_2 t^2 + f_3 t^3 \\ g(t) &= g_0 + g_1 t + g_2 g^2 + g_3 t^3 \\ h(t) &= h_0 + h_1 t + h_2 t^2 + h_3 t^3 \\ k(t) &= k_0 + k_1 t + k_2 t^2 + k_3 t^3 \end{aligned} \tag{3.39}$$

As noted prior to Proposition 3.21, under parameterisation by t the curve c has a projective geometric structure compatible with the space $\mathbb{P}^3$ containing c. The parameterisation of c by t is not unique, in that each bilinear transformation of t, $t \mapsto (at+b)/(ct+d)$ yields a new parameterisation. In order to fix a unique parameterisation it is required that $\mathbf{c}$ corresponds to $t = 0$, $\mathbf{n}$ corresponds to $t = \infty$ and one of the intersections of c with Π_∞ corresponds to $t = i$. These requirements ensure that the values of t at the remaining points of c are uniquely determined.

With this choice of parameterisation, the restriction of τ to c is given by $t \mapsto -t$, because this is the only non-trivial involution of c with fixed points at $t = 0$ and $t = \infty$. The plane Π_∞ is invariant under τ, thus the three points of $c \cap \Pi_\infty$ are given by $t = 0$, $t = i$, $t = -i$. The points given by $t = \pm i$ are not, in general, the points of contact of the tangents drawn from $\mathbf{n}$ to the absolute conic, Ω.

As a result of the above restrictions on c and the choice of parameterisation of c, the equations of (3.39) reduce to

$$\begin{aligned} f(t) &= f_1 t + f_3 t^3 \\ g(t) &= g_1 t + g_2 t^2 \\ h(t) &= g_1 t - g_2 t^2 \\ k(t) &= k_0(1+t^2) \end{aligned} \tag{3.40}$$

It follows from (3.40) that the points given by $t = \pm i$ at which c meets Π_∞ are

$$(i(f_1 - f_3), ig_1 - g_2, ig_1 + g_2, 0)^\top$$
$$(-i(f_1 - f_3), -ig_1 - g_2, -ig_1 + g_2, 0)^\top$$

In general, neither of these points is equal to the points of contact $\mathbf{i}_n$, $\mathbf{j}_n$ of the tangents drawn from $\mathbf{n}$ to Ω, thus the twisted cubic c is not, in general, a horopter curve.

Let c now be a horopter curve and let $t = i$ correspond to the point $\mathbf{i}_n = (0,1,i,0)^\top$. It follows that

$$(i(f_1 - f_3), ig_1 - g_2, ig_1 + g_2, 0)^\top = (0,1,i,0)^\top$$

thus $f_1 = f_3$ and $g_1 = -g_2$. The equations of (3.40) reduce to

$$\begin{aligned} f(t) &= f_1(t + t^3) \\ g(t) &= g_1(t - t^2) \\ h(t) &= g_1(t + t^2) \\ k(t) &= k_0(1+t^2) \end{aligned} \tag{3.41}$$

The coefficients f_1, g_1, k_0 of (3.41) can be further restricted by specifying the tangent line to c at $\mathbf{n}$. For example, if this line is required to be $\langle \mathbf{n}, (0,-1,1,1)^\top \rangle$, then $k_0 = g_1$.

3.3.5 Horopter Curves on Rectangular Quadrics

The horopter curves arising in the ambiguous case of reconstruction are contained within a critical surface. It is thus necessary to examine the relation between a horopter curve and the critical surface containing it. The main aim in this subsection is to show that the horopter curves contained in a rectangular quadric ψ and invariant under the skew involution τ_ψ form four one-parameter families.

The four families of horopter curves are defined. Let $\mathbf{m}$, $\mathbf{n}$ be the principal points of ψ and let $\mathcal{F}_1$, $\mathcal{F}_2$ be the two families of generators of ψ. Let $\mathcal{H}_1(\mathbf{n})$ be the set of horopter curves in ψ that are invariant under τ_ψ, that contain $\mathbf{n}$ and that meet each generator of $\mathcal{F}_1$ twice. The sets $\mathcal{H}_2(\mathbf{n})$, $\mathcal{H}_i(\mathbf{m})$ are defined *mutatis mutandis*. The sets $\mathcal{H}_i(\mathbf{m})$, $\mathcal{H}_i(\mathbf{n})$ contain all the horopter curves invariant under τ_ψ, because firstly each horopter curve c in ψ contains either $\mathbf{m}$ or $\mathbf{n}$ and secondly each horopter curve in ψ either meets each generator of $\mathcal{F}_1$ twice, or it meets each generator of $\mathcal{F}_2$ twice.

Proposition 3.25. *The intersection of any two of the four families $\mathcal{H}_i(\mathbf{m})$, $\mathcal{H}_i(\mathbf{n})$ of horopter curves contains only a finite number of curves. The curves in any such intersection are split.*

Proof. Let c be a curve in $\mathcal{H}_i(\mathbf{m}) \cap \mathcal{H}_j(\mathbf{n})$. Then $c \cap \Pi_\infty$ contains both $\mathbf{m}$, $\mathbf{n}$ and either the pair of points $\mathbf{i}_m$, $\mathbf{j}_m$ or the pair of points $\mathbf{i}_n$, $\mathbf{j}_n$. The curve c is of degree three. Thus, as Π_∞ contains four points of c, Π_∞ contains a component of c. The component is the conic $s_\infty = \psi \cap \Pi_\infty$, because c is contained in ψ. The remaining component of c is a generator of ψ invariant under τ_ψ. There are only four such generators. It follows that there are at most four such curves c, and that they all split.

To complete the proof, suppose if possible that c is a curve contained in $\mathcal{H}_1(\mathbf{m}) \cap \mathcal{H}_2(\mathbf{m})$ or in $\mathcal{H}_1(\mathbf{n}) \cap \mathcal{H}_2(\mathbf{n})$. Let g_1, g_2 be generators of ψ taken from different families $\mathcal{F}_1$, $\mathcal{F}_2$. Then in general, c meets the plane $\langle g_1, g_2 \rangle$ at four points. This is impossible, because c is of degree three. It follows from this contradiction that $\mathcal{H}_1(\mathbf{m}) \cap \mathcal{H}_2(\mathbf{m})$ and $\mathcal{H}_1(\mathbf{n}) \cap \mathcal{H}_2(\mathbf{n})$ are empty. □

Theorem 3.26. *The four families $\mathcal{H}_i(\mathbf{m})$, $\mathcal{H}_i(\mathbf{n})$ of horopter curves are each parameterised by $\mathbf{P}^1$.*

Proof. Let ψ be the non-singular rectangular quadric and let l_ψ be the line formed by the intersection of the tangent planes to ψ at the principal points $\mathbf{m}$ and $\mathbf{n}$ of ψ. Let $\mathbf{c}_1$, $\mathbf{c}_2$ be the points at which l_ψ intersects ψ. The generators $g_1(\mathbf{n})$, $g_2(\mathbf{n})$ of ψ passing through $\mathbf{n}$ meet l_ψ. After relabelling $\mathbf{c}_1$ and $\mathbf{c}_2$ if necessary, it is assumed that

$$
\begin{aligned}
g_1(\mathbf{n}) &= \langle \mathbf{n}, \mathbf{c}_1 \rangle \\
g_2(\mathbf{n}) &= \langle \mathbf{n}, \mathbf{c}_2 \rangle
\end{aligned}
$$

as illustrated in Fig. 3.7. It follows from the definition of τ_ψ, given in Sect. 3.2.2, that $\mathbf{c}_1$ and $\mathbf{c}_2$ are the only fixed points of τ_ψ contained in $\psi \setminus \Pi_\infty$. Thus any

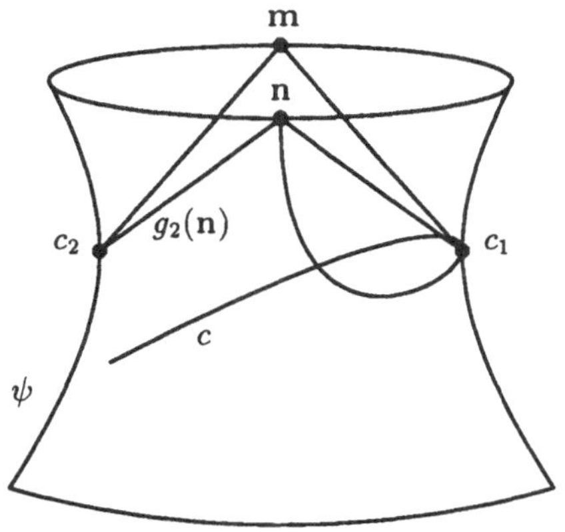

Fig. 3.7. Horopter curves on a rectangular quadric

horopter curve contained in ψ and invariant under τ_ψ has a centre either at $\mathbf{c}_1$ or at $\mathbf{c}_2$. Each horopter curve c in $\mathcal{H}_1(\mathbf{n})$ meets $g_1(\mathbf{n})$ twice and $g_2(\mathbf{n})$ once (at $\mathbf{n}$). The point $\mathbf{c}_2$ is thus not on c. Hence c has its centre at $\mathbf{c}_1$.

Let $\mathbf{p}$ be a general point of ψ. It is shown that $\mathcal{H}_1(\mathbf{n})$ is parameterised by the points of $g_2(\mathbf{p})$. Let c be a horopter curve in $\mathcal{H}_1(\mathbf{n})$ that meets $g_2(\mathbf{p})$ at $\mathbf{a}$. It follows from Theorem 3.24 that c is the only horopter curve in $\mathcal{H}_1(\mathbf{n})$ that contains $\mathbf{a}$. It is necessary to show that each point $\mathbf{a}$ of $g_2(\mathbf{p})$ lies on a horopter curve in $\mathcal{H}_1(\mathbf{n})$. Let ϕ be the unique cone with vertex $\mathbf{n}$ containing the points $S = \{\mathbf{n}, \mathbf{i}_n, \mathbf{j}_n, \mathbf{c}_1, \mathbf{a}, \tau_\psi(\mathbf{a})\}$. The set S is invariant under τ_ψ and $\mathbf{n}$ is fixed by τ_ψ, thus $\tau_\psi(\phi) = \phi$. The intersection $\phi \cap \psi$ contains a common generator of ϕ and ψ, namely $\langle \mathbf{n}, \mathbf{c}_1 \rangle$, thus

$$\phi \cap \psi = \langle \mathbf{n}, \mathbf{c}_1 \rangle \cup c$$

where c is a twisted cubic meeting $\langle \mathbf{n}, \mathbf{c}_1 \rangle$ twice. The curve c is invariant under τ_ψ because ϕ and ψ are both invariant under τ_ψ. The projection of c from $\mathbf{n}$ is a conic because c lies in ϕ. It follows that $\mathbf{n}$ is a point of c. The curve c also contains $\mathbf{i}_n$ and $\mathbf{j}_n$, thus by Theorem 3.21, c is a horopter curve. Hence c is in $\mathcal{H}_1(\mathbf{n})$, as required. It follows that $\mathcal{H}_1(\mathbf{n})$ is a one-parameter family of horopter curves, parameterised by the points of $g_2(\mathbf{p})$. Similar arguments establish that $\mathcal{H}_2(\mathbf{n})$ and the $\mathcal{H}_i(\mathbf{m})$ are also one-parameter families of horopter curves. □

The next result is useful for the classification of horopter curves.

Proposition 3.27. *Let k be any line tangent to ψ at the centre $\mathbf{c}_1$ of the horopter curves in $\mathcal{H}_1(\mathbf{n})$. Then there exists a unique horopter curve c in $\mathcal{H}_1(\mathbf{n})$ with tangent k at $\mathbf{c}_1$.*

Proof. Let k intersect Π_∞ at $\mathbf{k}$ and let s be the conic in Π_∞ containing $\mathbf{n}$, $\mathbf{i}_n$, $\mathbf{j}_n$, $\mathbf{k}$ that is invariant under τ_ψ. The conic s is unique because it contains the four points $\mathbf{n}$, $\mathbf{i}_n$, $\mathbf{j}_n$, $\mathbf{k}$ and because the tangent to s at $\mathbf{k}$ passes through the vertex of τ. Let ϕ be the cone with vertex $\mathbf{c}_1$ and base s. Then ϕ is invariant under τ_ψ. The intersection $\psi \cap \phi$ is a space curve of degree four which splits into the line $\langle \mathbf{c}_1, \mathbf{n} \rangle$ and a space curve c' of degree three. The surfaces ψ, ϕ and the line $\langle \mathbf{c}_1, \mathbf{n} \rangle$ are invariant under τ_ψ, thus c' is invariant under τ_ψ. The curve c' contains $\mathbf{i}_n$ and $\mathbf{j}_n$. The curve c' also contains $\mathbf{n}$ because the projection of c' from $\mathbf{n}$ is a conic. It follows from Theorem 3.21 that c' is a horopter curve. The line k is tangent to c' at $\mathbf{c}_1$ because k is contained in ϕ. □

The question arises of how different are the horopter curves in any one of the families $\mathcal{H}_i(\mathbf{m})$, $\mathcal{H}_i(\mathbf{n})$. In the remainder of this subsection it is shown that if any two horopter curves in $\mathcal{H}_1(\mathbf{n})$ are given then there is no Euclidean transformation which transforms the first horopter curve into the second horopter curve. In pursuit of this result it is convenient to make the following definition.

Definition 3.28. *Two horopter curves, c_1 and c_2 are conjugate if there exists a Euclidean transformation ω of $\mathbb{P}^3$ such that $\omega(c_1) = c_2$. Conjugacy between horopter curves is denoted by $c_1 \sim c_2$.*

It follows from Definition 3.28 that conjugacy of horopter curves is an equivalence relation: if c_1, c_2, c_3 are three horopter curves then i) $c_1 \sim c_1$; ii) if $c_1 \sim c_2$ then $c_2 \sim c_1$; and iii) if $c_1 \sim c_2$ and $c_2 \sim c_3$ then $c_1 \sim c_3$. Any set with an equivalence relation defined on it is partitioned into disjoint equivalence classes, such that the members of a given class are conjugate to one another and such that members of different classes are not conjugate.

The conjugacy classes of horopter curves are described. The first step is to show that two distinct horopter curves drawn from one of the families $\mathcal{H}_i(\mathbf{m})$, $\mathcal{H}_i(\mathbf{n})$ are not conjugate. This is done in Theorem 3.30. Proposition 3.29 is a useful preliminary.

Proposition 3.29. *Let ψ be a non-singular rectangular quadric with distinct principal points $\mathbf{m}$, $\mathbf{n}$. Then the horopter curves in $\mathcal{H}_1(\mathbf{n})$ or in $\mathcal{H}_2(\mathbf{n})$ all have the same real asymptotic line. Similarly, the horopter curves in $\mathcal{H}_1(\mathbf{m})$ or in $\mathcal{H}_2(\mathbf{m})$ have the same real asymptotic line.*

Proof. Let l be the real asymptotic line of a horopter curve c in $\mathcal{H}_1(\mathbf{n})$ or $\mathcal{H}_2(\mathbf{n})$. Then l is in the tangent plane Π to ψ at $\mathbf{n}$ and l is tangent to c at $\mathbf{n}$. The line l is invariant under the rigid skew involution τ_ψ of ψ, thus l is a common transversal of the lines h_τ, g_τ of fixed points of τ_ψ. Let h_τ be the line of fixed points of τ_ψ not contained in Π_∞. Then $l = \langle \mathbf{n}, h_\tau \cap \Pi \rangle$, thus l is independent of the choice of c. □

Theorem 3.30. *The horopter curves in $\mathcal{H}_1(\mathbf{n})$ are pairwise non-conjugate.*

Proof. Let c_1, c_2 be horopter curves in $\mathcal{H}_1(\mathbf{n})$ and let ω be a Euclidean transformation such that $\omega(c_1) = c_2$. It is shown that $c_1 = c_2$. Let h_τ, g_τ be the two skew lines of fixed points of τ_ψ, chosen such that g_τ is contained in the plane at infinity, Π_∞. The collineation $\omega^{-1}\tau_\psi\omega$ is a non-trivial rigid skew involution of c_1. Thus, by Theorem 3.22, $\omega^{-1}\tau_\psi\omega = \tau_\psi$. Let $\mathbf{p}$ be any point of h_τ. Then $\omega(\mathbf{p})$ is a fixed point of τ_ψ because $\omega^{-1}\tau_\psi\omega(\mathbf{p}) = \mathbf{p}$, thus $\omega(\mathbf{p})$ is a fixed point of τ_ψ. It follows that either $\omega(h_\tau) = h_\tau$ or $\omega(h_\tau) = g_\tau$. The plane Π_∞ is invariant under ω, thus $\omega(h_\tau) \neq g_\tau$. It follows that $\omega(h_\tau) = h_\tau$. A similar argument establishes that $\omega(g_\tau) = g_\tau$. The point $h_\tau \cap \Pi_\infty$ is fixed by ω. The point $\mathbf{n}$ is also fixed by ω because $\mathbf{n}$ is the only point of

$$c_1 \cap \Pi_\infty = c_2 \cap \Pi_\infty = \{\mathbf{n}, \mathbf{i}_n, \mathbf{j}_n\}$$

not contained in the absolute conic Ω. On composing ω with τ_ψ if necessary, it is supposed that $\mathbf{i}_n$ and $\mathbf{j}_n$ are fixed by ω. In consequence, the four points $\{h_\tau \cap \Pi_\infty, \mathbf{n}, \mathbf{i}_n, \mathbf{j}_n\}$ are fixed by ω. The restriction of ω to Π_∞ is thus the identity.

The common centre $\mathbf{c}$ of c_1 and c_2 is the unique point at which c_1 and c_2 meet h_τ. The line h_τ is invariant under ω, thus $\omega(\mathbf{c}) = \mathbf{c}$. The curves c_1, c_2 are in the same family $\mathcal{H}_1(\mathbf{n})$, thus by Proposition 3.29, c_1 and c_2 have the same real asymptotic line l. The line l is invariant under ω, thus the intersection $l \cap h_\tau$ is a fixed point of ω. The line h_τ thus contains three fixed points of ω, namely, $l \cap h_\tau$, $\mathbf{c}$ and $\Pi_\infty \cap h_\tau$. Hence, the restriction of ω to h_τ is the identity. It follows that ω is the identity, hence $c_1 = c_2$ as required. □

It follows from Theorem 3.30 that each horopter curve in $\mathcal{H}_1(\mathbf{n})$ is drawn from a different conjugacy class. In the next theorem it is shown that every conjugacy class of horopter curves contains a curve belonging to $\mathcal{H}_1(\mathbf{n})$. The final result is that $\mathcal{H}_1(\mathbf{n})$ contains a unique representative from each conjugacy class of horopter curves under the action of the Euclidean transformations.

Theorem 3.31. *Each non-singular horopter curve c is conjugate to a horopter curve in $\mathcal{H}_1(\mathbf{n})$.*

Proof. Let $\mathbf{c}_1$ be the common centre of the horopter curves in $\mathcal{H}_1(\mathbf{n})$ obtained in the proof of Theorem 3.26 and let l_n be the real asymptotic line common to the horopter curves in $\mathcal{H}_1(\mathbf{n})$ obtained in Proposition 3.29. Let the horopter curve c be invariant under the rigid skew involution τ_c, and let g_τ, h_τ be the lines of fixed points of τ_c, labelled such that g_τ is contained in Π_∞. Let $\mathbf{c}$ be the centre of c, and let l be the real asymptotic line of c. Let l_ψ be the axis of τ_ψ. Then there exists a Euclidean transformation ω such that

$$\begin{aligned} \omega(g_\tau) &= \langle \mathbf{m}, \mathbf{n} \rangle \\ \omega(h_\tau) &= l_\psi \\ \omega(l) &= l_n \\ \omega(\mathbf{c}) &= \mathbf{c}_1 \end{aligned}$$

It is straightforward to check that a suitable ω exists. There is a Euclidean transformation taking g_τ to $\langle \mathbf{m}, \mathbf{n} \rangle$ and h_τ to l_ψ, because g_τ is the polar of $h_\tau \cap \Pi_\infty$ with respect to Ω and $\langle \mathbf{m}, \mathbf{n} \rangle$ is the polar of $l_\psi \cap \Pi_\infty$ with respect to Ω. An additional translation ensures that $\mathbf{c}$ is taken to $\mathbf{c}_1$ and a uniform change of scale with $\mathbf{c}_1$ as origin ensures that l is taken to l_ψ. Let $c' = \omega(c)$. Then c' is invariant under the rigid skew involution $\omega\tau_c\omega^{-1}$. The involution $\omega\tau_c\omega^{-1}$ has the same two lines of fixed points as τ_ψ, thus $\omega\tau_c\omega^{-1} = \tau_\psi$.

It remains to show that c' is contained in $\mathcal{H}_1(\mathbf{n})$. Let Π be the tangent plane to ψ at $\mathbf{c}_1$ and let k be the tangent line of c' at $\mathbf{c}_1$. The line k is invariant under τ_ψ, thus k is contained in Π. By Proposition 3.27, each line tangent to ψ at $\mathbf{c}_1$ is also the tangent line to a horopter curve in $\mathcal{H}_1(\mathbf{n})$. It follows that there exists a curve c'' in $\mathcal{H}_1(\mathbf{n})$ with tangent line k at $\mathbf{c}_1$. It follows from Theorem 3.24 that $c' = c''$, thus c is conjugate to a horopter curve in $\mathcal{H}_1(\mathbf{n})$. □

3.4 Horopter Curves and Reconstruction

Horopter curves arise in the ambiguous case of reconstruction as one component of the intersection of a critical surface pair, ψ, ϕ. The surfaces ψ, ϕ intersect in a space curve which splits into a horopter curve c and a common generator g. The line g contains the optical centre $\mathbf{o}$ of the camera before the displacement and c contains the two possible optical centres $\mathbf{a}$, $\mathbf{b}$ of the camera after the displacement. In addition, ψ contains the line $\langle \mathbf{o}, \mathbf{b} \rangle$ and ϕ contains the line $\langle \mathbf{o}, \mathbf{a} \rangle$. In the first result of this section it is shown that ϕ can be chosen such that the unique non-trivial rigid involution of c is equal to the rigid involution τ_ψ of ψ described in Sect. 3.2.2.

Theorem 3.32. *Let ψ be a fixed non-singular critical surface and let $\mathbf{o}$, $\mathbf{a}$ be the optical centres of the camera before and after the displacement. Let ϕ be a critical surface such that ψ, ϕ form a critical surface pair. There is a one-parameter family of possible choices for ϕ. The critical surface ϕ can be chosen such that that ψ and the horopter curve contained in $\psi \cap \phi$ are both invariant under the same non-trivial rigid involution.*

Proof. Let $\mathbf{m}$, $\mathbf{n}$ be the principal points of ψ. Cartesian coordinates are chosen with the origin $(0,0,0,1)^\top$ at $\mathbf{o}$. Let $\mathbf{b}$ be the alternative to $\mathbf{a}$ for the optical centre of the camera after the displacement. The equation for ψ is recalled from (2.24),

$$(U\mathbf{x} \times \mathbf{x}).\mathbf{b} + (\mathbf{x} \times U\mathbf{a}).\mathbf{b} = 0 \tag{3.42}$$

where U is a real orthogonal collineation. It follows from (3.42) that $\mathbf{b}$ lies on a generator h of ψ through $\mathbf{o}$. The position of $\mathbf{b}$ on h is not uniquely determined by ψ. A different choice of $\mathbf{b}$ on h yields the same surface ψ, but a different surface ϕ such that ψ, ϕ are a critical surface pair. As $\mathbf{b}$ moves on h, a one-parameter family of critical surfaces ϕ_b is obtained, such that ψ, ϕ_b are a critical surface pair. The surfaces ϕ_b are given by (2.25). Each intersection $\psi \cap \phi_b$ contains a horopter curve c_b passing through $\mathbf{b}$. The curves c_b form a one parameter family indexed by the points $\mathbf{b}$ of h.

The point $\mathbf{b}$ is chosen on h such that the line $\langle \mathbf{a}, \mathbf{b} \rangle$ meets Π_∞ at a point of $\langle \mathbf{m}, \mathbf{n} \rangle$, as illustrated in Fig. 3.8. Let c be the horopter curve in $\psi \cap \phi_b$ arising from this choice of $\mathbf{b}$. It follows from Theorem 3.22 that c is invariant under a non-trivial rigid skew involution τ. Let g_τ and h_τ be the skew lines of fixed points of τ, chosen such that g_τ is contained in Π_∞. It follows from the

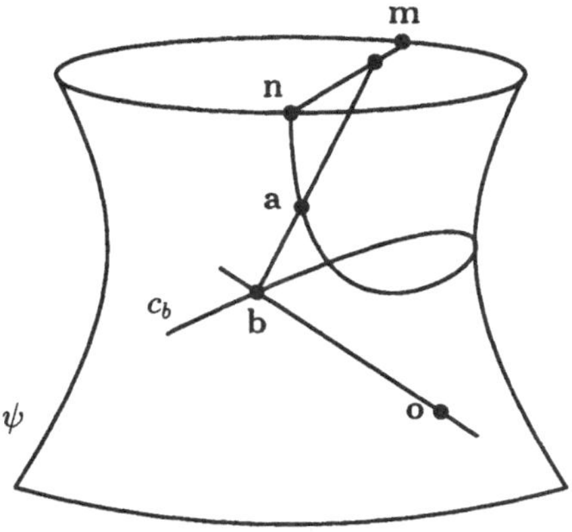

Fig. 3.8. Illustration to Theorem 3.32

explicit construction of τ, given in Theorem 3.22, that g_τ contains both $\mathbf{n}$ and $\langle \mathbf{a}, \mathbf{b} \rangle \cap \langle \mathbf{m}, \mathbf{n} \rangle$, thus $g_\tau = \langle \mathbf{m}, \mathbf{n} \rangle$.

Let $s_\infty = \psi \cap \Pi_\infty$. The intersection $\tau(s_\infty) \cap s_\infty$ contains the six points $\mathbf{m}$, $\mathbf{i}_m$, $\mathbf{j}_m$, $\mathbf{n}$, $\mathbf{i}_n$, $\mathbf{j}_n$. This set of points is invariant under τ, because $\mathbf{m}$, $\mathbf{n}$ are fixed by τ, and because τ leaves the absolute conic invariant. The conic s_∞ is thus invariant under τ, i.e. $\tau(s_\infty) = s_\infty$. It follows that $\tau(\psi) \cap \psi$ contains a (split) curve of degree five comprising the cubic space curve c and the conic s_∞. The intersection of two distinct quadrics is a space curve of degree four, thus $\tau(\psi)$ and ψ are not distinct. In other words $\tau(\psi) = \psi$. □

Corollary 1. The rigid involution τ is equal to the rigid involution τ_ψ defined in Sect. 3.2.2, because τ_ψ is the unique non-trivial rigid skew involution of ψ which fixes both $\mathbf{m}$ and $\mathbf{n}$.

Corollary 2. The involution τ_ψ takes $\mathbf{a}$ to a point of $\langle \mathbf{o}, \mathbf{b} \rangle$.

An algebraic proof that $\tau_\psi(\mathbf{a})$ is a point of $\langle \mathbf{o}, \mathbf{b} \rangle$ can be obtained from (3.42). It suffices to obtain the following three equations:

$$\begin{aligned} (\tau_\psi(\mathbf{a}) \times \mathbf{b}).\mathbf{m} &= 0 \\ (\tau_\psi(\mathbf{a}) \times \mathbf{b}).\mathbf{n} &= 0 \\ (\tau_\psi(\mathbf{a}) \times \mathbf{b}).(\mathbf{m} \times \mathbf{n}) &= 0 \end{aligned}$$

It then follows that $\tau_\psi(\mathbf{a}) \times \mathbf{b} = 0$. The details are omitted because they are lengthy and unenlightening.

Theorem 3.32 has a converse.

Theorem 3.33. *With the notation of Theorem 3.32, let* $\mathbf{o}$, $\mathbf{a}$, $\tau_\psi(\mathbf{a})$ *be distinct points of a non-singular rectangular quadric* ψ *with distinct principal points* $\mathbf{m}$, $\mathbf{n}$, *such that* $\tau_\psi(\mathbf{a})$ *is on a generator of* ψ *passing through* $\mathbf{o}$. *Then* ψ *is a critical surface, such that* $\mathbf{o}$ *is the optical centre of the camera prior to the rigid displace-*

ment and $\mathbf{a}$, $\tau_\psi(\mathbf{a})$ *are the two possible optical centres for the camera after the rigid displacement.*

Proof. Let $\mathbf{b}$ be the point defined by $\mathbf{b} = \tau_\psi(\mathbf{a})$. It follows from the proof of Theorem 3.26 that ψ contains a unique horopter curve c invariant under τ_ψ and containing $\mathbf{n}$ and $\mathbf{a}$, such that c intersects the generator $\langle \mathbf{o}, \mathbf{b} \rangle$ exactly once. The curve c contains $\mathbf{a}$, thus c contains $\tau_\psi(\mathbf{a}) = \mathbf{b}$. The curve c thus intersects $\langle \mathbf{o}, \mathbf{b} \rangle$ at $\mathbf{b}$ only. It follows from Corollary 2 to Proposition 3.21 that there exists an orthogonal collineation ω from the star of lines through $\mathbf{a}$ to the star of lines through $\mathbf{b}$ such that c is the locus of the intersections $k \cap \omega(k)$ as k varies through the lines containing $\mathbf{a}$.

Let $h = \omega^{-1}\langle \mathbf{o}, \mathbf{b} \rangle$ and let ψ' be the surface swept out by the lines $\Pi \cap \omega(\Pi)$ as Π varies through the pencil of planes containing h. It follows from the construction given in Sect. 2.3.2 that ψ' is a critical surface. The intersection, $\psi \cap \psi'$, contains the horopter curve c and the two lines $< \mathbf{o}, \mathbf{b} >$ and h. Each line $l = \Pi \cap \omega(\Pi)$ meets ψ at three points, namely, $l \cap h$, $l \cap \langle \mathbf{o}, \mathbf{b} \rangle$ and $l \cap c$. It follows that l is contained in ψ. The surfaces ψ and ψ' are thus identical. □

3.4.1 A Formula for τ_ψ

An algebraic expression is obtained for the rigid skew involution τ_ψ of a rectangular quadric ψ with distinct principal points $\mathbf{m}$, $\mathbf{n}$. With the notation of Theorem 3.32, let the principal points of ψ be $\mathbf{m}$, $\mathbf{n}$ and let Cartesian coordinates be chosen such that $\mathbf{o} = (0, 0, 0, 1)^\top$. The quadric ψ has an equation

$$\mathbf{x}^\top M \mathbf{x} + \mathbf{l}.\mathbf{x} = 0 \tag{3.43}$$

where $\mathbf{l}$ is a vector and M is a real 3×3 symmetric matrix. It follows from Theorem 3.13 that

$$M = \frac{1}{2}(\mathbf{m} \otimes \mathbf{n} + \mathbf{n} \otimes \mathbf{m}) - \mathbf{m}.\mathbf{n} I$$

It follows from the definition of τ_ψ given in Sect. 3.2.2 that the axis l_ψ of τ_ψ is the intersection of the tangent planes to ψ at $\mathbf{m}$ and $\mathbf{n}$. In order to obtain the equations of the two tangent planes the additional coordinate x_4 is introduced. Equation (3.43) is written in the homogeneous form

$$\mathbf{x}^\top M \mathbf{x} + x_4 \mathbf{l}.\mathbf{x} = 0 \tag{3.44}$$

The tangent plane to ψ at a point $\mathbf{p}$ consists of the points $(\mathbf{x}, x_4)^\top$ which satisfy

$$(\mathbf{x}, x_4)^\top.(\nabla\psi|_{\mathbf{p}}) = 0 \tag{3.45}$$

The differential of (3.44) yields

$$\nabla\psi = (2M\mathbf{x} + x_4 \mathbf{l},\ \mathbf{l}.\mathbf{x})^\top \tag{3.46}$$

It follows from (3.45) and (3.46) that the inhomogeneous equation of the tangent plane to ψ at $(\mathbf{m}, 0)^\top$ is

$$2\mathbf{m}^\top M \mathbf{x} + \mathbf{l}.\mathbf{m} = 0 \tag{3.47}$$

Similarly, the tangent plane to ψ at $(\mathbf{n}, 0)^{\top}$ is

$$2\mathbf{n}^{\top} M\mathbf{x} + \mathbf{n}.\mathbf{l} = 0 \tag{3.48}$$

The axis l_{ψ} of τ_{ψ} is the intersection of the planes (3.47) and (3.48). It follows from Proposition 3.15 that l_{ψ} has direction $\mathbf{m} \times \mathbf{n}$. Let $\mathbf{s}$ be a general point of l_{ψ}. The line l_{ψ} has a parameterisation of the form

$$t \mapsto \mathbf{s} + t\mathbf{m} \times \mathbf{n} \tag{3.49}$$

The point $\mathbf{s}$ is chosen such that $\mathbf{s}.(\mathbf{m} \times \mathbf{n}) = 0$. Let a, b be scalars such that

$$\mathbf{s} = a\mathbf{m} + b\mathbf{n} \tag{3.50}$$

The point $\mathbf{s}$ is on l_{ψ}, thus $\mathbf{s}$ satisfies (3.47) and (3.48). On substituting the right-hand side of (3.50) for $\mathbf{s}$ in (3.47) an equation for $\mathbf{b}$ is obtained,

$$2b\mathbf{m}^{\top} M\mathbf{n} + \mathbf{l}.\mathbf{m} = 0$$

from which it follows that

$$b = -\frac{1}{2}\left(\frac{\mathbf{l}.\mathbf{m}}{\mathbf{m}^{\top} M\mathbf{n}}\right) \tag{3.51}$$

Similarly, the substitution of the right-hand side of (3.50) for $\mathbf{s}$ in (3.48) yields

$$a = -\frac{1}{2}\left(\frac{\mathbf{l}.\mathbf{n}}{\mathbf{m}^{\top} M\mathbf{n}}\right) \tag{3.52}$$

Let $\mathbf{r}$ be the vector defined by $\mathbf{r} = \mathbf{m} \times \mathbf{n}$. The evaluation of $2\mathbf{m}^{\top} M\mathbf{n}$ yields

$$2\mathbf{m}^{\top} M\mathbf{n} = \|\mathbf{m}\|^2\|\mathbf{n}\|^2 - (\mathbf{m}.\mathbf{n})^2 = \|\mathbf{r}\|^2 \tag{3.53}$$

It follows from (3.50), (3.51), (3.52) and (3.53) that

$$\mathbf{s} = -[(\mathbf{l}.\mathbf{n})\mathbf{m} + (\mathbf{l}.\mathbf{m})\mathbf{n}]/\|\mathbf{r}\|^2$$

On substituting for $\mathbf{s}$ in (3.49) the following parameterisation of l_{ψ} is obtained,

$$t \mapsto -\frac{(\mathbf{n}.\mathbf{l})\mathbf{m} + (\mathbf{m}.\mathbf{l})\mathbf{n}}{\|\mathbf{r}\|^2} + t\,\mathbf{r} \tag{3.54}$$

The involution τ_{ψ} is a rotation of 180° about the axis l_{ψ}. Let $\mathbf{x}$ be any point of $\mathbb{R}^3$. The mid-point of the line segment $[\mathbf{x}, \tau_{\psi}(\mathbf{x})]$ is on l_{ψ}, thus there exists a value of t, depending on $\mathbf{x}$, such that

$$\frac{1}{2}[\mathbf{x} + \tau_{\psi}(\mathbf{x})] = \mathbf{s} + t\,\mathbf{r} \tag{3.55}$$

The line $\langle \mathbf{x}, \tau_\psi(\mathbf{x})\rangle$ is orthogonal to l_ψ thus

$$(\mathbf{x} - \tau_\psi(\mathbf{x})).\mathbf{r} = 0 \tag{3.56}$$

Let $\hat{\mathbf{r}}$ be the unit vector defined by $\hat{\mathbf{r}} = \mathbf{r}/\|\mathbf{r}\|$. The scalar product of (3.55) with $\hat{\mathbf{r}}$ yields

$$\frac{1}{2}[\mathbf{x}.\hat{\mathbf{r}} + \tau_\psi(\mathbf{x}).\hat{\mathbf{r}}] = \mathbf{s}.\hat{\mathbf{r}} + t\|\mathbf{r}\| \tag{3.57}$$

It follows from (3.56) and (3.57) that $t = \|\mathbf{r}\|^{-1}\mathbf{x}.\hat{\mathbf{r}}$. On substituting for t and $\mathbf{s}$ in (3.55) and rearranging terms an expression for $\tau_\psi(\mathbf{x})$ is obtained,

$$\tau_\psi(\mathbf{x}) = -\frac{2[(\mathbf{n}.\mathbf{l})\mathbf{m} + (\mathbf{m}.\mathbf{l})\mathbf{n}]}{\|\mathbf{r}\|^2} + 2(\mathbf{x}.\hat{\mathbf{r}})\hat{\mathbf{r}} - \mathbf{x} \tag{3.58}$$

3.4.2 Two Cubic Constraints on Critical Surfaces

In Corollary 2 to Theorem 3.32 it is shown that τ_ψ interchanges the two possible optical centres for the camera after the displacement. This result leads to a cubic constraint on the critical surface ψ. The constraint is obtained in two steps. In the first step, Theorem 3.34, the constraint is expressed in terms of the vectors $\mathbf{l}$, $\mathbf{m}$, $\mathbf{n}$, $\mathbf{r}$ employed in Sect. 3.4.1. In the second step, Proposition 3.35, the constraint is shown to be a cubic polynomial in the coefficients M, $\mathbf{l}$ of the equation (3.43) for ψ.

Theorem 3.34. *With the notation of Sect. 3.4.1, let ψ be a critical surface with principal points $\mathbf{m}$, $\mathbf{n}$, let $\mathbf{r}$ be the vector defined by $\mathbf{r} = \mathbf{m} \times \mathbf{n}$ and let $\mathbf{l}$ be a vector normal to the tangent plane to ψ at $\mathbf{o}$. Let $\mathbf{a}$ be a possible optical centre for the camera after the displacement not lying on a generator through $\mathbf{o}$. Let Cartesian coordinates be chosen with origin at $\mathbf{o}$. Then the following equation holds,*

$$-4(\mathbf{m}.\mathbf{l})(\mathbf{n}.\mathbf{l}) + 2(\mathbf{a}.\mathbf{r})(\mathbf{l}.\mathbf{r}) - (\mathbf{a}.\mathbf{l})(\mathbf{r}.\mathbf{r}) = 0 \tag{3.59}$$

Proof. It follows from Corollary 2 of Theorem 3.32 that $\tau_\psi(\mathbf{a})$ lies in the tangent plane to ψ at $\mathbf{o}$. This gives the constraint $\mathbf{l}.\tau_\psi(\mathbf{a}) = 0$. Equation (3.59) follows on substituting the expression for $\tau_\psi(\mathbf{a})$ given by (3.58) into the equation $\mathbf{l}.\tau_\psi(\mathbf{a}) = 0$. □

The next result is a preliminary to showing that (3.59) is a cubic constraint on ψ.

Proposition 3.35. *Let $\mathbf{m}$, $\mathbf{n}$, $\mathbf{e}_1$, $\mathbf{e}_2$, $\mathbf{e}_3$ be vectors such that*

$$\frac{1}{2}(\mathbf{m} \otimes \mathbf{n} + \mathbf{n} \otimes \mathbf{m}) = \begin{pmatrix} \mathbf{e}_1^\top \\ \mathbf{e}_2^\top \\ \mathbf{e}_3^\top \end{pmatrix}$$

then

$$\frac{1}{4}(\mathbf{m} \times \mathbf{n}) \otimes (\mathbf{m} \times \mathbf{n}) = \begin{pmatrix} \mathbf{e}_3^\top \times \mathbf{e}_2^\top \\ \mathbf{e}_1^\top \times \mathbf{e}_3^\top \\ \mathbf{e}_2^\top \times \mathbf{e}_1^\top \end{pmatrix} \tag{3.60}$$

Proof. It follows from the hypotheses that

$$\begin{aligned} 2\mathbf{e}_1 &= m_1\mathbf{n} + n_1\mathbf{m} \\ 2\mathbf{e}_2 &= m_2\mathbf{n} + n_2\mathbf{m} \\ 2\mathbf{e}_3 &= m_3\mathbf{n} + n_3\mathbf{m} \end{aligned}$$

thus

$$\begin{aligned} 4\mathbf{e}_3 \times \mathbf{e}_2 &= (\mathbf{m} \times \mathbf{n})_1\mathbf{m} \times \mathbf{n} \\ 4\mathbf{e}_1 \times \mathbf{e}_3 &= (\mathbf{m} \times \mathbf{n})_2\mathbf{m} \times \mathbf{n} \\ 4\mathbf{e}_2 \times \mathbf{e}_1 &= (\mathbf{m} \times \mathbf{n})_3\mathbf{m} \times \mathbf{n} \end{aligned} \tag{3.61}$$

Equation (3.60) follows from (3.61). □

Theorem 3.36. *Let two images of a set of points be obtained by projections with optical centres at the points* **o**, **a** *respectively. If reconstruction from corresponding pairs of image points is ambiguous then the associated critical surface satisfies a cubic polynomial constraint.*

Proof. Cartesian coordinates are chosen with origin at **o**. Let ψ be a critical surface obtained by reconstruction and let **m**, **n** be the principal points of ψ. Then ψ has an equation of the form

$$\mathbf{x}^\top M\mathbf{x} + \mathbf{l}.\mathbf{x} = 0 \tag{3.62}$$

where **l** is a vector and where M is the symmetric matrix defined by

$$M = \frac{1}{2}(\mathbf{m} \otimes \mathbf{n} + \mathbf{n} \otimes \mathbf{m}) - \mathbf{m}.\mathbf{n}I$$

Let N, L be the matrices defined by

$$N = \frac{1}{2}(\mathbf{m} \otimes \mathbf{n} + \mathbf{n} \otimes \mathbf{m}) \tag{3.63}$$

$$L = (\mathbf{m} \times \mathbf{n}) \otimes (\mathbf{m} \times \mathbf{n}) \tag{3.64}$$

The coefficients of N are linear functions of the coefficients of M because

$$N = M - \frac{1}{2}\mathrm{tr}(M)I$$

It follows from Proposition 3.35 that the coefficients of L are quadratic functions of the coefficients of N, and hence quadratic functions of the coefficients of M. The theorem is proved by expressing (3.59) in terms of **a**, **l**, L and N.

The term $-4(\mathbf{l.m})(\mathbf{l.n})$ on the left-hand side of (3.59) has the form

$$-4(\mathbf{l.m})(\mathbf{l.n}) = -4\,\mathbf{l}^{\mathsf{T}} N\,\mathbf{l} \tag{3.65}$$

The remaining two terms on the left-hand side of (3.59) have the form

$$2(\mathbf{a.r})(\mathbf{l.r}) - (\mathbf{a.l})(\mathbf{r.r}) = 2\mathbf{a}^{\mathsf{T}} L\,\mathbf{l} - (\mathbf{a.l})\mathrm{tr}(L) \tag{3.66}$$

It follows from (3.65) and (3.66) that (3.59) has the form

$$-4\mathbf{l}^{\mathsf{T}} N\mathbf{l} + 2\mathbf{a}^{\mathsf{T}} L\mathbf{l} - (\mathbf{a.l})\mathrm{tr}(L) = 0 \tag{3.67}$$

Equation (3.67) is the required constraint on the coefficients M, $\mathbf{l}$ of (3.62). □

Corollary. Let $\mathbf{e}_1^{\mathsf{T}}$, $\mathbf{e}_2^{\mathsf{T}}$, $\mathbf{e}_3^{\mathsf{T}}$ be the rows of N. Then (3.67) is equivalent to

$$-\mathbf{l}^{\mathsf{T}}\begin{pmatrix}\mathbf{e}_1^{\mathsf{T}}\\ \mathbf{e}_2^{\mathsf{T}}\\ \mathbf{e}_3^{\mathsf{T}}\end{pmatrix}\mathbf{l} + 2\mathbf{a}^{\mathsf{T}}\begin{pmatrix}\mathbf{e}_3^{\mathsf{T}}\times\mathbf{e}_2^{\mathsf{T}}\\ \mathbf{e}_1^{\mathsf{T}}\times\mathbf{e}_3^{\mathsf{T}}\\ \mathbf{e}_2^{\mathsf{T}}\times\mathbf{e}_1^{\mathsf{T}}\end{pmatrix}\mathbf{l} - (\mathbf{a.l})\mathrm{tr}\begin{pmatrix}\mathbf{e}_3^{\mathsf{T}}\times\mathbf{e}_2^{\mathsf{T}}\\ \mathbf{e}_1^{\mathsf{T}}\times\mathbf{e}_3^{\mathsf{T}}\\ \mathbf{e}_2^{\mathsf{T}}\times\mathbf{e}_1^{\mathsf{T}}\end{pmatrix} = 0 \tag{3.68}$$

There is second cubic constraint on ψ arising from the fact that N has rank two,

$$\det(N) = 0 \tag{3.69}$$

Equation (3.69) is identical to

$$\det(M - \frac{1}{2}\mathrm{tr}(M)I) = 0 \tag{3.70}$$

Equation (3.70) states that $(1/2)\mathrm{tr}(M)$ is an eigenvalue of M. Let λ_1, λ_2, λ_3 be the eigenvalues of M, labelled such that $\lambda_3 = (1/2)\mathrm{tr}(M)$. It is shown in Sect. 2.2.3 that (3.70) holds if and only if the λ_i satisfy $\lambda_1 + \lambda_2 = \lambda_3$.

3.4.3 An Example

The full version (3.68) of the cubic constraint on critical surfaces appears to be complicated. One route to a better understanding of the constraint is to study special cases in which it simplifies. It is shown that the constraint takes a simpler form on certain one-dimensional spaces of rectangular quadrics. It reduces to the condition that three variable lines intersect at a single point.

Let ψ_1, ψ_2 be two non-singular rectangular quadrics with principal points $\mathbf{m}$, $\mathbf{n}_1$ and $\mathbf{m}$, $\mathbf{n}_2$ respectively, such that the three points $\mathbf{m}$, $\mathbf{n}_1$, $\mathbf{n}_2$ lie on a fixed line g in Π_∞. Let $\mathbf{o}$, $\mathbf{a}$ be distinct points of $\mathbb{R}^3$ contained in both ψ_1 and ψ_2. Let $\langle\psi_1, \psi_2\rangle$ be the one dimensional space of quadrics generated by ψ_1 and ψ_2. A general quadric ψ in $\langle\psi_1, \psi_2\rangle$ is given by

$$\psi = \lambda_1\psi_1 + \lambda_2\psi_2$$

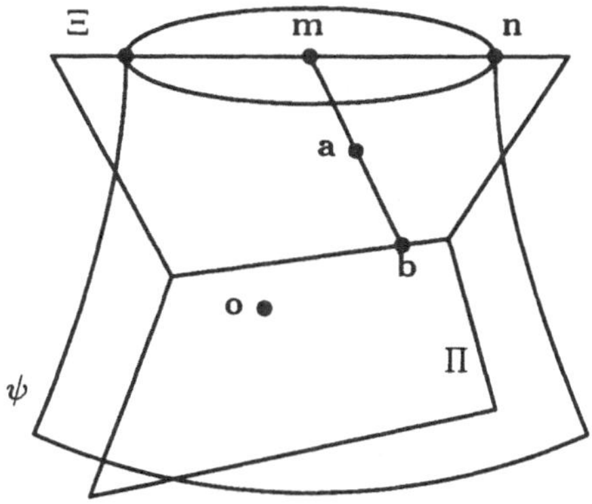

Fig. 3.9. The geometry of a pencil of rectangular quadrics

where (λ_1, λ_2) is a point of $\mathbb{P}^1$ depending on ψ. The quadric ψ meets Π_∞ in the conic $\mathbf{x}^\top M \mathbf{x} = 0$, where M is the symmetric matrix defined by

$$M = \frac{1}{2}[\mathbf{m} \otimes (\lambda_1 \mathbf{n}_1 + \lambda_2 \mathbf{n}_2) + (\lambda_1 \mathbf{n}_1 + \lambda_2 \mathbf{n}_2) \otimes \mathbf{m}] - \mathbf{m}.(\lambda_1 \mathbf{n}_1 + \lambda_2 \mathbf{n}_2) I \quad (3.71)$$

It follows from (3.71) that ψ satisfies (3.69) for all choices of (λ_1, λ_2). The principal points of ψ are $\mathbf{m}$ and the point $\mathbf{n}$ defined by $\mathbf{n} = \lambda_1 \mathbf{n}_1 + \lambda_2 \mathbf{n}_2$.

The quadric ψ is illustrated in Fig. 3.9. The plane Π shown in Fig. 3.9 is defined to be the tangent plane to ψ at $\mathbf{o}$. The quadric ψ contains $\mathbf{o}$ and $\mathbf{a}$ because ψ_1 and ψ_2 both contain $\mathbf{o}$ and $\mathbf{a}$. Let τ_ψ be the unique non-trivial rigid involution of ψ that fixes both $\mathbf{m}$ and $\mathbf{n}$. Then g is one of the lines of fixed points of τ_ψ. The plane $\Xi = \langle g, \mathbf{a} \rangle$ is invariant under τ_ψ because Ξ contains g.

Let $\mathbf{b}$ be the point defined by $\mathbf{b} = \tau_\psi(\mathbf{a})$. Then $\mathbf{b}$ is in Ξ, because $\mathbf{a}$ is in Ξ and because Ξ is invariant under τ_ψ. It follows from (3.58) that

$$\mathbf{b} = -\frac{2[(\mathbf{n}.\mathbf{l})\mathbf{m} + (\mathbf{m}.\mathbf{l})\mathbf{n}]}{\|\mathbf{m} \times \mathbf{n}\|^2} + \frac{2[\mathbf{a}.(\mathbf{m} \times \mathbf{n})]\mathbf{m} \times \mathbf{n}}{\|\mathbf{m} \times \mathbf{n}\|^2} - \mathbf{a}$$

Let N, L be the matrices defined by (3.63) and (3.64) respectively. As a point of $\mathbb{R}^3$, $\mathbf{b}$ is given by

$$\mathbf{b} = \frac{-4N\mathbf{l} + 2L\mathbf{a}}{\mathrm{tr}(L)} - \mathbf{a}$$

As a point of $\mathbb{P}^3$, $\mathbf{b}$ is given by

$$\mathbf{b} = (-4N\mathbf{l} + 2L\mathbf{a} - \mathrm{tr}(L)\mathbf{a}, \mathrm{tr}(L))^\top \quad (3.72)$$

The entries of the matrix N and the components of the vector $\mathbf{l}$ are linear functions of ψ, and therefore linear functions of $(\lambda_1, \lambda_2)^\top$. Similarly, the entries of L are quadratic functions of $(\lambda_1, \lambda_2)^\top$. It thus follows from (3.72) that $\mathbf{b}$ is a quadratic function of $(\lambda_1, \lambda_2)^\top$. As ψ varies through the space $\langle \psi_1, \psi_2 \rangle$, $\mathbf{b}$ varies on a conic in Ξ. Let this conic be s and let $\mathbf{r}$, $\mathbf{s}$ be two arbitrary points chosen on

s. Let $\mathbf{g}_\psi$ be the coefficients of the line $\langle \mathbf{b}, \mathbf{r} \rangle$ in a convenient coordinate system on Ξ. Thus a point $\mathbf{x}$ of Ξ is on $\langle \mathbf{b}, \mathbf{r} \rangle$ if and only if $\mathbf{g}_\psi.\mathbf{x} = 0$. Similarly, let $\mathbf{h}_\psi$ be the coefficients of the line $\langle \mathbf{b}, \mathbf{s} \rangle$. It follows from Steiner's theorem that $\mathbf{g}_\psi$ and $\mathbf{h}_\psi$ both depend linearly on $(\lambda_1, \lambda_2)^\top$.

Equation (3.68) is satisfied if and only if $\mathbf{b}$ lies on $\Pi \cap \Xi$, where Π is the tangent plane to ψ at $\mathbf{o}$. Let $\mathbf{k}_\psi$ be the coefficients of the line $\Pi \cap \Xi$. The plane Π depends linearly on ψ, and Ξ is fixed independently of ψ, thus $\mathbf{k}_\psi$ depends linearly on $(\lambda_1, \lambda_2)^\top$. The condition that $\mathbf{b}$ lies on both s and on the line $\mathbf{k}_\psi.\mathbf{x} = 0$ is

$$\det(\mathbf{g}_\psi | \mathbf{h}_\psi | \mathbf{k}_\psi) = 0 \tag{3.73}$$

Equation (3.73) is the cubic constraint on ψ obtained by applying (3.67) to this special case.

3.5 Reconstruction up to a Collineation

In reconstruction up to a collineation the ambiguous case arises when there are two or more sets of epipoles which are compatible with the same set of image correspondences. It is shown in Sect. 2.4 that the critical surfaces in reconstruction up to a collineation are hyperboloids of one sheet. The critical surfaces are not, in general, rectangular hyperboloids. In the first result of this section a description is given of the triples of points on a hyperboloid of one sheet that can serve as optical centres in the ambiguous case.

Proposition 3.37. *Let ψ be a non-singular hyperboloid of one sheet and let* $\mathbf{o}$, $\mathbf{a}$, $\mathbf{b}$ *be points of ψ such that $\langle \mathbf{o}, \mathbf{b} \rangle$ is a generator of ψ and* $\mathbf{a}$ *is not a point of $\langle \mathbf{o}, \mathbf{b} \rangle$. Then ψ is a critical surface for reconstruction up to a collineation. In the reconstruction* $\mathbf{o}$ *is the optical centre of the camera before the displacement and* $\mathbf{a}$, $\mathbf{b}$ *are two possible positions for the optical centre after the displacement.*

Proof. The result follows from the observations made in Sect. 2.4. Let h be the generator of ψ through $\mathbf{a}$ skew to $\langle \mathbf{o}, \mathbf{b} \rangle$ and let ω define a homographic correspondence between the pencil of planes through h and the pencil of planes through $\langle \mathbf{o}, \mathbf{b} \rangle$, such that ψ is swept out by the lines $\Pi \cap \omega(\Pi)$ as Π varies through the pencil of planes containing h. The homography ω is induced by a collineation ω from the lines through $\mathbf{a}$ to the lines through $\mathbf{b}$. Let $\mathbf{p}$ be a general point of ψ and let l be the generator of ψ through $\mathbf{p}$ that intersects h and $\langle \mathbf{o}, \mathbf{b} \rangle$. Then $\omega\langle \mathbf{a}, \mathbf{p} \rangle$ is a line through $\mathbf{b}$ contained in the plane $\langle l, \mathbf{o}, \mathbf{b} \rangle$. It follows that $\omega\langle \mathbf{a}, \mathbf{p} \rangle$ intersects $\langle \mathbf{o}, \mathbf{p} \rangle$. Then $\langle \mathbf{o}, \mathbf{p} \rangle \leftrightarrow \langle \mathbf{a}, \mathbf{p} \rangle$ defines a correspondence between the image obtained by projection to $\mathbf{o}$ and the image obtained by projection to $\mathbf{a}$ such that ψ is the reconstructed surface. The fact that $\omega\langle \mathbf{a}, \mathbf{p} \rangle$ intersects $\langle \mathbf{o}, \mathbf{p} \rangle$ ensures that reconstruction is possible when the camera is at $\mathbf{b}$. □

The involutions of ψ that interchange $\mathbf{a}$ and $\mathbf{b}$ are described and the twisted cubics contained in ψ that pass through $\mathbf{a}$ and $\mathbf{b}$ are also described.

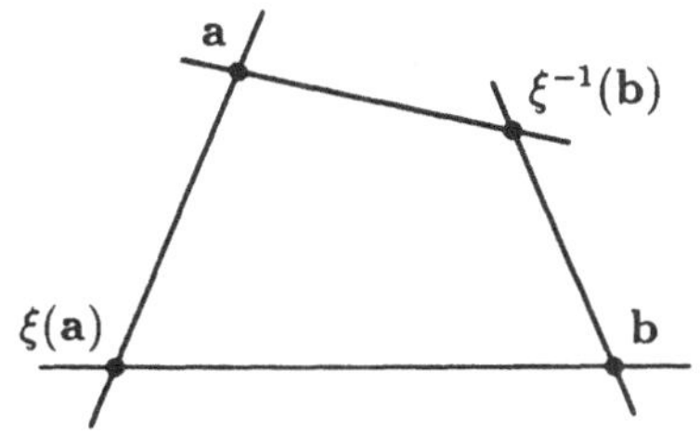

Fig. 3.10. A skew quadrilateral of generators of ψ

Proposition 3.38. *Let ψ be a non-singular quadric and let* **a**, **b** *be two points of ψ not contained in the same generator of ψ. Then the skew involutions of ψ which interchange* **a** *and* **b** *form a non-linear two-parameter family.*

Proof. Let $g_1(\mathbf{a})$ be the generator of ψ through **a** and let $g_1(\mathbf{b})$ be a generator of ψ through **b** skew to $g_1(\mathbf{a})$. Let ξ be a homography between the points of $g_1(\mathbf{a})$ and the points of $g_1(\mathbf{b})$ such that $\langle \mathbf{p}, \xi(\mathbf{p})\rangle$ is a generator of ψ for each point **p** in $g_1(\mathbf{a})$. The action of ξ is illustrated in Fig. 3.10. Let τ be a skew involution of ψ that interchanges **a** and **b**. It follows from Proposition 3.17 that $\tau(g_1(\mathbf{a})) = g_1(\mathbf{b})$ and $\tau(g_2(\mathbf{a})) = g_2(\mathbf{b})$, where $g_2(\mathbf{a})$, $g_2(\mathbf{b})$ are the generators of ψ through **a** and **b** distinct from $g_1(\mathbf{a})$ and $g_1(\mathbf{b})$. The involution τ thus satisfies the constraints

$$\begin{aligned} \tau(\xi^{-1}(\mathbf{b})) &= \xi(\mathbf{a}) \\ \tau(\mathbf{a}) &= \mathbf{b} \end{aligned} \tag{3.74}$$

The first equation of (3.74) follows from the observation $\xi^{-1}(\mathbf{b}) = g_1(\mathbf{a}) \cap g_2(\mathbf{b})$. Next, let τ be any involution of $\mathbb{P}^3$ satisfying (3.74). Let q be the skew quadrilateral constructed from the two generators of ψ through **a** and the two generators of ψ through **b**, as illustrated in Fig. 3.10. Then q is invariant under τ, and each of the two diagonals of q is separately invariant under τ. The intersection $\tau(\psi) \cap \psi$ contains q. No point of q is a fixed point of τ. This observation is sufficient to ensure that τ is a skew involution, because a harmonic homology has at least one fixed point on every line. The quadric ψ intersects the set of fixed points of τ thus $\tau(\psi) \cap \psi$ is strictly greater than q. Two distinct quadrics intersect in a space curve of degree four. It follows that $\tau(\psi)$ and ψ are not distinct. In other words, $\tau(\psi) = \psi$.

In order to prove the proposition it is enough to describe the involutions τ which satisfy (3.74). Let $\mathbf{p}_1$ be an arbitrary point of $\langle \mathbf{a}, \mathbf{b}\rangle$ and let $\mathbf{p}_2$ be an arbitrary point of $\langle \xi(\mathbf{a}), \xi^{-1}(\mathbf{b})\rangle$. There is a unique involution σ_1 of $\langle \mathbf{a}, \mathbf{b}\rangle$ such that $\sigma_1(\mathbf{a}) = \mathbf{b}$, $\sigma_1(\mathbf{p}_1) = \mathbf{p}_1$. Let $\mathbf{q}_1$ be the fixed point of σ_1 distinct from $\mathbf{p}_1$. There is a unique involution σ_2 of $\langle \xi(\mathbf{a}), \xi^{-1}(\mathbf{b})\rangle$ such that $\sigma_2(\xi(\mathbf{a})) = \xi^{-1}(\mathbf{b})$, $\sigma_2(\mathbf{p}_2) = \mathbf{p}_2$. Let $\mathbf{q}_2$ be the fixed point of σ_2 distinct from $\mathbf{p}_2$. Let τ be the unique involution of $\mathbb{P}^3$ with the two skew lines of fixed points $\langle \mathbf{p}_1, \mathbf{p}_2\rangle$ and $\langle \mathbf{q}_1, \mathbf{q}_2\rangle$. It follows from the definition of τ that $\tau(\mathbf{a}) = \mathbf{b}$, $\tau(\xi^{-1}(\mathbf{b})) = \xi(\mathbf{a})$. As a result,

$\tau(q) = q$ and ψ is invariant under τ. The involution τ is uniquely defined by the pair $\mathbf{p}_1$, $\mathbf{p}_2$. The possible choices of $\mathbf{p}_1$, $\mathbf{p}_2$ are parameterised by the space $\langle \mathbf{a}, \mathbf{b} \rangle \times \langle \xi(\mathbf{a}), \xi^{-1}(\mathbf{b}) \rangle$. There is a collineation from $\langle \mathbf{a}, \mathbf{b} \rangle \times \langle \xi(\mathbf{a}), \xi^{-1}(\mathbf{b}) \rangle$ to a non-singular quadric embedded in $\mathbb{P}^3$. □

Proposition 3.39. *Let ψ be a non-singular quadric, let $\mathbf{a}$, $\mathbf{b}$ be distinct points of ψ not on the same generator of ψ and let τ be a skew involution of ψ that interchanges $\mathbf{a}$ and $\mathbf{b}$. Then there are four one-parameter families of twisted cubics that contain $\mathbf{a}$, $\mathbf{b}$ and that are invariant under τ.*

Proof. Let g, h be the two skew lines of fixed points of τ and let c be a twisted cubic in ψ that contains $\mathbf{a}$ and $\mathbf{b}$, and that is invariant under τ. Then c contains two fixed points of τ. Suppose, if possible, that both fixed points of τ on c are in g. Let $\mathbf{p}$ be any point of c. The plane $\langle g, \mathbf{p} \rangle$ is invariant under τ, thus the set $\langle g, \mathbf{p} \rangle \cap c$ is invariant under τ. The intersection $\langle g, \mathbf{p} \rangle \cap c$ contains three points, namely $\mathbf{p}$ and two points on g. It follows that $\mathbf{p}$ is fixed by τ. Then τ is the identity contrary to hypothesis, because $\mathbf{p}$ is any point of c. It follows from this contradiction that one of the fixed points of τ on c is in $\psi \cap g$ and the other is in $\psi \cap h$. The set $\psi \cap g$ contains two points and $\psi \cap h$ contains two points, thus there are four different ways of choosing the fixed points of τ on c. It is shown that each choice gives rise to a one parameter family of twisted cubics. Let $\mathbf{p}_1$ be a point of $\psi \cap g$ and let $\mathbf{p}_2$ be a point of $\psi \cap h$. The surface ψ is projected from $\mathbf{p}_1$ to a plane Π. The involution τ induces an involution σ of Π.

Let $\mathbf{v}$ be the vertex of σ and let l be the line of fixed points of σ. The point $\mathbf{v}$ is the projection of g and l is the projection of h. The conics in Π that contain the projections of $\mathbf{a}$, $\mathbf{b}$, $\mathbf{p}_2$ form a two-parameter family. A conic in this family is invariant under σ if and only if it is tangent to $\langle \mathbf{v}, \mathbf{p}_2 \rangle$ at $\mathbf{p}_2$. Thus the conics invariant under σ that contain the projections of $\mathbf{a}$, $\mathbf{b}$, $\mathbf{p}_2$ form a one-parameter family. Each such conic is the projection of a twisted cubic contained in ψ that is invariant under τ and that contains $\mathbf{a}$ and $\mathbf{b}$. □

Theorem 3.40. *Let ψ be a fixed critical surface arising from reconstruction up to collineation. Let $\mathbf{o}$ be the optical center for the first camera and let g_1, g_2 be the two generators of ψ through $\mathbf{o}$. Let $\mathbf{a}$ be a possible optical centre for the second camera not on g_1 or g_2. Let ϕ_1, ϕ_2 be critical surfaces chosen such that ψ, ϕ_1, ϕ_2 are compatible with the same set of image correspondences and such that the epipoles associated with ψ, ϕ_1 and ϕ_2 are all different. Then ϕ_1 and ϕ_2 can be labelled such that ϕ_1 contains g_1 and ϕ_2 contains g_2. The surfaces that can be chosen for ϕ_1 are in one-to-one correspondence with the twisted cubics contained in ψ that pass through $\mathbf{a}$ and that meet g_2 exactly once. Similarly, the surfaces ϕ_2 are in one-to-one correspondence with the twisted cubics contained in ψ that pass through $\mathbf{a}$ and that meet g_1 exactly once.*

Proof. Let ϕ be a second critical surface associated with ψ. It follows from Proposition 2.34 that the intersection $\psi \cap \phi$ splits into a twisted cubic c passing through $\mathbf{o}$ and a common generator of ψ and ϕ. The curve c meets the common generator twice. The curve c is either a $(1,2)$ curve or a $(2,1)$ curve of ψ, thus either c meets g_1 once and g_2 twice, or c meets g_2 once and g_1 twice. In the first case $\phi = \phi_2$ and in the second case $\phi = \phi_1$.

Conversely, let c be a twisted cubic contained in ψ that passes through $\mathbf{a}$ and that meets g_2 exactly once, at a point $\mathbf{b}$. Let ω be the unique collineation from the star of lines through $\mathbf{a}$ to the star of lines through $\mathbf{b}$ such that c is the locus of the intersections of corresponding lines. Let $k = \omega^{-1}(g_2)$ and let ψ' be the surface swept out by the line $l = \Pi \cap \omega(\Pi)$ as Π varies through the pencil of planes containing k. Each line l meets c twice and l also meets g_2. It follows that l meets ψ three times, thus l is contained in ψ. The surfaces ψ' and ψ are thus identical.

Let ϕ_1 be the surface swept out by the line $l = \Pi \cap \omega(\Pi)$ as Π varies through the pencil of planes containing $\langle \mathbf{o}, \mathbf{a} \rangle$. Then ψ, ϕ_1 are a critical surface pair and $\psi \cap \phi_1$ contains c. The twisted cubic c and the choice of $\mathbf{a}$ and $\mathbf{b}$ uniquely determine ω, hence they uniquely determine ϕ_i. □

Two different arguments are given to show that the surfaces ϕ_1 in the proof of Theorem 3.40 form a linear family of dimension three. The subscript 1 of ϕ_1 is omitted. Let ω be a fixed collineation from the star of lines through $\mathbf{a}$ to the star of lines through $\mathbf{b}$, with an associated critical surface ϕ and let ω' be any self-collineation of the star of lines through $\mathbf{a}$ such that ω' leaves invariant all planes containing the line $h = \omega^{-1}\langle \mathbf{o}, \mathbf{b} \rangle$. The collineation $\omega\omega'$ yields a critical surface ϕ' complementary to ψ, but in general different from ϕ. All critical surfaces ϕ' complementary to ψ that meet g_2 exactly once arise in this way. It is shown that the space of possible ω' has dimension three. Coordinates are chosen for the star of lines through $\mathbf{a}$ such that $h = (0,0,1)^\top$. Let the entries of the matrix A' of ω' be given by

$$A' = \begin{pmatrix} a_{11} & a_{12} & a_{13} \\ a_{21} & a_{22} & a_{23} \\ a_{31} & a_{32} & a_{33} \end{pmatrix}$$

The condition $\omega(h) = h$ is satisfied if and only if $a_{13} = a_{23} = 0$. Let $\mathbf{q} = (x, y, z)^\top$ be coordinates of an arbitrary line q through $\mathbf{a}$ and let Π_q be the plane in $\mathbf{P}^3$ containing h and q. The condition $\omega'(\Pi_q) = \Pi_q$ holds if and only if

$$(\mathbf{q} \times \mathbf{h}).\omega(\mathbf{q}) = 0 \tag{3.75}$$

identically in x, y, z. The expansion of (3.75) yields

$$(a_{11} - a_{22})xy + a_{12}y^2 - a_{21}x^2 = 0$$

It follows that $a_{21} = a_{12} = 0$ and $a_{11} = a_{22}$. The matrix A' thus reduces to

$$A' = \begin{pmatrix} a_{11} & 0 & 0 \\ 0 & a_{11} & 0 \\ a_{31} & a_{32} & a_{33} \end{pmatrix} \tag{3.76}$$

It follows from (3.76) that the space of collineations ω' is parameterised by the points $(a_{11}, a_{31}, a_{32}, a_{33})^{\top}$ of $\mathbf{P}^3$.

A second parameterisation of the same space of critical surfaces ϕ is obtained using the results of Theorem 3.40. The critical surface ϕ is uniquely determined by the twisted cubic contained in $\psi \cap \phi$. Let $\mathcal{S}$ be the space of twisted cubics contained in ψ that contain $\mathbf{a}$ and that meet the generator g_2 at one point only, namely the fixed point $\mathbf{b}$. It is shown that $\mathcal{S}$ is a linear space of dimension three. Let k be the generator of ψ passing through $\mathbf{b}$, but distinct from g. The curves of $\mathcal{S}$ are projected from $\mathbf{b}$ to a fixed plane. Let k project to a point $\mathbf{k}$, and let $\langle \mathbf{a}, \mathbf{b} \rangle$ project to a point $\mathbf{p}$. The curves of $\mathcal{S}$ project to conics containing $\mathbf{k}$ and $\mathbf{p}$. Conversely, each conic containing $\mathbf{k}$ and $\mathbf{p}$ is the projection of a unique twisted cubic contained in $\mathcal{S}$. The space of conics contained in a fixed plane and constrained to pass through two fixed points is linear and of dimension three as required.

References

Helmholtz H. von 1925 *Physiological Optics*, vol. **3**. (Ed. J.P.C. Southall Optical Soc. America.

Hofmann W. 1950 Das Problem der "Gefährlichen Flächen in Theorie und Praxis". *Dissertation, Fakultät für Bauwesen der Technischen Hochschule München, München, FR Germany. Published in Reihe C, No. 3 der Deutschen Geodatischen Kommission bei der Bayerischen Akademie der Wissenschaften, München 1953.*

Maybank S.J. 1990 The projective geometry of ambiguous surfaces. *Phil. Trans. Royal Soc. London, Series A* **332**, 1-47.

Semple J.G. & Kneebone G.T. 1953 *Algebraic Projective Geometry.* Oxford: Clarendon Press (reprinted 1979).

Wunderlich W. 1942 Zur Eindeutigkeitsfrage der Hauptaufgabe der Photogrammetrie. *Monatsch. Math. Physik* **50**, 151-164.

4 Reconstruction from Image Velocities

The velocities of points in an image taken by a moving camera depend on the motion of the camera relative to the environment and on the distances to the object points from which they arise. It is these dependencies that allow the reconstruction of the camera velocity and the shape of the scene from the image velocities. If the full camera calibration is known then reconstruction yields the angular velocity of the camera, the direction of the translational velocity and the shape of the scene up to a single unknown scale factor. Reconstruction from image velocities is a limiting case of reconstruction from image correspondences, as the distances between corresponding points become small. In the limit the underlying equations are simplified, but many of the properties of reconstruction from image correspondences are retained, most notably in the ambiguous case.

Image velocities are important both theoretically and practically. They provide a method of reconstruction which is an approximation to the more complicated reconstruction from image correspondences. Reconstruction from image velocities can be analysed mathematically in more detail. Image velocities are important practically because in many applications it is necessary to estimate the velocities and positions of objects relative to the camera in a short space of time, for example to avoid collisions. If a vehicle equipped with a vision system moves at a walking pace in an indoor environment, then it is necessary to locate obstacles within about 1 s of their appearance in the image. The velocities of points in the image can be estimated quickly and reconstruction from image velocities is algebraically simpler than reconstruction from image correspondences.

The main disadvantage of image velocities is that they are susceptible to noise. The variance of the errors in locating points in an image is to a first approximation independent of image velocity. The image changes from which the velocities are estimated are usually small, thus the errors have a large proportional effect on the estimates.

Recent work, for example in Grzywacz & Hildreth (1987), indicates that the effects of image noise on reconstruction from image velocities are severe in some cases. Grzywacz and Hildreth show that the stability of reconstruction from an image sequence is determined by the amount of difference between the images involved, to the extent that the accuracy of the reconstruction depends more on the difference between the first and the last image than it does on the number of intervening images. Jerian & Jain (1990) produce additional evidence that the total change in the appearance of the image is the most important factor. The accuracy and stability of reconstruction from image velocities can be improved by using additional information about the camera motion or the shape of the scene. For example, Murray & Buxton (1990) describe a system for estimating the shape and velocity of a polyhedron from image velocities.

The contents of this chapter are as follows. In Sect. 4.1 the vector equation relating the camera velocity to the image velocities and to the shape of the rigid surface in the field of view is obtained. If enough image velocity vectors are available then reconstruction based on this equation is in general unique up to a single unknown scale factor. In certain ambiguous cases two or more reconstructions are obtained. It is shown in Sect. 4.2 that ambiguity is possible only if the scene points giving rise to the image velocity vectors lie on critical surfaces which, as in the case of reconstruction from image correspondences, are rectangular hyperboloids. Each ambiguous case of reconstruction yields a pair of critical surfaces, which intersect in a space curve of degree four. The space curve splits into a straight line and a horopter curve. Each critical surface is subject to two cubic constraints arising from the known camera calibration. At the end of Sect. 4.2 it is shown that there are at most three reconstructions compatible with a dense image velocity field.

In Sect. 4.3 it is shown that four image velocity vectors constrain the translational velocity of the camera such that it lies on a quartic plane curve. If the image velocity field is irregular then the leading order terms of the quartic split into two linear factors and a quadratic factor. Only one of these three factors contains information about the velocity of the surface. The significance of this result for reconstruction is elaborated in Chap. 6. The linear factor can be obtained easily in practice and it is stable in the presence of noise. If the four image velocity vectors are varied whilst holding the base points fixed then a linear system of quartic plane curves is obtained. It is shown in Sect. 4.4 that the linear system has dimension four and that it is characterised by ten base points, in that a quartic is contained in the linear system if and only if it passes through the ten points.

In Sect. 4.5 expressions are obtained for the spatial derivatives of the image velocity field to second order. From these derivatives three homogeneous cubic polynomial constraints on the translational velocity of the camera are obtained. An expression for the time to contact of the camera with an obstacle is derived.

4.1 Framework

It is assumed that the image velocities arise from the motion of a single rigid surface with respect to the camera. It is always assumed that the camera calibration is known. The velocity of the surface relative to the camera is described by two vectors, the translational velocity $\mathbf{v}$ and the angular velocity $\mathbf{w}$. The vectors $\mathbf{v}$, $\mathbf{w}$ do not uniquely specify the velocity of the surface relative to the camera. In order to ensure uniqueness it is necessary to give a point in space about which to measure $\mathbf{w}$ (Sokolnikoff & Redheffer 1966).

The effect of changing the point about which $\mathbf{w}$ is measured is described. Let $\{\mathbf{v}, \mathbf{w}\}$ specify the velocity of the surface relative to the camera when the angular velocity is measured about a point $\mathbf{p}$ and let $\{\mathbf{v}', \mathbf{w}'\}$ specify the velocity of the surface relative to the camera when the angular velocity is measured about a point $\mathbf{p}'$. The velocity $\dot{\mathbf{x}}$ of a point $\mathbf{x}$ on the surface is given in terms of $\{\mathbf{v}, \mathbf{w}\}$ by

$$\dot{\mathbf{x}} = \mathbf{v} + \mathbf{w} \times (\mathbf{x} - \mathbf{p}) \tag{4.1}$$

The same velocity $\dot{\mathbf{x}}$ is given in terms of $\{\mathbf{v}', \mathbf{w}'\}$ by

$$\dot{\mathbf{x}} = \mathbf{v}' + \mathbf{w}' \times (\mathbf{x} - \mathbf{p}') \tag{4.2}$$

The velocity $\dot{\mathbf{x}}$ of the surface is the same independently of the method chosen to describe it. The right-hand sides of (4.1) and (4.2) are thus equated to obtain

$$\mathbf{v} + \mathbf{w} \times (\mathbf{x} - \mathbf{p}) = \mathbf{v}' + \mathbf{w}' \times (\mathbf{x} - \mathbf{p}')$$

from which it follows that

$$\begin{aligned} \mathbf{w}' &= \mathbf{w} \\ \mathbf{v}' &= \mathbf{v} + \mathbf{w} \times (\mathbf{p}' - \mathbf{p}) \end{aligned}$$

In this book the angular velocity is always taken about the optical centre of the camera. This choice is adequate for theoretical purposes, but in some applications a different choice may be more natural. For example, if the camera tracks a top which is simultaneously spinning and translating then it is more convenient to take the angular velocity about a point fixed in the top.

Cartesian coordinates are chosen with origin $\mathbf{o}$ at the optical centre of the camera. Let $\mathbf{x}$ be a point on the rigid surface which projects to a point $\mathbf{q}$ on the unit sphere. It follows from the choice of origin that the velocity $\dot{\mathbf{x}}$ of $\mathbf{x}$ relative to the camera is given by

$$\dot{\mathbf{x}} = \mathbf{v} + \mathbf{w} \times \mathbf{x} \tag{4.3}$$

The point $\mathbf{q}$ on the projection sphere is given by

$$\mathbf{q} = \frac{\mathbf{x}}{\|\mathbf{x}\|} = \frac{\mathbf{x}}{\sqrt{\mathbf{x}.\mathbf{x}}} \tag{4.4}$$

The derivative of (4.4) with respect to time is

$$\begin{aligned} \dot{\mathbf{q}} &= \frac{\dot{\mathbf{x}}}{\|\mathbf{x}\|} - \frac{(\mathbf{x}.\dot{\mathbf{x}})\mathbf{x}}{\|\mathbf{x}\|^3} \\ &= [\dot{\mathbf{x}} - (\mathbf{q}.\dot{\mathbf{x}})\mathbf{q}]\|\mathbf{x}\|^{-1} \end{aligned} \tag{4.5}$$

The point $\mathbf{q}$ is called the base point of the image velocity vector $\dot{\mathbf{q}}$. The inverse distance K is defined by $K = \|\mathbf{x}\|^{-1}$. The notation $K(\mathbf{q}_i)$ or K_i is used to indicate the dependence of K on the base point $\mathbf{q}_i$. On using (4.3) to substitute in $\dot{\mathbf{x}}$ in (4.5) an equation giving $\dot{\mathbf{q}}$ in terms of the velocity and shape of the surface is obtained,

$$\dot{\mathbf{q}} = [\mathbf{v} - (\mathbf{v}.\mathbf{q})\mathbf{q}]K + \mathbf{w} \times \mathbf{q} \tag{4.6}$$

Equation (4.6) is a vector equation, but only two of the three components yield any information. The component of (4.6) in the direction $\mathbf{q}$ yields no information because it holds identically, regardless of the values of $\mathbf{v}$, $\mathbf{w}$ and K. The scalar product of (4.6) with $\mathbf{q}$ yields $\dot{\mathbf{q}}.\mathbf{q} = 0$. The equation $\dot{\mathbf{q}}.\mathbf{q} = 0$ can be obtained directly by differentiating the identity $\mathbf{q}.\mathbf{q} = 1$.

It follows from (4.6) that $\dot{\mathbf{q}}$ is the sum of two components,

$$\dot{\mathbf{q}} = \dot{\mathbf{q}}_v + \dot{\mathbf{q}}_w$$

where $\dot{\mathbf{q}}_v$ depends on the translational velocity of the camera and $\dot{\mathbf{q}}_w$ depends on the angular velocity of the camera. The two components $\dot{\mathbf{q}}_v$, $\dot{\mathbf{q}}_w$ are defined by

$$\begin{aligned} \dot{\mathbf{q}}_v &= [\mathbf{v} - (\mathbf{v}.\mathbf{q})\mathbf{q}]K \\ \dot{\mathbf{q}}_w &= \mathbf{w} \times \mathbf{q} \end{aligned}$$

The component $\dot{\mathbf{q}}_w$ is independent of the variable K describing the shape of the rigid surface.

Equation (4.6) is the basis of the following formulation of reconstruction from image velocities: given n image velocity vectors $\dot{\mathbf{q}}_i$ with base points $\mathbf{q}_i$, solve the n simultaneous equations of the form (4.6) for the inverse distances $K(\mathbf{q}_i)$ and for the velocity $\{\mathbf{v}, \mathbf{w}\}$ of the surface with respect to the camera. It is apparent from (4.6) that reconstruction from image velocities is possible only up to an ambiguity in scale. The ambiguity arises because $\mathbf{v}$ occurs in (4.6) only as a product with K. Thus, if $\{\mathbf{v}, \mathbf{w}\}$ is a solution to (4.6), then $\{\lambda\mathbf{v}, \mathbf{w}\}$ is also a solution to (4.6) for any non-zero scalar λ. If K is an inverse distance at a base point $\mathbf{q}$ compatible with the velocity $\{\mathbf{v}, \mathbf{w}\}$ then $\lambda^{-1}K$ is the inverse distance at the same base point compatible with $\{\lambda\mathbf{v}, \mathbf{w}\}$. The scaling ambiguity cannot be removed simply by increasing the number n of image velocity vectors. It is sometimes resolved artificially by imposing on $\mathbf{v}$ an additional condition such as $\|\mathbf{v}\| = 1$. In order for a reconstruction to be physically realisable it is usually necessary to have all the inverse distances K greater than zero. This condition arises because the moving surface is known to be in front of the camera rather than behind it.

A count of the number of unknown variables shows that at least five equations (4.6) are needed in order to obtain by reconstruction at most a finite number of camera velocities up to the unknown scale factor. In detail, the number of unknown variables in n equations is $n + 5$, where the inverse distances K_i contribute n, the components of $\mathbf{w}$ contribute three and $\mathbf{v}$ contributes two. Only two unknown variables are obtained from $\mathbf{v}$, because changes in the magnitude of $\mathbf{v}$ only affect the scale of the reconstruction. Each equation (4.6) yields two constraints. The number of scalar equations constraining $\mathbf{v}$, $\mathbf{w}$ and the K_i obtained from n equations of the form (4.6) is thus $2n$. It follows that in order to obtain only a finite number of solutions it is necessary to have $2n \geq n+5$ or equivalently $n \geq 5$. The fact that five image velocity vectors are sufficient to restrict the possible reconstructions to a finite number is established in Sect. 5.4, where it is shown that in general five velocity vectors are compatible with exactly ten different reconstructions.

Two useful equations are obtained from (4.6). The vector product of (4.6) with $\mathbf{q}$ yields

$$\begin{aligned} \dot{\mathbf{q}} \times \mathbf{q} &= (\mathbf{v} \times \mathbf{q})K + (\mathbf{w} \times \mathbf{q}) \times \mathbf{q} \\ &= (\mathbf{v} \times \mathbf{q})K + (\mathbf{w}.\mathbf{q})\mathbf{q} - \mathbf{w} \end{aligned} \tag{4.7}$$

Equation (4.7) is equivalent to (4.6), in that the vector product of (4.7) with $\mathbf{q}$ yields (4.6) up to sign. The inverse distance K is eliminated from (4.7) by taking the scalar product with $\mathbf{v}$,

$$(\dot{\mathbf{q}} \times \mathbf{q}).\mathbf{v} = (\mathbf{w}.\mathbf{q})(\mathbf{v}.\mathbf{q}) - \mathbf{w}.\mathbf{v} \tag{4.8}$$

Equation (4.8) depends on $\dot{\mathbf{q}}$, $\mathbf{q}$ and the camera velocity $\{\mathbf{v}, \mathbf{w}\}$, but not on the inverse distances K. It plays a role in reconstruction from image velocities analogous to that played by (2.8) in the theory of reconstruction from image correspondences.

4.2 Ambiguity

It is shown in Sect. 4.1 that reconstruction from image velocities is always accompanied by a scaling ambiguity. The magnitude of the translational velocity and the absolute sizes of the objects in the field of view cannot be obtained using image velocities alone. For most image velocity fields this is the only ambiguity that can arise. However, certain special sets of image velocities yield a further ambiguity, in that they are compatible with two or more reconstructions that are not related by a simple change of scale. In order to exclude the scaling ambiguity the definition of an ambiguous image velocity field is formulated as follows.

Definition. *An image velocity field* $\dot{\mathbf{q}}$ *is ambiguous if it has two decompositions*

$$\dot{\mathbf{q}} = \dot{\mathbf{q}}_{v_1} + \dot{\mathbf{q}}_{w_1} = \dot{\mathbf{q}}_{v_2} + \dot{\mathbf{q}}_{w_2} \tag{4.9}$$

such that $\dot{\mathbf{q}}_{v_1} \neq \dot{\mathbf{q}}_{v_2}$ *and* $\dot{\mathbf{q}}_{w_1} \neq \dot{\mathbf{q}}_{w_2}$.

When referring to an image velocity field as ambiguous it is always in the sense of Definition 4.1. Ambiguous image velocity fields are discussed by Horn (1987), Longuet-Higgins (1984) and Maybank (1985,1987b,1990).

If the image velocity field contains only a few vectors then ambiguity is likely simply because of the lack of data. It is often convenient to exclude this effect by specifying a dense image velocity field. A dense image velocity field is one defined at all points within a non-empty open set of the unit sphere. In order to fix ideas, the open set can be the interior of a circle of non-zero radius drawn on the unit sphere.

4.2.1 Preliminary Results

It is shown that ambiguity does not arise if the image velocities are due to angular velocity alone. It is also shown that ambiguity can only arise if the two translational velocities $\mathbf{v}_1$ and $\mathbf{v}_2$ of (4.9) are not parallel. The proofs of these results depend on equations in $\mathbf{q}$ similar to

$$(\mathbf{l}.\mathbf{q})(\mathbf{m}.\mathbf{q}) - (\mathbf{l}.\mathbf{m})(\mathbf{q}.\mathbf{q}) = (\mathbf{l}'.\mathbf{q})(\mathbf{m}'.\mathbf{q}) - (\mathbf{l}'.\mathbf{m}')(\mathbf{q}.\mathbf{q}) \tag{4.10}$$

The $\mathbf{l}$, $\mathbf{m}$, $\mathbf{l}'$, $\mathbf{m}'$ are constant three dimensional vectors and the equation holds for all $\mathbf{q}$ on the unit sphere. Let $\sim$ be an equivalence relation between pairs of vectors, $(\mathbf{l},\mathbf{m}) \sim (\mathbf{l}',\mathbf{m}')$, defined by

- One of $\mathbf{l}$, $\mathbf{m}$ and one of $\mathbf{l}'$, $\mathbf{m}'$ is zero; or
- There exists a non-zero scalar λ such that either $\mathbf{l} = \lambda\mathbf{l}'$, $\mathbf{m} = \lambda^{-1}\mathbf{m}'$ or $\mathbf{l} = \lambda\mathbf{m}'$, $\mathbf{m} = \lambda^{-1}\mathbf{l}'$.

Proposition 4.1. *Let* $\mathbf{l}$, $\mathbf{m}$, $\mathbf{l}'$, $\mathbf{m}'$ *be four vectors each of dimension three. Then* $(\mathbf{l},\mathbf{m}) \sim (\mathbf{l}',\mathbf{m}')$ *if and only if (4.10) holds for all* $\mathbf{q}$.

Proof. If $(\mathbf{l},\mathbf{m}) \sim (\mathbf{l}',\mathbf{m}')$ then (4.10) holds for all $\mathbf{q}$. Conversely, let (4.10) hold for all $\mathbf{q}$. It follows that

$$\mathbf{l} \otimes \mathbf{m} + \mathbf{m} \otimes \mathbf{l} - 2(\mathbf{l}.\mathbf{m})I = \mathbf{l}' \otimes \mathbf{m}' + \mathbf{m}' \otimes \mathbf{l}' - 2(\mathbf{l}'.\mathbf{m}')I \tag{4.11}$$

The trace of (4.11) yields $\mathbf{l}.\mathbf{m} = \mathbf{l}'.\mathbf{m}'$. It follows that

$$\mathbf{l} \otimes \mathbf{m} + \mathbf{m} \otimes \mathbf{l} = \mathbf{l}' \otimes \mathbf{m}' + \mathbf{m}' \otimes \mathbf{l}' \tag{4.12}$$

Let $\mathbf{l}$ and $\mathbf{l}'$ be linearly independent and let A be any invertible matrix such that $A\mathbf{l} = (1,0,0)^\top$ and $A\mathbf{l}' = (0,1,0)^\top$. It follows from (4.12) that

$$A(\mathbf{l} \otimes \mathbf{m})A^\top + A(\mathbf{m} \otimes \mathbf{l})A^\top = A(\mathbf{l}' \otimes \mathbf{m}')A^\top + A(\mathbf{m}' \otimes \mathbf{l}')A^\top \tag{4.13}$$

Let $\mathbf{a}$ and $\mathbf{a}'$ be the vectors defined by $\mathbf{a} = A\mathbf{m}$ and $\mathbf{a}' = A\mathbf{m}'$. Equation (4.13) yields

$$\begin{pmatrix} 2a_1 & a_2 & a_3 \\ a_2 & 0 & 0 \\ a_3 & 0 & 0 \end{pmatrix} = \begin{pmatrix} 0 & a_1' & 0 \\ a_1' & 2a_2' & 0 \\ 0 & a_3' & 0 \end{pmatrix}$$

It follows that $a_1 = a_3 = 0$ and $a_2' = a_3' = 0$. The vector $A\mathbf{m}$ is parallel to $A\mathbf{l}'$ and $A\mathbf{m}'$ is parallel to $A\mathbf{l}$. The result $(\mathbf{l},\mathbf{m}) \sim (\mathbf{l}',\mathbf{m}')$ follows.

It remains to deal with the case in which $\mathbf{l}$ and $\mathbf{l}'$ are linearly dependent. If $\mathbf{l}$ and $\mathbf{l}'$ are both zero then the result follows. Let one of $\mathbf{l}$ and $\mathbf{l}'$ be non-zero, say $\mathbf{l} \neq 0$, and let α be a scalar such that $\mathbf{l}' = \alpha\mathbf{l}$. Then (4.12) yields

$$\mathbf{l} \otimes (\mathbf{m} - \alpha\mathbf{m}') + (\mathbf{m} - \alpha\mathbf{m}') \otimes \mathbf{l} = 0 \tag{4.14}$$

Let $\mathbf{a}$ be any vector such that $\mathbf{a}.(\mathbf{m} - \alpha\mathbf{m}') = 0$ and $\mathbf{a}.\mathbf{l} \neq 1$. On applying the left-hand side of (4.14) to $\mathbf{a}$ it follows that $\mathbf{m} = \alpha\mathbf{m}'$. The result follows. □

Proposition 4.2. *A dense image velocity field which arises from angular velocity alone is not ambiguous.*

Proof. Let $\dot{\mathbf{q}}$ be a dense image velocity field such that

$$\dot{\mathbf{q}} = \dot{\mathbf{q}}_{w_1} = \dot{\mathbf{q}}_{v_2} + \dot{\mathbf{q}}_{w_2} \tag{4.15}$$

It is shown that $\dot{\mathbf{q}}_{v_2} = 0$. It follows from (4.15) that

$$\dot{\mathbf{q}}_{v_2} = \dot{\mathbf{q}}_{w_1} - \dot{\mathbf{q}}_{w_2}$$

Let $\mathbf{w}$ be the angular velocity such that

$$\dot{\mathbf{q}}_{w_1} - \dot{\mathbf{q}}_{w_2} = \mathbf{w} \times \mathbf{q} \tag{4.16}$$

Let $\mathbf{v}_2$ be a translational velocity and let $K_2 = K_2(\mathbf{q})$ be an inverse distance such that

$$\dot{\mathbf{q}}_{v_2} = [\mathbf{v}_2 - (\mathbf{v}_2.\mathbf{q})\mathbf{q}]K_2 \tag{4.17}$$

It follows from (4.15), (4.16) and (4.17) that

$$\mathbf{w} \times \mathbf{q} = [\mathbf{v}_2 - (\mathbf{v}_2.\mathbf{q})\mathbf{q}]K_2$$

The scalar product of (4.33) with $\mathbf{v}_2 \times \mathbf{q}$ yields

$$(\mathbf{w} \times \mathbf{q}).(\mathbf{v}_2 \times \mathbf{q}) = 0 \tag{4.18}$$

It follows from (4.18) that $(\mathbf{v}_2, \mathbf{w}) \sim (0, 0)$. Thus, either $\mathbf{v}_2 = 0$ or $\mathbf{w} = 0$. In both cases (4.15) yields $\dot{\mathbf{q}} = \dot{\mathbf{q}}_{w_1} = \dot{\mathbf{q}}_{w_2}$. □

Theorem 4.3. *Let $\dot{\mathbf{q}}$ be an ambiguous dense image velocity field and let*

$$\dot{\mathbf{q}} = \dot{\mathbf{q}}_{v_1} + \dot{\mathbf{q}}_{w_1} = \dot{\mathbf{q}}_{v_2} + \dot{\mathbf{q}}_{w_2} \tag{4.19}$$

be two decompositions of $\dot{\mathbf{q}}$ into components due to translational velocity and angular velocity. Let $\mathbf{v}_1$ be the translational velocity associated with the first decomposition and let $\mathbf{v}_2$ be the translational velocity associated with the second decomposition. Then $\mathbf{v}_1$ and $\mathbf{v}_2$ are non-zero and non-parallel.

Proof. It follows from Proposition 4.2 that $\mathbf{v}_1 \neq 0$ and $\mathbf{v}_2 \neq 0$. It is assumed that $\mathbf{v}_1 = \lambda \mathbf{v}_2$ for a non-zero scalar λ and a contradiction is obtained. The assumption yields

$$\begin{aligned} \dot{\mathbf{q}}_{v_1} - \dot{\mathbf{q}}_{v_2} &= [\mathbf{v}_1 - (\mathbf{v}_1.\mathbf{q})\mathbf{q}]K_1 - [\mathbf{v}_2 - (\mathbf{v}_2.\mathbf{q})\mathbf{q}]K_2 \\ &= [\mathbf{v}_2 - (\mathbf{v}_2.\mathbf{q})\mathbf{q}](\lambda K_1 - K_2) \end{aligned} \tag{4.20}$$

Let $\dot{\mathbf{q}}_v$ be defined by $\dot{\mathbf{q}}_v = \dot{\mathbf{q}}_{v_1} - \dot{\mathbf{q}}_{v_2}$. It follows from (4.20) that $\dot{\mathbf{q}}_v$ has the same form as an image velocity field arising from translational velocity alone. The application of Proposition 4.2 to

$$\dot{\mathbf{q}} = \dot{\mathbf{q}}_v + \dot{\mathbf{q}}_{w_1} = \dot{\mathbf{q}}_{w_2} \tag{4.21}$$

yields $\mathbf{w}_1 = \mathbf{w}_2$. This contradicts the hypothesis that the two decompositions of $\dot{\mathbf{q}}$ in (4.19) are different. □

4.2.2 Critical Surfaces

An image velocity field is ambiguous only if it arises from motion of the camera relative to a particular type of surface known as a critical surface. As in the case of reconstruction from image correspondences the critical surfaces are rectangular hyperboloids of one sheet.

Theorem 4.4. *Let ψ be a rigid surface moving with velocity $\{\mathbf{v}_1, \mathbf{w}_1\}$ relative to the camera. Let $\{\mathbf{v}_2, \mathbf{w}_2\}$ be a rigid velocity compatible with the image velocity field arising from the motion of ψ, such that $\mathbf{v}_1 \times \mathbf{v}_2 \neq 0$. Let $\mathbf{w}$ be the vector defined by $\mathbf{w} = \mathbf{w}_2 - \mathbf{w}_1$. If Cartesian coordinates are chosen with origin at the optical centre $\mathbf{o}$ of the camera then ψ has the equation*

$$(\mathbf{w} \times \mathbf{x}).(\mathbf{v}_2 \times \mathbf{x}) + (\mathbf{v}_2 \times \mathbf{v}_1).\mathbf{x} = 0 \tag{4.22}$$

Proof. Let $\dot{\mathbf{q}}$ be the image velocity field arising from the motion of ψ relative to the camera and let ϕ be a rigid surface moving with velocity $\{\mathbf{v}_2, \mathbf{w}_2\}$ that also gives rise to $\dot{\mathbf{q}}$. Let $K_1(\mathbf{q})$, $K_2(\mathbf{q})$ be the inverse distances to ψ, ϕ respectively, measured in the direction $\mathbf{q}$. Then (4.6) yields

$$\begin{aligned} \dot{\mathbf{q}} &= [\mathbf{v}_1 - (\mathbf{v}_1.\mathbf{q})\mathbf{q}]K_1 + \mathbf{w}_1 \times \mathbf{q} \\ &= [\mathbf{v}_2 - (\mathbf{v}_2.\mathbf{q})\mathbf{q}]K_2 + \mathbf{w}_2 \times \mathbf{q} \end{aligned} \tag{4.23}$$

The equating of the two expressions for $\dot{\mathbf{q}}$ in (4.23) yields

$$[\mathbf{v}_1 - (\mathbf{v}_1.\mathbf{q})\mathbf{q}]K_1 - [\mathbf{v}_2 - (\mathbf{v}_2.\mathbf{q})\mathbf{q}]K_2 = \mathbf{w} \times \mathbf{q} \tag{4.24}$$

The scalar product of (4.24) with $\mathbf{v}_2 \times \mathbf{q}$ yields

$$[\mathbf{v}_1.(\mathbf{v}_2 \times \mathbf{q})]K_1 = (\mathbf{w} \times \mathbf{q}).(\mathbf{v}_2 \times \mathbf{q}) \tag{4.25}$$

Let $\mathbf{x}$ be the point of ψ projecting to $\mathbf{q}$. The point $\mathbf{x}$ is given by $\mathbf{x} = K_1\mathbf{q}$. With this definition of $\mathbf{x}$, (4.22) follows from (4.25). □

Corollary. The surface ψ is a hyperboloid of one sheet which passes through the optical centre of the camera. To prove the corollary note first that it follows from (4.22) that ψ is a surface of degree two which contains $\mathbf{o}$. The surface ψ is a hyperboloid of one sheet because it contains the real line $\langle \mathbf{o}, \mathbf{v}_2 \rangle$ and because it intersects the plane at infinity in a non-singular conic with an infinity of real points.

Proposition 4.5. *The critical surfaces arising in reconstruction from image velocities are rectangular.*

Proof. It follows from (4.22) that the intersection of ψ with the plane at infinity is the conic s_∞ given by the equation

$$(\mathbf{w} \times \mathbf{x}).(\mathbf{v}_2 \times \mathbf{x}) = 0 \tag{4.26}$$

Equation (4.26) has the form $\mathbf{x}^\top M \mathbf{x} = 0$, where M is the symmetric 3×3 matrix defined by

$$M = \frac{1}{2}(\mathbf{w} \otimes \mathbf{v}_2 + \mathbf{v}_2 \otimes \mathbf{w}) - (\mathbf{w}.\mathbf{v}_2)I$$

The result follows from Theorem 3.13. □

Corollary 1. The principal points of ψ are $(\mathbf{w}, 0)^\top$ and $(\mathbf{v}_2, 0)^\top$.

Corollary 2. The line $\langle \mathbf{o}, \mathbf{v}_2 \rangle$ is a generator of ψ which contains both the optical centre $\mathbf{o}$ of the camera and the principal point $\mathbf{v}_2$ of ψ.

In stating Corollary 2 to Proposition 4.5 the notation is simplified by writing $\mathbf{v}_2$ for the point $(\mathbf{v}_2, 0)^\top$ in the plane at infinity. The line $\langle \mathbf{o}, \mathbf{v}_2 \rangle$ is thus the line through $\mathbf{o}$ with direction $\mathbf{v}_2$. Corollary 2 does not hold for the critical surfaces arising in reconstruction from image correspondences. It leads to simplifications in the theory of ambiguous reconstructions from image velocities. These simplifications are apparent in the investigation of singular critical surfaces carried out in Sect. 4.2.3.

For the next theorem it is recalled from Sect. 3.2.2 that a main generator of a rectangular quadric is any generator which passes through one of the principal points of the quadric.

Theorem 4.6 *Let ψ be a rectangular hyperboloid of one sheet with distinct principal points* $\mathbf{m}$, $\mathbf{n}$, *and let* $\mathbf{o}$ *be any point on one of the main generators of ψ. Then there exists a rigid velocity of ψ such that the image velocities obtained on projecting the points of ψ from* $\mathbf{o}$ *are ambiguous. The surface ψ is obtained by reconstruction from this image velocity field.*

Proof. Cartesian coordinates are chosen with origin at $\mathbf{o}$. Let $\mathbf{l}$ be the direction of the normal to ψ at $\mathbf{o}$. The equation for ψ is

$$\mathbf{x}^\top[\frac{1}{2}(\mathbf{m} \otimes \mathbf{n} + \mathbf{n} \otimes \mathbf{m}) - (\mathbf{m}.\mathbf{n})I]\mathbf{x} + \mathbf{l}.\mathbf{x} = 0$$

which is equivalent to

$$-(\mathbf{m}\times\mathbf{x}).(\mathbf{n}\times\mathbf{x})+\mathbf{l}.\mathbf{x}=0 \tag{4.27}$$

Without loss of generality, let $\langle \mathbf{o}, \mathbf{n}\rangle$ be the main generator of ψ referred to in the statement of the theorem. It follows from (4.27) that $\mathbf{n}.\mathbf{l} = 0$, thus there exists a vector $\mathbf{v}$ such that $\mathbf{l} = \mathbf{v}\times\mathbf{n}$. The substitution of $\mathbf{v}\times\mathbf{n}$ for $\mathbf{l}$ in (4.27) yields

$$(\mathbf{m}\times\mathbf{x}).(\mathbf{n}\times\mathbf{x})+(\mathbf{n}\times\mathbf{v}).\mathbf{x}=0 \tag{4.28}$$

The result follows from (4.28) and (4.22) after making the substitutions $\mathbf{m} = \mathbf{w}$, $\mathbf{n} = \mathbf{v}_2$ and $\mathbf{v} = \mathbf{v}_1$. □

4.2.3 Singular Critical Surfaces

The critical surfaces are in general non-singular quadrics, but singular critical surfaces can arise in special cases. In this subsection a necessary and sufficient condition for a critical surface to have a singular point is obtained and an expression for the singular point is given. Under a general collineation there are three classes of singular quadric in $\mathbb{P}^3$, the irreducible cones, the plane pairs and the coincident plane pairs. If a quadric contains only one singular point then it is a cone and the singular point is the vertex. If the quadric contains more than one singular point then it splits into a plane pair and every point in the intersection of the two planes is a singular point. Both the cone and the plane pair have the property that any line contained in the surface passes through a singular point.

Theorem 4.7. *Let Cartesian coordinates be chosen with origin* $\mathbf{o}$ *at the optical centre of the camera and let* ψ *be a critical surface with the equation*

$$(\mathbf{w}\times\mathbf{x}).(\mathbf{v}_2\times\mathbf{x})+(\mathbf{v}_2\times\mathbf{v}_1).\mathbf{x}=0 \tag{4.29}$$

Then ψ *contains a singular point if and only if there exist scalars* a, b *such that*

$$\mathbf{w}=a(\mathbf{v}_2\times\mathbf{v}_1)+b\mathbf{v}_2 \tag{4.30}$$

If (4.30) holds, then $(\mathbf{v}_2, a)^\top$ *is a singular point of* ψ.

Proof. Let $F(\mathbf{x}, x_4)$ be the homogeneous polynomial defined by

$$F(\mathbf{x},x_4)=(\mathbf{w}\times\mathbf{x}).(\mathbf{v}_2\times\mathbf{x})+x_4(\mathbf{v}_2\times\mathbf{v}_1).\mathbf{x}$$

The inhomogeneous equation (4.29) for ψ is obtained from $F(\mathbf{x}, x_4) = 0$ on setting $x_4 = 1$. The surface ψ has a singular point if and only if there exists a point $(\mathbf{x}, x_4)^\top$ at which $\nabla F = 0$. It follows that $\mathbf{x}$ is a singular point of ψ if and only if

$$\begin{aligned}2(\mathbf{w}.\mathbf{v}_2)\mathbf{x}-(\mathbf{w}.\mathbf{x})\mathbf{v}_2-(\mathbf{v}_2.\mathbf{x})\mathbf{w}+x_4(\mathbf{v}_2\times\mathbf{v}_1) &= 0\\ (\mathbf{v}_2\times\mathbf{v}_1).\mathbf{x} &= 0\end{aligned} \tag{4.31}$$

It follows from the Corollary to Theorem 4.4 that ψ contains the line g through $\mathbf{o}$ parameterised by $t \mapsto t\mathbf{v}_2$. If ψ has a singular point then every line of ψ contains a singular point of ψ. Thus a solution $\mathbf{x}$ to (4.31) of the form $\mathbf{x} = t\mathbf{v}_2$ is sought. With this choice of $\mathbf{x}$, the second equation of (4.31) is satisfied for all t. The substitution of $t\mathbf{v}_2$ for $\mathbf{x}$ in the first equation of (4.31) yields

$$t(\mathbf{w}.\mathbf{v}_2)\mathbf{v}_2 - t(\mathbf{v}_2.\mathbf{v}_2)\mathbf{w} + x_4(\mathbf{v}_2 \times \mathbf{v}_1) = 0 \tag{4.32}$$

If $t = 0$ then (4.32) yields $x_4 = 0$. This case does not arise because $(t\mathbf{v}_2, x_4)^\top$ is required to be a point of $\mathbb{P}^3$. It follows that $t \neq 0$. The cancellation of t from (4.32) yields

$$\mathbf{w} = (\mathbf{v}_2.\mathbf{v}_2)^{-1}[(\mathbf{w}.\mathbf{v}_2)\mathbf{v}_2 + t^{-1}x_4(\mathbf{v}_2 \times \mathbf{v}_1)] \tag{4.33}$$

It follows from (4.33) that $\mathbf{w}$ is a linear combination of $\mathbf{v}_2$ and $\mathbf{v}_2 \times \mathbf{v}_1$ as required.

Conversely, let a, b be scalars such (4.30) holds and let $\mathbf{x} = t\mathbf{v}_2$. A value of t is found such that (4.31) holds. If $t = 0$ then the first equation of (4.31) yields $x_4 = 0$. This case does not arise because $(t\mathbf{v}_2, x_4)^\top$ is required to be a point of $\mathbb{P}^3$. The second equation of (4.31) holds whenever $\mathbf{x} = t\mathbf{v}_2$. On substituting for $\mathbf{w}$ and $\mathbf{x}$ on the right-hand side of the first equation of (4.31) it follows that

$$tb(\mathbf{v}_2.\mathbf{v}_2)\mathbf{v}_2 - t(\mathbf{v}_2.\mathbf{v}_2)[a(\mathbf{v}_2 \times \mathbf{v}_1) + b\mathbf{v}_2] + x_4(\mathbf{v}_2 \times \mathbf{v}_1) \tag{4.34}$$

The expression (4.34) is zero if $x_4 = t\,a(\mathbf{v}_2.\mathbf{v}_2)$. Thus ψ has a singular point at

$$(t\mathbf{v}_2, x_4)^\top = (\mathbf{v}_2, t^{-1}x_4)^\top = (\mathbf{v}_2, a(\mathbf{v}_2.\mathbf{v}_2))^\top \qquad \square$$

It is shown in Theorem 4.7 that if ψ has a singular point anywhere, then it has a singular point at $(\mathbf{v}_2, a)^\top$, where a is given by (4.30). The scalar product of (4.30) with $\mathbf{v}_2 \times \mathbf{v}_1$ yields

$$a = \frac{\mathbf{w}.(\mathbf{v}_2 \times \mathbf{v}_1)}{\|\mathbf{v}_2 \times \mathbf{v}_1\|^2}$$

It follows that if ψ has a singular point, then it has a singular point at

$$(\mathbf{x}, x_4)^\top = (\|\mathbf{v}_2 \times \mathbf{v}_1\|^2\mathbf{v}_2, \mathbf{w}.(\mathbf{v}_2 \times \mathbf{v}_1))^\top$$

4.2.4 Critical Surface Pairs

In the ambiguous case each possible reconstruction yields a critical surface. A critical surface pair is thus obtained in which both of the surfaces are compatible with the same image velocity field. Let $\dot{\mathbf{q}}$ be an ambiguous image velocity field yielding a critical surface pair ψ, ϕ such that ψ has a velocity $\{\mathbf{v}_1, \mathbf{w}_1\}$ relative to the camera and ϕ has a velocity $\{\mathbf{v}_2, \mathbf{w}_2\}$ relative to the camera. Cartesian coordinates are chosen with origin at the optical centre $\mathbf{o}$ of the camera and the vector $\mathbf{w}$ is defined by $\mathbf{w} = \mathbf{w}_2 - \mathbf{w}_1$. It follows from (4.22) that the equations for ψ and ϕ are

$$(\mathbf{w}.\mathbf{x})(\mathbf{v}_2.\mathbf{x}) - (\mathbf{w}.\mathbf{v}_2)(\mathbf{x}.\mathbf{x}) + (\mathbf{v}_1 \times \mathbf{v}_2).\mathbf{x} = 0 \quad (4.35)$$

$$(\mathbf{w}.\mathbf{x})(\mathbf{v}_1.\mathbf{x}) - (\mathbf{w}.\mathbf{v}_1)(\mathbf{x}.\mathbf{x}) + (\mathbf{v}_1 \times \mathbf{v}_2).\mathbf{x} = 0 \quad (4.36)$$

The critical surface pairs arising in reconstruction from image velocities have some properties in common with the critical surfaces arising in reconstruction from point correspondences, as shown in the following theorem.

Theorem 4.8. *Let ψ, ϕ be a critical surface pair. Then $\psi \cap \phi$ is a space curve of degree four which splits into a line g and a horopter curve c. Both c and g pass through the optical centre of the camera.*

Proof. Cartesian coordinates are chosen such that ψ and ϕ are given by (4.35) and (4.36) respectively. The point $\mathbf{o}$ is then the optical centre of the camera. It follows that ψ and ϕ have the same tangent plane at $\mathbf{o}$, namely the plane with normal $\mathbf{v}_1 \times \mathbf{v}_2$. A short calculation shows that ψ and ϕ contain a common generator, g, parameterised by

$$t \mapsto t[a\mathbf{v}_1 - b\mathbf{v}_2] \quad (4.37)$$

where a and b are defined by

$$\begin{aligned} a &= (\mathbf{w} \times \mathbf{v}_2).(\mathbf{v}_2 \times \mathbf{v}_1) \\ b &= (\mathbf{w} \times \mathbf{v}_1).(\mathbf{v}_2 \times \mathbf{v}_1) \end{aligned}$$

The surfaces ψ and ϕ are each of degree two, thus the intersection $\psi \cap \phi$ is a space curve of degree four. It follows that $\psi \cap \phi$ splits into the line g and a space curve c of degree three. The curve c meets the plane at infinity at $\mathbf{w}$, $\mathbf{i}_w$, $\mathbf{j}_w$, where $\mathbf{i}_w$, $\mathbf{j}_w$ are the points of contact of the tangents drawn from $\mathbf{w}$ to the absolute conic. The points $\mathbf{w}$, $\mathbf{i}_w$, $\mathbf{j}_w$ are in c because they are contained in $\psi \cap \phi$, but they are not in general points of g. It follows from Proposition 3.21 that c is a horopter curve.

It is clear from (4.37) that g passes through $\mathbf{o}$. To show that c passes through $\mathbf{o}$ (4.36) is subtracted from (4.35) to obtain

$$(\mathbf{w}.\mathbf{x})(\mathbf{v}_2.\mathbf{x} - \mathbf{v}_1.\mathbf{x}) - (\mathbf{w}.\mathbf{v}_2 - \mathbf{w}.\mathbf{v}_1)(\mathbf{x}.\mathbf{x}) = 0 \quad (4.38)$$

The points $\mathbf{x}$ of $\psi \cap \phi$ satisfy (4.38). This equation is homogeneous of degree two in $\mathbf{x}$. It follows that the projection of $\psi \cap \phi$ from $\mathbf{o}$ is a conic, s. The projection of g from $\mathbf{o}$ is a single point, thus the projection of c from $\mathbf{o}$ is s. The projection of a horopter curve from a point is a conic if and only if the point is on the horopter curve. It follows that $\mathbf{o}$ is a point of c. □

It is shown in Theorem 3.32 that in the ambiguous case of reconstruction from image correspondences the critical surface pair ψ, ϕ can be chosen such that ψ

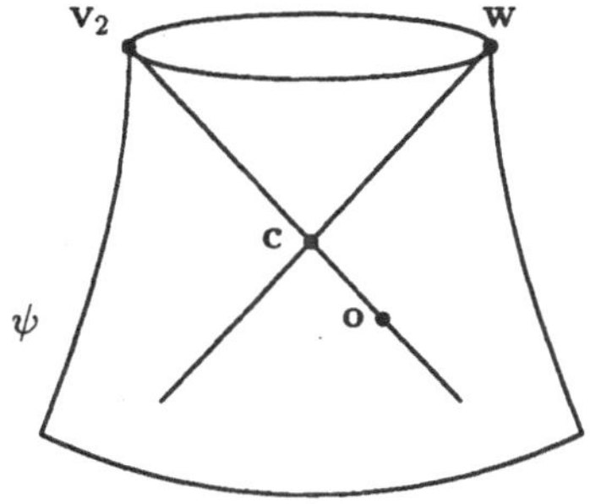

Fig. 4.1. Illustration to Theorem 4.9

and the horopter curve contained in $\psi \cap \phi$ are invariant under the same non-trivial rigid involution. This prompts the analogous question for image velocities: given ψ, can ϕ be chosen such that ψ and the horopter curve contained in $\psi \cap \phi$ are invariant under the same non-trivial rigid involution? In the next theorem it is shown that the answer to the question is no.

Theorem 4.9. *Let ψ be a fixed non-singular critical surface and let ϕ be a variable critical surface compatible with the same image velocity field. Let τ_ψ be the unique non-trivial rigid skew involution of ψ that fixes the principal points of ψ and let τ_c be the unique non-trivial rigid skew involution of the horopter curve c contained in $\psi \cap \phi$. Let $\mathbf{v}_2$ be the translational velocity of ϕ relative to the camera. As the magnitude of $\mathbf{v}_2$ varies, ϕ and c vary, but ψ remains fixed. Then for all possible choices of the magnitude of $\mathbf{v}_2$, $\tau_\psi \neq \tau_c$.*

Proof. Cartesian coordinates are chosen with origin at the optical centre $\mathbf{o}$ of the camera. Let the equations for ψ and ϕ be given by (4.35) and (4.36), respectively, where $\{\mathbf{v}_1, \mathbf{w}_1\}$ is the velocity of ψ relative to the camera, $\{\mathbf{v}_2, \mathbf{w}_2\}$ is the velocity of ϕ relative to the camera and $\mathbf{w} = \mathbf{w}_2 - \mathbf{w}_1$. It is assumed that for some choice of ϕ, $\tau_\psi = \tau_c$, and a contradiction is obtained.

Let $\mathbf{c}$ be the centre of c, as illustrated in Fig. 4.1. Then $\mathbf{c}$ is fixed by τ_c, thus $\mathbf{c}$ is fixed by τ_ψ. It follows from the definition of τ_ψ in Sect. 3.2.2 that $\langle \mathbf{c}, \mathbf{v}_2 \rangle$ and $\langle \mathbf{c}, \mathbf{w} \rangle$ are two of the main generators of ψ. It follows from Theorem 4.8 that c contains $\mathbf{o}$. If $\mathbf{c} = \mathbf{o}$ then the normal to ψ at $\mathbf{o}$ is perpendicular to $\mathbf{w}$. It follows that

$$\mathbf{w}.(\mathbf{v}_1 \times \mathbf{v}_2) = 0 \qquad (4.39)$$

In general, (4.39) does not hold, thus $\mathbf{c} \neq \mathbf{o}$. The main generator $\langle \mathbf{c}, \mathbf{w} \rangle$ intersects c twice. It follows that $\langle \mathbf{c}, \mathbf{v}_2 \rangle$ intersects c once only (at $\mathbf{c}$), because c is a $(2,1)$ curve on ψ. The line $\langle \mathbf{o}, \mathbf{v}_2 \rangle$ is also a generator of ψ through $\mathbf{v}_2$ which intersects c once only (at $\mathbf{o}$). It follows that $\langle \mathbf{o}, \mathbf{v}_2 \rangle = \langle \mathbf{c}, \mathbf{v}_2 \rangle$ and that $\mathbf{c} = \mathbf{o}$, thus contradicting the result $\mathbf{c} \neq \mathbf{o}$. The inequality $\tau_\psi \neq \tau_c$ thus holds for all choices of $\mathbf{v}_2$. □

4.2.5 Cubic Polynomial Constraints on Critical Surfaces

Two cubic polynomial constraints on critical surfaces are obtained. These constraints are limiting cases of two constraints on the critical surfaces obtained in Sect. 3.4.2 for the ambiguous case of reconstruction from corresponding points.

Theorem 4.10. *The critical surfaces arising in the ambiguous case of reconstruction from image velocities are subject to two cubic polynomial constraints.*

Proof. Cartesian coordinates are chosen with origin at the optical centre $\mathbf{o}$ of the camera. Let ψ be a critical surface arising from a set of image velocities. It follows from (4.22) that the equation for ψ is

$$\mathbf{x}^{\top} M\mathbf{x} + \mathbf{l}.\mathbf{x} = 0$$

where M is the symmetric 3×3 matrix defined by

$$M = \frac{1}{2}(\mathbf{w} \otimes \mathbf{v}_2 + \mathbf{v}_2 \otimes \mathbf{w}) - (\mathbf{w}.\mathbf{v}_2)I$$

and $\mathbf{l}$ is the vector defined by $\mathbf{l} = \mathbf{v}_2 \times \mathbf{v}_1$. Let N be the matrix defined by

$$N = M - \frac{1}{2}\mathrm{tr}(M)I = \frac{1}{2}(\mathbf{w} \otimes \mathbf{v}_2 + \mathbf{v}_2 \otimes \mathbf{w}) \tag{4.40}$$

It follows from (4.40) and the definition of $\mathbf{l}$ that

$$\det(N) = 0 \tag{4.41}$$

$$\mathbf{l}^{\top} N\mathbf{l} = 0 \tag{4.42}$$

Equations (4.41) and (4.42) are the required cubic constraints on ψ. □

Equation (4.41) is the limit of (3.69) and (4.42) is the limit of (3.68), as the distances between corresponding points become small.

A converse to Theorem 4.10 is obtained.

Theorem 4.11. *Let ψ be a hyperboloid of one sheet with equation*

$$\mathbf{x}^{\top} M\mathbf{x} + \mathbf{l}.\mathbf{x} = 0$$

where M is a symmetric 3×3 matrix and $\mathbf{l}$ is a vector. Let N be the matrix defined by

$$N = M - (1/2)\mathrm{tr}(M)I$$

If (4.41) and (4.42) hold, then ψ is a critical surface for reconstruction from image velocities, where the optical centre $\mathbf{o}$ of the camera is at the origin of coordinates.

Proof. It follows from (4.41) that N has rank two or less. In addition, N is symmetric because M is symmetric. Thus, there exist vectors $\mathbf{w}$, $\mathbf{v}_2$ such that

$$N = \frac{1}{2}(\mathbf{w} \otimes \mathbf{v}_2 + \mathbf{v}_2 \otimes \mathbf{w})$$

The constraint (4.42) on N yields $(\mathbf{l}.\mathbf{w})(\mathbf{l}.\mathbf{v}_2) = 0$. It follows that either $\mathbf{l}.\mathbf{w} = 0$ or $\mathbf{l}.\mathbf{v}_2 = 0$. Without loss of generality, $\mathbf{w}$ and $\mathbf{v}_2$ are chosen such that $\mathbf{l}.\mathbf{v}_2 = 0$. It follows that there exists a vector $\mathbf{v}_1$ such that

$$\mathbf{l} = \mathbf{v}_2 \times \mathbf{v}_1$$

The surface ψ thus has an equation of the form (4.22). □

4.2.6 The Maximum Number of Reconstructions

It is shown that in the worst case there are at most three reconstructions compatible with a dense image velocity field such that no two of the reconstructions are related by a change of scale. This is in contrast with the general case, in which reconstruction is unique up to a single unknown scale factor. If the surface giving rise to the image velocity field is a plane then there is in general a two-fold ambiguity (Longuet-Higgins 1984). In proving that at most three reconstructions are possible the planar case is treated separately, as it differs significantly from the general case. An analogous result for image correspondences, that there is at worst a three-fold ambiguity, is obtained in Sect. 2.1.3.

Theorem 4.12. *There are in general exactly two reconstructions compatible with a dense image velocity field arising from a moving rigid planar surface. The image velocity field cannot also arise from a moving rigid non-planar surface.*

Proof. Cartesian coordinates are chosen with origin $\mathbf{o}$ at the optical centre of the camera. Let the plane move with rigid velocity $\{\mathbf{v}_1, \mathbf{w}_1\}$, let $\mathbf{u}$ be the unit normal of the plane and let k^{-1} be the distance from $\mathbf{o}$ to the plane. It follows that the distance p to the plane from $\mathbf{o}$ in the direction $\mathbf{q}$ is given by $p = k^{-1}(\mathbf{q}.\mathbf{u})^{-1}$.

Let the image velocity field also arise from a second rigid surface moving with rigid velocity $\{\mathbf{v}_2, \mathbf{w}_2\}$. Equation (4.6) yields the following two expressions for $\dot{\mathbf{q}}$

$$\begin{aligned} \dot{\mathbf{q}} &= [\mathbf{v}_1 - (\mathbf{v}_1.\mathbf{q})\mathbf{q}]k(\mathbf{q}.\mathbf{u}) + \mathbf{w}_1 \times \mathbf{q} \\ &= [\mathbf{v}_2 - (\mathbf{v}_2.\mathbf{q})\mathbf{q}]K_2 + \mathbf{w}_2 \times \mathbf{q} \end{aligned} \tag{4.43}$$

where K_2 is the inverse distance to the second surface in the direction $\mathbf{q}$. On equating the two expressions in (4.43) for $\dot{\mathbf{q}}$ and then taking the scalar product of the resulting equation with $\mathbf{v}_2 \times \mathbf{q}$ it follows that

$$[\mathbf{v}_1.(\mathbf{v}_2 \times \mathbf{q})]k(\mathbf{q}.\mathbf{u}) + (\mathbf{w}_1 \times \mathbf{q}).(\mathbf{v}_2 \times \mathbf{q}) = (\mathbf{w}_2 \times \mathbf{q}).(\mathbf{v}_2 \times \mathbf{q}) \tag{4.44}$$

The vector $\mathbf{w}$ is defined by $\mathbf{w} = \mathbf{w}_2 - \mathbf{w}_1$. A rearrangement of (4.44) yields

$$\begin{aligned} k(\mathbf{u}.\mathbf{q})[(\mathbf{v}_1 \times \mathbf{v}_2).\mathbf{q}] &= (\mathbf{w} \times \mathbf{q}).(\mathbf{v}_2 \times \mathbf{q}) \\ &= (\mathbf{v}_2.\mathbf{w})(\mathbf{q}.\mathbf{q}) - (\mathbf{v}_2.\mathbf{q})(\mathbf{w}.\mathbf{q}) \end{aligned} \tag{4.45}$$

It follows from (4.45) that

$$\begin{aligned} \mathbf{w}.\mathbf{v}_2 &= 0 \\ (\mathbf{w}, \mathbf{v}_2) &\sim (\mathbf{v}_2 \times \mathbf{v}_1, k\mathbf{u}) \end{aligned} \tag{4.46}$$

where $\sim$ is the equivalence relation defined in Sect. 4.2.1. It is shown in Proposition 4.2 that $\mathbf{v}_2$ and $\mathbf{v}_2 \times \mathbf{v}_1$ are both non-zero and non-parallel. Thus it follows from (4.46) that there exists a scalar c such that

$$\mathbf{v}_2 = ck\mathbf{u} \tag{4.47}$$

$$\mathbf{w} = c^{-1}(\mathbf{v}_2 \times \mathbf{v}_1) \tag{4.48}$$

On eliminating $\mathbf{v}_2$ from (4.47) and (4.48) it follows that $\mathbf{w} = k(\mathbf{u} \times \mathbf{v}_1)$. This yields

$$\mathbf{w}_2 = \mathbf{w}_1 + k(\mathbf{u} \times \mathbf{v}_1)$$

The angular velocity $\mathbf{w}_2$ is thus uniquely determined by the plane and the rigid velocity $\{\mathbf{v}_1, \mathbf{w}_1\}$. The vector $\mathbf{v}_2$ is parallel to $\mathbf{u}$. It follows that if $\{\mathbf{v}_2, \mathbf{w}_2\}$ is compatible with the image velocity field then it is determined up to a single unknown scale factor, the magnitude of $\mathbf{v}_2$. There are thus at most two reconstructions.

To complete the proof, it is shown that $\{\mathbf{v}_2, \mathbf{w}_2\}$ does indeed yield a reconstruction compatible with $\dot{\mathbf{q}}$ and that the surface obtained in this second reconstruction is a plane. Define the image velocity field $\dot{\mathbf{q}}'$ by

$$\dot{\mathbf{q}}' = [\mathbf{v}_2 - (\mathbf{v}_2.\mathbf{q})\mathbf{q}]K_2 + \mathbf{w}_2 \times \mathbf{q} \tag{4.49}$$

It follows from (4.43), (4.49) and the definition of $\mathbf{w}$ that

$$(\dot{\mathbf{q}} - \dot{\mathbf{q}}') \times \mathbf{q} = (\mathbf{v}_1 \times \mathbf{q})k(\mathbf{q}.\mathbf{u}) - (\mathbf{v}_2 \times \mathbf{q})K_2 - (\mathbf{w} \times \mathbf{q}) \times \mathbf{q} \tag{4.50}$$

On using (4.47) and (4.48) to substitute for $\mathbf{v}_2$ and $\mathbf{w}$ in (4.50) the following equation is obtained,

$$\begin{aligned}(\dot{\mathbf{q}} - \dot{\mathbf{q}}') \times \mathbf{q} &= (\mathbf{v}_1 \times \mathbf{q})k(\mathbf{q}.\mathbf{u}) - ck(\mathbf{u} \times \mathbf{q})K_2 - k[(\mathbf{u}.\mathbf{q})\mathbf{v}_1 - (\mathbf{v}_1.\mathbf{q})\mathbf{u}] \times \mathbf{q} \\ &= (\mathbf{u} \times \mathbf{q})k[(\mathbf{v}_1.\mathbf{q}) - cK_2]\end{aligned} \tag{4.51}$$

It follows from (4.51) that $\dot{\mathbf{q}} = \dot{\mathbf{q}}'$ if and only if $K_2 = c^{-1}(\mathbf{v}_1.\mathbf{q})$. The image velocity field $\dot{\mathbf{q}}$ can thus arise from a planar surface with normal $\mathbf{v}_1$ moving with velocity $\{\mathbf{v}_2, \mathbf{w}_2\}$. □

Theorem 4.13. *There are at most three essentially different reconstructions compatible with a dense image velocity field.*

Proof. In view of Theorem 4.12 it suffices to consider the case in which the image velocity field $\dot{\mathbf{q}}$ arises from a moving rigid non-planar surface. Let three essentially different reconstructions be obtained from $\dot{\mathbf{q}}$. Then $\dot{\mathbf{q}}$ has three distinct decompositions into components arising from translational and angular velocities, as follows,

$$\dot{\mathbf{q}} = \dot{\mathbf{q}}_{v_1} + \dot{\mathbf{q}}_{w_1} = \dot{\mathbf{q}}_{v_2} + \dot{\mathbf{q}}_{w_2} = \dot{\mathbf{q}}_{v_3} + \dot{\mathbf{q}}_{w_3}$$

Let $\mathbf{v}_1$, $\mathbf{v}_2$, $\mathbf{v}_3$ be the translational velocities associated with $\dot{\mathbf{q}}_{v_1}$, $\dot{\mathbf{q}}_{v_2}$, $\dot{\mathbf{q}}_{v_3}$, respectively, and let $\mathbf{w}_1$, $\mathbf{w}_2$, $\mathbf{w}_3$ be the angular velocities associated with $\dot{\mathbf{q}}_{w_1}$, $\dot{\mathbf{q}}_{w_2}$,

$\dot{\mathbf{q}}_{w_3}$, respectively. It follows from Proposition 4.2 that $\mathbf{v}_1$, $\mathbf{v}_2$, $\mathbf{v}_3$ are non-zero and pairwise non-parallel. The vectors $\mathbf{w}$, $\mathbf{w}'$ are defined by

$$\begin{aligned}\mathbf{w} &= \mathbf{w}_2 - \mathbf{w}_1 \\ \mathbf{w}' &= \mathbf{w}_3 - \mathbf{w}_1\end{aligned}$$

Let $K(\mathbf{q})$ be the inverse distance to the rigid surface associated with the decomposition $\dot{\mathbf{q}} = \dot{\mathbf{q}}_{v_1} + \dot{\mathbf{q}}_{w_1}$. On pairing $\dot{\mathbf{q}}_{v_1}$, $\dot{\mathbf{q}}_{w_1}$ with $\dot{\mathbf{q}}_{v_2}$, $\dot{\mathbf{q}}_{w_2}$, and then with $\dot{\mathbf{q}}_{v_3}$, $\dot{\mathbf{q}}_{w_3}$, (4.25) yields the following two expressions for K:

$$\begin{aligned}K &= \frac{(\mathbf{w}.\mathbf{q})(\mathbf{v}_2.\mathbf{q}) - \mathbf{w}.\mathbf{v}_2}{(\mathbf{v}_2 \times \mathbf{v}_1).\mathbf{q}} \\ K &= \frac{(\mathbf{w}'.\mathbf{q})(\mathbf{v}_3.\mathbf{q}) - \mathbf{w}'.\mathbf{v}_3}{(\mathbf{v}_3 \times \mathbf{v}_1).\mathbf{q}}\end{aligned} \tag{4.52}$$

On equating the two expressions for K given in (4.52) and then multiplying out the denominators the following equation is obtained,

$$[(\mathbf{v}_2 \times \mathbf{v}_1).\mathbf{q}][(\mathbf{w}'.\mathbf{q})(\mathbf{v}_3.\mathbf{q}) - \mathbf{w}'.\mathbf{v}_3] = [(\mathbf{v}_3 \times \mathbf{v}_1).\mathbf{q}][(\mathbf{w}.\mathbf{q})(\mathbf{v}_2.\mathbf{q}) - \mathbf{w}.\mathbf{v}_2] \tag{4.53}$$

Equation (4.53) is written homogeneously in $\mathbf{q}$,

$$(\mathbf{v}_2 \times \mathbf{v}_1).\mathbf{q}[(\mathbf{w}'.\mathbf{q})(\mathbf{v}_3.\mathbf{q}) - (\mathbf{w}'.\mathbf{v}_3)(\mathbf{q}.\mathbf{q})] = (\mathbf{v}_3 \times \mathbf{v}_1).\mathbf{q}[(\mathbf{w}.\mathbf{q})(\mathbf{v}_2.\mathbf{q}) - (\mathbf{w}.\mathbf{v}_2)(\mathbf{q}.\mathbf{q})] \tag{4.54}$$

It follows from the hypothesis of a dense image velocity field that (4.54) holds for all $\mathbf{q}$ in an open set on the projection sphere. Equation (4.54) is polynomial in $\mathbf{q}$, thus it holds for all $\mathbf{q}$ on the projection sphere.

The polynomial $(\mathbf{v}_2 \times \mathbf{v}_1).\mathbf{q}$ is of total degree one in the components x, y, z of $\mathbf{q}$, thus it divides a product of two polynomials if and only if it divides at least one of the polynomials. In particular, $(\mathbf{v}_2 \times \mathbf{v}_1).\mathbf{q}$ divides one of the factors on the right-hand side of (4.54). If $(\mathbf{v}_2 \times \mathbf{v}_1).\mathbf{q}$ divides

$$(\mathbf{w}.\mathbf{q})(\mathbf{v}_2.\mathbf{q}) - (\mathbf{w}.\mathbf{v}_2)(\mathbf{q}.\mathbf{q}) \tag{4.55}$$

then it follows from (4.52) that K is equal to a linear expression in x, y, z. The image velocity field can then be generated by a moving planar surface, contrary to hypothesis. It follows that $(\mathbf{v}_2 \times \mathbf{v}_1).\mathbf{q}$ does not divide (4.55). It follows from (4.54) that $(\mathbf{v}_2 \times \mathbf{v}_1).\mathbf{q}$ divides $(\mathbf{v}_3 \times \mathbf{v}_1).\mathbf{q}$. The polynomials $(\mathbf{v}_2 \times \mathbf{v}_1).\mathbf{q}$ and $(\mathbf{v}_3 \times \mathbf{v}_1).\mathbf{q}$ are both of degree one. Thus, there exists a non-zero scalar c such that

$$(\mathbf{v}_2 \times \mathbf{v}_1).\mathbf{q} = c(\mathbf{v}_3 \times \mathbf{v}_1).\mathbf{q} \tag{4.56}$$

Equation (4.56) is used to substitute for $(\mathbf{v}_2 \times \mathbf{v}_1).\mathbf{q}$ in (4.54). The cancellation of $(\mathbf{v}_3 \times \mathbf{v}_1).\mathbf{q}$ from the resulting equation yields

$$c[(\mathbf{w}'.\mathbf{q})(\mathbf{v}_3.\mathbf{q}) - (\mathbf{w}'.\mathbf{v}_3)(\mathbf{q}.\mathbf{q})] = (\mathbf{w}.\mathbf{q})(\mathbf{v}_2.\mathbf{q}) - (\mathbf{w}.\mathbf{v}_2)(\mathbf{q}.\mathbf{q})$$

It follows that

$$(c\mathbf{w}', \mathbf{v}_3) \sim (\mathbf{w}, \mathbf{v}_2) \tag{4.57}$$

None of the vectors $\mathbf{w}$, $\mathbf{v}_2$, $\mathbf{w}'$, $\mathbf{v}_3$ are zero and, by Proposition 4.2, $\mathbf{v}_2$ and $\mathbf{v}_3$ are not parallel. It thus follows from (4.57) that $\mathbf{v}_3$ is parallel to $\mathbf{w}_2$. If there existed a fourth reconstruction from $\dot{\mathbf{q}}$, giving rise to a decomposition $\dot{\mathbf{q}} = \dot{\mathbf{q}}_{v_4} + \dot{\mathbf{q}}_{w_4}$, then the above argument establishes that $\mathbf{v}_4$ would be parallel to $\mathbf{w}$. Hence $\mathbf{v}_4$ would be parallel to $\mathbf{v}_3$, in contradiction with Proposition 4.2. Thus $\dot{\mathbf{q}}$ yields at most three different reconstructions. □

The ambiguous image velocity fields are parameterised by the subset of $\mathbf{P}^2 \times \mathbf{P}^2 \times \mathbf{R}^3$ consisting of triples $(\mathbf{v}_1, \mathbf{v}_2, \mathbf{w})$ such that $\mathbf{v}_1$ and $\mathbf{v}_2$ are not parallel. The inverse distances to the critical surfaces reconstructed from a triple $\{\mathbf{v}_1, \mathbf{v}_2, \mathbf{w}\}$ are obtained from (4.52). If $\mathbf{w}$ is coplanar with $\mathbf{v}_1$, $\mathbf{v}_2$, but $\mathbf{w}$ is not parallel to $\mathbf{v}_1$ or $\mathbf{v}_2$ then a unique pair of vectors $(\mathbf{v}_3, \mathbf{w}')$ can be found such that $\mathbf{v}_3$ is parallel to $\mathbf{w}$ and such that for some non-zero scalar c,

$$(c\mathbf{w}', \mathbf{v}_3) \sim (\mathbf{w}, \mathbf{v}_2) \qquad \text{and} \qquad c\mathbf{v}_3 \times \mathbf{v} = \mathbf{v}_2 \times \mathbf{v}$$

In this case a third reconstruction is possible.

4.3 Algebraic Properties of Four Image Velocity Vectors

The algebraic nature of reconstruction is particularly apparent when the image velocity field contains only a small number of vectors. Four is the least number of vectors sufficient to impose a non-trivial constraint on the translational velocity of a rigid surface. The constraint takes the form of a quartic polynomial in the projective plane of possible directions for the translational velocity. The quartic is independent of the angular velocity and it determines the four image velocity vectors up to a component due to angular velocity alone. If the four image velocity vectors are irregular, in that the points in space giving rise to them vary significantly in depth, then the quartic is simplified because the leading order terms split into a quadratic factor and two linear factors. In practical cases of reconstruction the special properties of irregular image velocity fields are particularly important (Maybank 1987b; Murray & Buxton 1990). Reconstruction is easier than in the case of smooth image velocity fields, and the results obtained are less affected by small errors in estimates of the image velocity vectors.

For a general smooth image velocity field containing four vectors the leading order terms of the quartic constraint on the translational velocity are complicated. However, the leading order terms are simplified if the base points are placed symmetrically on the unit sphere, for example at the vertices of a square, or at the vertices and over the centroid of an equilateral triangle (tripod). The quartics arising from a square of base points are qualitatively different from the quartics arising from a tripod of base points.

The qualitative difference between the quartics arising from a square and from a tripod suggests the problem of describing the general quartic arising from a particular fixed choice of base points. This problem remains unsolved. As a possible first step towards a solution an examination is made of the vector space L of all quartics associated with a fixed set of base points. In algebraic geometry L is an example of a linear system. It is shown that there are exactly ten points through which all the quartics in L pass. These points are referred to as the base points of L. The converse result is also obtained. Any quartic passing through the ten base points is in L.

The order notation $O()$ of Hardy (1967) is used to indicate the magnitudes of quantities. Let f, g be two real valued functions defined on $\mathbb{R}^n$. Then $f = O(g)$ in a subset S of $\mathbb{R}^n$ if and only if there exists a constant c such that $|f(\mathbf{x})| \leq c|g(\mathbf{x})|$ for all $\mathbf{x}$ in S. Many of the equations given in the remainder of this chapter and in Chap. 6 contain terms such as $O(d^n)$. The term $O(d^n)$ includes quantities which are not written out explicitly, and which decrease like d^n as d tends towards zero. The order notation is particularly useful for equations which contain a few large terms and a great number of much smaller terms. The smaller terms are included in a single $O(d^n)$.

4.3.1 The Quartic Polynomial Constraint

The translational velocity $\mathbf{v}$ compatible with a given set of four image velocity vectors is subject to a quartic polynomial constraint q. Let $\dot{\mathbf{q}}_i$, $1 \leq i \leq 4$, be a set of image velocity vectors on the projection sphere with base points $\mathbf{q}_i$, $1 \leq i \leq 4$. Let A be the 4×3 matrix with ith row equal to

$$(\mathbf{v}.\mathbf{q}_i)\mathbf{q}_i^\top - \mathbf{v}^\top$$

Let $\mathbf{r}_i$ be the vector defined by $\mathbf{r}_i = \dot{\mathbf{q}}_i \times \mathbf{q}_i$ and let $\mathbf{l}$ be the four-dimensional vector with ith entry equal to $\mathbf{r}_i.\mathbf{v}$. The quartic polynomial q is defined by

$$q(\mathbf{v}) \equiv \det(A|\mathbf{l}) = \det \begin{pmatrix} (\mathbf{v}.\mathbf{q}_1)\mathbf{q}_1^\top - \mathbf{v}^\top & \mathbf{r}_1.\mathbf{v} \\ (\mathbf{v}.\mathbf{q}_2)\mathbf{q}_2^\top - \mathbf{v}^\top & \mathbf{r}_2.\mathbf{v} \\ (\mathbf{v}.\mathbf{q}_3)\mathbf{q}_3^\top - \mathbf{v}^\top & \mathbf{r}_3.\mathbf{v} \\ (\mathbf{v}.\mathbf{q}_4)\mathbf{q}_4^\top - \mathbf{v}^\top & \mathbf{r}_4.\mathbf{v} \end{pmatrix} \tag{4.58}$$

The definition of q arises from the condition that four equations of the form (4.8),

$$(\dot{\mathbf{q}} \times \mathbf{q}).\mathbf{v} = (\mathbf{w}.\mathbf{q})(\mathbf{v}.\mathbf{q}) - \mathbf{w}.\mathbf{v}$$

have a solution for $\mathbf{w}$ at a fixed value of $\mathbf{v}$. If $\hat{\mathbf{v}}$ is the translational velocity of the rigid surface giving rise to the image velocity field $\dot{\mathbf{q}}_i$ then $q(\hat{\mathbf{v}}) = 0$. Conversely, if $\mathbf{v}$ is any non-zero vector such that $q(\mathbf{v}) = 0$ and such that $A(\mathbf{v})$ has rank three, then there exists a rigid surface moving with translational velocity $\mathbf{v}$ that gives rise to the four image velocity vectors $\dot{\mathbf{q}}_i$.

The notation $\langle\dot{\mathbf{q}}\rangle$ is a short way of referring to the image velocity field $\dot{\mathbf{q}}_i$, $1 \leq i \leq 4$. Where necessary to avoid ambiguity, the dependence of q on $\langle\dot{\mathbf{q}}\rangle$ is indicated by the notation $\mathbf{v} \mapsto q(\langle\dot{\mathbf{q}}\rangle, \mathbf{v})$.

Proposition 4.14. *Let* $\langle\dot{\mathbf{q}}\rangle$, $\langle\dot{\mathbf{q}}'\rangle$ *be two image velocity fields with the same set of base points and let* $\mathbf{w}$ *be a vector such that*

$$\dot{\mathbf{q}}_i' = \dot{\mathbf{q}}_i + \mathbf{w} \times \mathbf{q}_i \qquad (1 \le i \le 4) \qquad (4.59)$$

Then the equation $q(\langle\dot{\mathbf{q}}'\rangle, \mathbf{v}) = q(\langle\dot{\mathbf{q}}\rangle, \mathbf{v})$ *holds for all* $\mathbf{v}$.

Proof. Let $\mathbf{l}$, $\mathbf{l}'$ be the four-dimensional vectors such that $l_i = (\dot{\mathbf{q}}_i \times \mathbf{q}_i).\mathbf{v}$ and $l_i' = (\dot{\mathbf{q}}_i' \times \mathbf{q}_i).\mathbf{v}$. It follows from (4.59) that

$$\begin{aligned} l_i' &= (\dot{\mathbf{q}}_i \times \mathbf{q}_i).\mathbf{v} + [(\mathbf{w} \times \mathbf{q}_i) \times \mathbf{q}_i].\mathbf{v} \\ &= l_i + [(\mathbf{w}.\mathbf{q}_i)\mathbf{q}_i - \mathbf{w}].\mathbf{v} \qquad (1 \le i \le 4) \end{aligned} \qquad (4.60)$$

It follows from (4.60) and the definition of A that

$$l_i' = l_i + (A\mathbf{w})_i \qquad (1 \le i \le 4) \qquad (4.61)$$

Equation (4.61) and the definition (4.58) of q yield

$$\begin{aligned} q(\langle\dot{\mathbf{q}}'\rangle, \mathbf{v}) &= \det(A|\mathbf{l}') \\ &= \det(A|\mathbf{l} + A\mathbf{w}) \\ &= \det(A|\mathbf{l}) \\ &= q(\langle\dot{\mathbf{q}}\rangle, \mathbf{v}) \end{aligned}$$

as required. □

The converse to Proposition 4.14 is obtained below in Theorem 4.26: if $q(\langle\dot{\mathbf{q}}'\rangle, \mathbf{v}) = q(\langle\dot{\mathbf{q}}\rangle, \mathbf{v})$ for all $\mathbf{v}$, then (4.59) holds for some choice of $\mathbf{w}$.

4.3.2 Irregular Image Velocity Fields

An irregular image velocity field is one in which the surface points giving rise to the image velocity vectors show a wide variation in depth over a small part of the image. The variation in depth ensures that there are relatively large differences in the length or orientation of image velocity vectors that are based at nearby points in the image. At first sight irregular image velocity fields appear to be more difficult to work with than smooth image velocity fields. However, first impressions are misleading. The irregularity leads to simplifications in the algebra and at the same time it ensures that the estimates of the camera velocity are more resistant to the effects of image noise. The simplifying effects of irregularity on algorithms for estimating the camera velocity from the image velocity field are described in Sect. 6.2.

The first results in this subsection deal with an extreme form of irregularity in which two of the base points coincide, but the associated image velocity vectors are different. The results are extended to a more general case in which the two base points are close together, but not necessarily coincident.

Proposition 4.15. *Let $\dot{\mathbf{q}}_i$ for $1 \le i \le 4$, be a set of four flow vectors with base points $\mathbf{q}_i$, such that $\mathbf{q}_1 = \mathbf{q}_2$, but $\dot{\mathbf{q}}_1 \ne \dot{\mathbf{q}}_2$. Then q splits into a linear factor and a cubic factor.*

Proof. Let $\mathbf{r}_i$ be the vector defined by $\mathbf{r}_i = \dot{\mathbf{q}}_i \times \mathbf{q}_i$ for $1 \le i \le 4$ and let $\Delta\mathbf{r}$ be defined by $\Delta\mathbf{r} = \mathbf{r}_2 - \mathbf{r}_1$. The subtraction of the second row of the determinant in (4.58) from the first row yields

$$\begin{aligned} q(\mathbf{v}) &= \det\begin{pmatrix} 0 & \Delta\mathbf{r}.\mathbf{v} \\ (\mathbf{v}.\mathbf{q}_2)\mathbf{q}_2^\top - \mathbf{v}^\top & \mathbf{r}_2.\mathbf{v} \\ (\mathbf{v}.\mathbf{q}_3)\mathbf{q}_3^\top - \mathbf{v}^\top & \mathbf{r}_3.\mathbf{v} \\ (\mathbf{v}.\mathbf{q}_4)\mathbf{q}_4^\top - \mathbf{v}^\top & \mathbf{r}_4.\mathbf{v} \end{pmatrix} \\ &= -(\Delta\mathbf{r}.\mathbf{v})\det\begin{pmatrix} (\mathbf{v}.\mathbf{q}_2)\mathbf{q}_2^\top - \mathbf{v}^\top \\ (\mathbf{v}.\mathbf{q}_3)\mathbf{q}_3^\top - \mathbf{v}^\top \\ (\mathbf{v}.\mathbf{q}_4)\mathbf{q}_4^\top - \mathbf{v}^\top \end{pmatrix} \end{aligned} \qquad (4.62)$$

□

Let f_{ijk} be the homogeneous cubic polynomial in $\mathbf{v}$ defined by

$$f_{ijk}(\mathbf{v}) = \det\begin{pmatrix} (\mathbf{v}.\mathbf{q}_i)\mathbf{q}_i^\top - \mathbf{v}^\top \\ (\mathbf{v}.\mathbf{q}_j)\mathbf{q}_j^\top - \mathbf{v}^\top \\ (\mathbf{v}.\mathbf{q}_k)\mathbf{q}_k^\top - \mathbf{v}^\top \end{pmatrix} \qquad (1 \le i, j, k \le 4)$$

If all the base points $\mathbf{q}_i$ are close together then the cubic factor f_{234} on the right-hand side of (4.62) can be approximated as follows.

Theorem 4.16. *With the notation of Proposition 4.15, let $\mathbf{q}$ be the vector of unit length parallel to $\mathbf{q}_2 + \mathbf{q}_3 + \mathbf{q}_4$, let d be defined by*

$$d = \max\{\|\mathbf{q}_i - \mathbf{q}\| \mid i = 2, 3, 4\}$$

and let Δ be the area of the plane triangle with vertices $\mathbf{q}_2$, $\mathbf{q}_3$, $\mathbf{q}_4$. Then the factor f_{234} on the right-hand side of (4.62) can be approximated by

$$f_{234}(\mathbf{v}) = 2\Delta(\mathbf{v}.\mathbf{q})\|\mathbf{v} \times \mathbf{q}\|^2 + O(d^3) \qquad (4.63)$$

Proof. Cartesian coordinates are chosen such that

$$\begin{aligned} \mathbf{q}_2 &= (0, 0, 1)^\top \\ \mathbf{q}_3 &= (x_3, 0, z_3)^\top \\ \mathbf{q}_4 &= (x_4, y_4, z_4)^\top \end{aligned}$$

The area Δ is given by

$$\begin{aligned} \Delta &= \frac{1}{2}\|(\mathbf{q}_3 - \mathbf{q}_2) \times (\mathbf{q}_4 - \mathbf{q}_2)\| \\ &= \frac{1}{2}x_3 y_4 + O(d^4) \end{aligned}$$

The expansion of the determinant on the left-hand side of (4.63) yields

$$\begin{aligned}
f_{234}(\mathbf{v}) &= \det\begin{pmatrix} -v_1 & -v_2 & 0 \\ (\mathbf{v}.\mathbf{q}_3)x_3 - v_1 & -v_2 & (\mathbf{v}.\mathbf{q}_3)z_3 - v_3 \\ (\mathbf{v}.\mathbf{q}_4)x_4 - v_1 & (\mathbf{v}.\mathbf{q}_4)y_4 - v_2 & (\mathbf{v}.\mathbf{q}_4)z_4 - v_3 \end{pmatrix} \\
&= \det\begin{pmatrix} -v_1 & -v_2 & 0 \\ (v_1x_3 + v_3z_3)x_3 & 0 & (v_1z_3 - v_3x_3)x_3 \\ (\mathbf{v}.\mathbf{q}_4)x_4 & (\mathbf{v}.\mathbf{q}_4)y_4 & (\mathbf{v}.\mathbf{q}_4)z_4 - v_3 \end{pmatrix} \\
&= x_3\det\begin{pmatrix} -v_1 & -v_2 & 0 \\ v_1x_3 + v_3z_3 & 0 & v_1z_3 - v_3x_3 \\ (\mathbf{v}.\mathbf{q}_4)x_4 & (\mathbf{v}.\mathbf{q}_4)y_4 & (\mathbf{v}.\mathbf{q}_4)z_4 - v_3 \end{pmatrix} \\
&= x_3\det\begin{pmatrix} -v_1 + v_2x_4/y_4 & -v_2 & 0 \\ v_1x_3 + v_3z_3 & 0 & v_1z_3 - v_3x_3 \\ 0 & (\mathbf{v}.\mathbf{q}_4)y_4 & (\mathbf{v}.\mathbf{q}_4)z_4 - v_3 \end{pmatrix} \\
&= x_3[(v_1y_4 - v_2x_4)(\mathbf{v}.\mathbf{q}_4)(v_1z_3 - v_3x_3) + v_2(v_1x_3 + v_3z_3)((\mathbf{v}.\mathbf{q}_4)z_4 - v_3)] \\
&= x_3(v_1^2v_3y_4 + v_3v_2^2y_4) + O(d^3) \\
&= 2\,\Delta v_3(v_1^2 + v_2^2) + O(d^3) \\
&= 2\,\Delta(\mathbf{v}.\mathbf{q})\|\mathbf{v}\times\mathbf{q}\|^2 + O(d^3)
\end{aligned}$$

The result follows. □

The drawback of Proposition 4.15 is that it requires two image velocity vectors with coincident base points. In practice it is rare to find two velocity vectors placed so conveniently together in the image. Even if such a case occurs, there remains the problem of separating the two image velocity fields. It is more usual to obtain image velocity vectors which differ significantly but which have their base points close together. The differences in the image velocity vectors arise from a large changes in the depth of the surface giving rise to the image velocity vectors. In this situation a weaker but more elaborate version of Proposition 4.15 holds.

Theorem 4.17. *Let $\dot{\mathbf{q}}_i$, $1 \le i \le 4$, be an irregular image velocity field and let Cartesian coordinates be chosen such that the coordinates $(x_i, y_i, z_i)^\top$ of the base points $\mathbf{q}_i$ satisfy*

$$\sum_{i=1}^{4} x_i = 0 \qquad \sum_{i=1}^{4} y_i = 0 \qquad \sum_{i=1}^{4} x_iy_i = 0 \tag{4.64}$$

Let $\mathbf{l}$ be the four-dimensional vector with ith entry

$$l_i = (\dot{\mathbf{q}}_i \times \mathbf{q}_i).\mathbf{v} \qquad (1 \le i \le 4) \tag{4.65}$$

Let $\mathbf{e}$, $\mathbf{x}$, $\mathbf{y}$ be four-dimensional vectors defined by

$$\begin{aligned}
\mathbf{e} &= (1, 1, 1, 1)^\top \\
\mathbf{x} &= (x_1, x_2, x_3, x_4)^\top \\
\mathbf{y} &= (y_1, y_2, y_3, y_4)^\top
\end{aligned}$$

and let $\hat{\mathbf{s}}$ *be a unit vector in* $\mathbb{R}^4$ *normal to* $\mathbf{e}$, $\mathbf{x}$ *and* $\mathbf{y}$. *Then the quartic polynomial constraint* q *on* $\mathbf{v}$ *defined by (4.58) reduces to*

$$q(\mathbf{v}) = 2\|\mathbf{x}\|\,\|\mathbf{y}\|v_3(v_1^2 + v_2^2)(\hat{\mathbf{s}}.\mathbf{l}) + O(d^3) \tag{4.66}$$

where d *is defined by*

$$d = \max\{\sqrt{x_i^2 + y_i^2} \;\; | 1 \le i \le 4\} \tag{4.67}$$

Proof. Let $\hat{\mathbf{e}}$, $\hat{\mathbf{x}}$, $\hat{\mathbf{y}}$ be the unit vectors defined by

$$\hat{\mathbf{e}} = \|\mathbf{e}\|^{-1}\mathbf{e} \qquad \hat{\mathbf{x}} = \|\mathbf{x}\|^{-1}\mathbf{x} \qquad \hat{\mathbf{y}} = \|\mathbf{y}\|^{-1}\mathbf{y} \tag{4.68}$$

The vector $\hat{\mathbf{s}}$ is chosen such that the orthogonal matrix U defined by $U^\top = (\hat{\mathbf{e}}|\hat{\mathbf{x}}|\hat{\mathbf{y}}|\hat{\mathbf{s}})$ has determinant $+1$. Let A be the 4×3 matrix with ith row $(\mathbf{v}.\mathbf{q}_i)\mathbf{q}_i^\top - \mathbf{v}^\top$. It follows from (4.58) that

$$q(\mathbf{v}) = \det(A|\mathbf{l}) = \det(UA|U\mathbf{l}) \tag{4.69}$$

On expanding A in powers of x_i, y_i and applying U the following equations are obtained

$$\begin{aligned}
UA &= \begin{pmatrix} \hat{\mathbf{e}}^\top \\ \hat{\mathbf{x}}^\top \\ \hat{\mathbf{y}}^\top \\ \hat{\mathbf{s}}^\top \end{pmatrix} \begin{pmatrix} -v_1 + v_3x_1 & -v_2 + v_3y_1 & v_1x_1 + v_2y_1 \\ -v_1 + v_3x_2 & -v_2 + v_3y_2 & v_1x_2 + v_2y_2 \\ -v_1 + v_3x_3 & -v_2 + v_3y_3 & v_1x_3 + v_2y_3 \\ -v_1 + v_3x_4 & -v_2 + v_3y_4 & v_1x_4 + v_2y_4 \end{pmatrix} + O(d^2) \\
&= \begin{pmatrix} \hat{\mathbf{e}}^\top \\ \hat{\mathbf{x}}^\top \\ \hat{\mathbf{y}}^\top \\ \hat{\mathbf{s}}^\top \end{pmatrix} [-v_1\mathbf{e} + v_3\mathbf{x}| - v_2\mathbf{e} + v_3\mathbf{y}|v_1\mathbf{x} + v_2\mathbf{y}] + O(d^2) \\
&= \begin{pmatrix} -2v_1 & -2v_2 & 0 \\ v_3\|\mathbf{x}\| & 0 & v_1\|\mathbf{x}\| \\ 0 & v_3\|\mathbf{y}\| & v_2\|\mathbf{y}\| \\ & & \end{pmatrix} + O(d^2)
\end{aligned} \tag{4.70}$$

On using (4.70) to substitute for UA in (4.69) it follows that

$$q(\mathbf{v}) = \det \begin{pmatrix} -2v_1 & -2v_2 & 0 & \hat{\mathbf{e}}.\mathbf{l} \\ v_3\|\mathbf{x}\| & 0 & v_1\|\mathbf{x}\| & \hat{\mathbf{x}}.\mathbf{l} \\ 0 & v_3\|\mathbf{y}\| & v_2\|\mathbf{y}\| & \hat{\mathbf{y}}.\mathbf{l} \\ 0 & 0 & 0 & \hat{\mathbf{s}}.\mathbf{l} \end{pmatrix} + O(d^3) \tag{4.71}$$

Equation (4.66) follows on expanding the determinant on the right-hand side of (4.71). □

Corollary. If the image velocity field is irregular then the leading order terms of q split into three factors. To prove the corollary, note that in the irregular case it follows from (4.66) that the leading order terms of q are

$$2\,\|\mathbf{x}\|\,\|\mathbf{y}\|v_3(v_1^2 + v_2^2)(\hat{\mathbf{s}}.\mathbf{l})$$

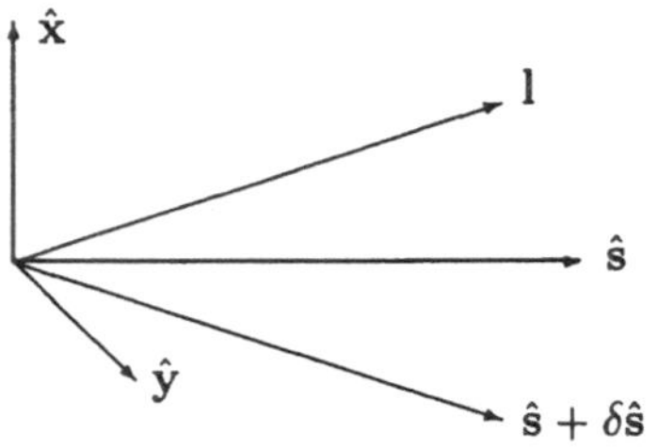

Fig. 4.2. A small perturbation of $\hat{\mathbf{s}}$

The two factors v_3 and $\hat{\mathbf{s}}.\mathbf{l}$ are linear and the remaining factor $v_1^2+v_2^2$ is quadratic.

If the image velocity field is smooth then Theorem 4.17 ceases to hold because the first term on the right-hand side of (4.66) is not necessarily the leading order term of $q(\mathbf{v})$.

The factor $\hat{\mathbf{s}}.\mathbf{l}$ is the most important of the three factors obtained in the Corollary to Theorem 4.17 because it is the only one that contains information about the velocity and shape of the rigid surface. The factor $\hat{\mathbf{s}}.\mathbf{l}$ is expanded as follows. It is assumed that the image velocity field $\langle\dot{\mathbf{q}}\rangle$ used in (4.65) to define the components l_i of $\mathbf{l}$ arises from a surface moving with zero angular velocity. There is no loss of generality in making this assumption because q is independent of angular velocity. It follows that

$$\begin{aligned}\hat{\mathbf{s}}.\mathbf{l} = \sum_{i=1}^{n}\hat{s}_i l_i &= \sum_{i=1}^{n}\hat{s}_i[(\dot{\mathbf{q}}_i \times \mathbf{q}_i).\mathbf{v}] \\ &= \sum_{i=1}^{n}\hat{s}_i[(\mathbf{v} \times \mathbf{q}_i).\hat{\mathbf{v}}]K_i \\ &= \left(\sum_{i=1}^{4}\hat{s}_i K_i(\hat{\mathbf{v}} \times \mathbf{q}_i)\right).\mathbf{v}\end{aligned}$$

Experiments with computer generated image velocity fields show that the approximation (4.66) to q obtained in Theorem 4.17 is accurate even if the image velocity field has only a slight irregularity. If the irregularity is pronounced then the $O(d^3)$ term in (4.66) is much smaller than the leading order terms and the approximation to q is excellent.

4.3.3 The Effects of Small Perturbations

The effects on q of small perturbations in the image velocities are assessed. In the irregular case a small perturbation causes only a slight shift in the linear factor of the leading order terms found in Theorem 4.17. In the case of a smooth image velocity field q is severely affected by small perturbations. The irregular case is described first.

Theorem 4.18. *Let $\dot{\mathbf{q}}_i$, $1 \le i \le 4$, be an irregular image velocity field. Let the base points $\mathbf{q}_i$ have coordinates $\mathbf{q}_i = (x_i, y_i, z_i)^\top$ and let coordinate axes be* *chosen such that (4.64) holds. Let d be defined by (4.67). Let the base points $\mathbf{q}_i$ and the image velocity vectors $\dot{\mathbf{q}}_i$ be subject to an $O(d^2)$ perturbation. Then the linear factor $\hat{\mathbf{s}}.\mathbf{l}$ of $q(\mathbf{v})$ obtained in the Corollary to Theorem 4.7 undergoes an $O(d)$ perturbation.*

Proof. The perturbations of $\hat{\mathbf{x}}$ and $\hat{\mathbf{y}}$ are $O(d)$ and $\hat{\mathbf{s}}$ is defined to be the unit vector normal to $\hat{\mathbf{e}}$, $\hat{\mathbf{x}}$ and $\hat{\mathbf{y}}$. Thus the perturbation $\delta\hat{\mathbf{s}}$ of $\hat{\mathbf{s}}$ is $O(d)$, as illustrated in Fig. 4.2. If $\hat{\mathbf{s}}.\mathbf{l} = O(d^2)$ then $\mathbf{l}$ would be closely approximated by a linear combination of $\hat{\mathbf{e}}$, $\hat{\mathbf{x}}$ and $\hat{\mathbf{y}}$. This would contradict the hypothesis that the image velocity field is irregular. It follows that $\hat{\mathbf{s}}.\mathbf{l} = O(1)$, in other words $\mathbf{l}$ is not near orthogonal to $\hat{\mathbf{s}}$. The $O(d)$ perturbation in $\hat{\mathbf{s}}$ produces an $O(d)$ perturbation in $\hat{\mathbf{s}}.\mathbf{l}$. □

If the image velocity field is very irregular then the vector $\mathbf{l}$ in the proof of Theorem 4.18 tends to be near parallel to $\hat{\mathbf{s}}$. If the angle between $\hat{\mathbf{s}}$ and $\mathbf{l}$ is $O(d)$ then an $O(d^2)$ peturbation in the base points and the image velocity vectors produces an $O(d^2)$ perturbation in $\hat{\mathbf{s}}.\mathbf{l}$.

In the case of a smooth image velocity field the effects of small perturbations are much more severe.

Theorem 4.19. *If the image velocity field $\dot{\mathbf{q}}_i$, $1 \le i \le 4$, is smooth and if $\mathbf{v}$ is any unit vector satisfying $|v_3|(v_1^2 + v_2^2) \gg d$, where d is defined by (4.67), then there is an image velocity field $\langle \dot{\mathbf{q}}' \rangle$ with the same base points as $\langle \dot{\mathbf{q}} \rangle$ such that*

$$\|\dot{\mathbf{q}}_i' - \dot{\mathbf{q}}_i\| = O(d^2) \qquad (1 \le i \le 4)$$

and such that $q(\langle \dot{\mathbf{q}}' \rangle, \mathbf{v}) = 0$, where q is the quartic polynomial defined by (4.58).

Proof. The definition of q is recalled from (4.58),

$$q(\mathbf{v}) = \det \begin{pmatrix} (\mathbf{v}.\mathbf{q}_1)\mathbf{q}_1^\top - \mathbf{v}^\top & \mathbf{r}_1.\mathbf{v} \\ (\mathbf{v}.\mathbf{q}_2)\mathbf{q}_2^\top - \mathbf{v}^\top & \mathbf{r}_2.\mathbf{v} \\ (\mathbf{v}.\mathbf{q}_3)\mathbf{q}_3^\top - \mathbf{v}^\top & \mathbf{r}_3.\mathbf{v} \\ (\mathbf{v}.\mathbf{q}_4)\mathbf{q}_4^\top - \mathbf{v}^\top & \mathbf{r}_4.\mathbf{v} \end{pmatrix} \qquad (4.72)$$

where $\mathbf{r}_i = \dot{\mathbf{q}}_i \times \mathbf{q}_i$ for $1 \le i \le 4$. Cartesian coordinates are chosen such that the coordinates x_i, y_i, z_i of the base points $\mathbf{q}_i$ satisfy

$$\sum_{i=1}^{4} x_i = 0 \qquad \sum_{i=1}^{4} y_i = 0 \qquad \sum_{i=1}^{4} x_i y_i = 0$$

By hypothesis the image velocity field is smooth. It follows that there exist three linear functions of $\mathbf{v}$, $a(\mathbf{v})$, $b(\mathbf{v})$, $c(\mathbf{v})$ such that $\mathbf{r}_i.\mathbf{v}$ has a good approximation linear in x_i, y_i,

$$\mathbf{r}_i.\mathbf{v} = a + bx_i + cy_i + O(d^2) \qquad (1 \le i \le 4) \qquad (4.73)$$

Let A be the 4×3 matrix formed by the first three columns of the matrix on the right-hand side of (4.72). Let $\hat{\mathbf{e}}$, $\hat{\mathbf{x}}$, $\hat{\mathbf{y}}$ be the unit vectors defined by (4.68). Let $\mathbf{w}$ be the vector defined by

$$v_3(v_1^2 + v_2^2)\mathbf{w} =$$

$$\left(-c\,v_1v_2 - a\,v_1v_3 + b\,v_2^2, c\,v_1^2 - b\,v_1v_2 - a\,v_2v_3, v_3(b\,v_1 + c\,v_2 + a\,v_3)\right)^{\mathsf{T}} \quad (4.74)$$

The expansion of A to first order in powers of the coordinates x_i, y_i of $\mathbf{q}_i$ is given in the proof of Theorem 4.17. It follows from this expansion and from the hypothesis $|v_3|(v_1^2 + v_2^2) \gg d$ that

$$\begin{aligned} A\mathbf{w} &= a\,\mathbf{e} + b\,\mathbf{x} + c\mathbf{y} + O(d^2 v_3^{-1}(v_1^2 + v_2^2)^{-1}) \\ &= a\mathbf{e} + b\mathbf{x} + c\mathbf{y} + O(d^2) \end{aligned} \quad (4.75)$$

The definition (4.74) of $\mathbf{w}$ is chosen to ensure that (4.75) holds. It follows from (4.73) and (4.75) that

$$\begin{pmatrix} \mathbf{r}_1.\mathbf{v} \\ \mathbf{r}_2.\mathbf{v} \\ \mathbf{r}_3.\mathbf{v} \\ \mathbf{r}_4.\mathbf{v} \end{pmatrix} - A\mathbf{w} = O(d^2)$$

Let $\delta\mathbf{r}_i$ be perturbations of the $\mathbf{r}_i$ of size $O(d^2)$ defined such that $\delta\mathbf{r}_i.\mathbf{q}_i = 0$ and

$$\delta\mathbf{r}_i.\mathbf{v} = \mathbf{r}_i.\mathbf{v} - (A\mathbf{w})_i \qquad (1 \le i \le 4) \qquad (4.76)$$

Let the $\dot{\mathbf{q}}_i'$ be defined by

$$\dot{\mathbf{q}}_i' = \dot{\mathbf{q}}_i + \delta\mathbf{r}_i \times \mathbf{q}_i \qquad (1 \le i \le 4)$$

It follows from (4.72) and (4.76) that $q(\langle\dot{\mathbf{q}}'\rangle, \mathbf{v}) = 0$. □

An example described in Sect. 4.3.4 shows that the condition $|v_3| \gg d$ subsumed in the condition $|v_3|(v_1^2 + v_2^2) \gg d$ of Theorem 4.19 is necessary. In the example the base points are at the vertices of a square. The positions of the zeros of q close to the line $v_3 = 0$ are then stable under small perturbations, even when the image velocity field is smooth.

4.3.4 Symmetric Arrangements of Base Points: The Square

In this subsection and the next the leading order terms of q are obtained for two highly symmetric arrangements of base points, the square and the tripod. The symmetry ensures that the calculations required to obtain the leading order terms of q are manageable. The quartics thus obtained illustrate Theorem 4.19, in that they are unstable when the image velocity field is smooth. The quartics also show clearly the way in which irregularity causes the leading order terms to split, to yield the linear factor $\hat{\mathbf{s}}.\mathbf{l}$ described in the Corollary to Theorem 4.17.

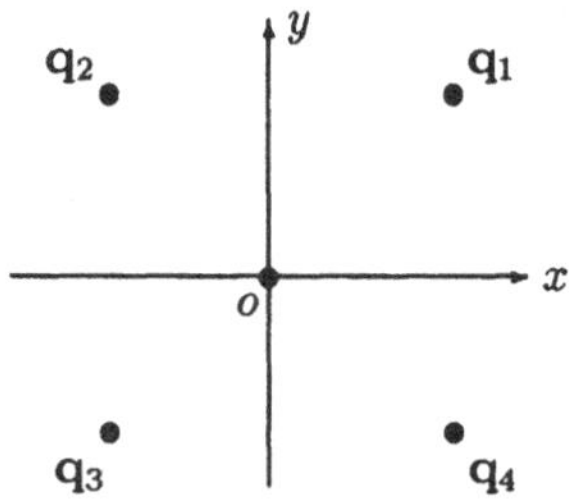

Fig. 4.3. The square of base points

Let d be a small quantity. The square of base points is defined by

$$\begin{aligned} \mathbf{q}_1 &= (d/\sqrt{2}, d/\sqrt{2}, \sqrt{1-d^2})^\top \\ \mathbf{q}_2 &= (-d/\sqrt{2}, d/\sqrt{2}, \sqrt{1-d^2})^\top \\ \mathbf{q}_3 &= (-d/\sqrt{2}, -d/\sqrt{2}, \sqrt{1-d^2})^\top \\ \mathbf{q}_4 &= (d/\sqrt{2}, -d/\sqrt{2}, \sqrt{1-d^2})^\top \end{aligned} \tag{4.77}$$

The square is illustrated in Fig. 4.3. In order to simplify the calculation of the leading order terms of q the following orthogonal matrix is required,

$$U = \begin{pmatrix} 1/2 & -1/2 & 1/2 & -1/2 \\ 1/\sqrt{2} & 0 & -1/\sqrt{2} & 0 \\ 0 & 1/\sqrt{2} & 0 & -1/\sqrt{2} \\ 1/2 & 1/2 & 1/2 & 1/2 \end{pmatrix}$$

It follows from the properties of the determinant that $q(\mathbf{v}) = \det(UA|U\mathbf{l})$, where as usual A is the 4×3 matrix formed from the first three columns of the matrix on the right-hand side of (4.72), and $\mathbf{l}$ is the vector with ith entry $l_i = \mathbf{r}_i.\mathbf{v}$. Let $\mathbf{q}$ be the vector defined by $\mathbf{q} = (0,0,1)^\top$ and let $\mathbf{\Delta}$, $\mathbf{\Psi}$ be the vectors defined by

$$\begin{aligned} \mathbf{q}_1 &= \mathbf{q} + \mathbf{\Delta} \\ \mathbf{q}_2 &= \mathbf{q} + \mathbf{\Psi} \\ \mathbf{q}_3 &= \mathbf{q} - \mathbf{\Delta} \\ \mathbf{q}_4 &= \mathbf{q} - \mathbf{\Psi} \end{aligned} \tag{4.78}$$

It follows from (4.77) and (4.78) that

$$\begin{aligned} \mathbf{\Delta} &= (d/\sqrt{2}, d/\sqrt{2}, -d^2/2)^\top + O(d^3) \\ \mathbf{\Psi} &= (-d/\sqrt{2}, d/\sqrt{2}, -d^2/2)^\top + O(d^3) \end{aligned}$$

The rows of UA are obtained in order,

$$\begin{aligned} (UA)_1 &= [(\mathbf{v}.\mathbf{q}_1)\mathbf{q}_1^\top - (\mathbf{v}.\mathbf{q}_2)\mathbf{q}_2^\top + (\mathbf{v}.\mathbf{q}_3)\mathbf{q}_3^\top - (\mathbf{v}.\mathbf{q}_4)\mathbf{q}_4^\top]/2 \\ (UA)_2 &= [(\mathbf{v}.\mathbf{q}_1)\mathbf{q}_1^\top - (\mathbf{v}.\mathbf{q}_3)\mathbf{q}_3^\top]/\sqrt{2} \\ (UA)_3 &= [(\mathbf{v}.\mathbf{q}_2)\mathbf{q}_2^\top - (\mathbf{v}.\mathbf{q}_4)\mathbf{q}_4^\top]/\sqrt{2} \\ (UA)_4 &= -2\mathbf{v} + [(\mathbf{v}.\mathbf{q}_1)\mathbf{q}_1^\top + (\mathbf{v}.\mathbf{q}_2)\mathbf{q}_2^\top + (\mathbf{v}.\mathbf{q}_3)\mathbf{q}_3^\top + (\mathbf{v}.\mathbf{q}_4)\mathbf{q}_4^\top]/2 \end{aligned} \tag{4.79}$$

Equation (4.78) is used to substitute for the $\mathbf{q}_i$ in (4.79),

$$\begin{aligned}
(UA)_1 &= (\mathbf{v}.\mathbf{\Delta})\mathbf{\Delta}^\top - (\mathbf{v}.\mathbf{\Psi})\mathbf{\Psi}^\top \\
&= d^2(v_2, v_1, 0)^\top + O(d^3) \\
(UA)_2 &= d(v_3, v_3, v_1 + v_2)^\top + O(d^2) \\
(UA)_3 &= -d(v_3, -v_3, v_1 - v_2)^\top + O(d^2) \\
(UA)_4 &= -2(v_1, v_2, 0)^\top + O(d^2)
\end{aligned} \tag{4.80}$$

The expansion of $U\mathbf{l}$ yields

$$U\mathbf{l} =$$
$$\frac{1}{2}\Big((\mathbf{r}_1 - \mathbf{r}_2 + \mathbf{r}_3 - \mathbf{r}_4).\mathbf{v}, \sqrt{2}(\mathbf{r}_1 - \mathbf{r}_3).\mathbf{v}, \sqrt{2}(\mathbf{r}_2 - \mathbf{r}_4).\mathbf{v}, (\mathbf{r}_1 + \mathbf{r}_2 + \mathbf{r}_3 + \mathbf{r}_4).\mathbf{v}\Big)^\top$$

Let the vectors $\mathbf{t}_i$ for $1 \le i \le 4$ be defined by

$$(\mathbf{t}_1.\mathbf{v}, \mathbf{t}_2.\mathbf{v}, \mathbf{t}_3.\mathbf{v}, \mathbf{t}_4.\mathbf{v})^\top = U\mathbf{l} \tag{4.81}$$

It follows that

$$\begin{aligned}
\mathbf{t}_1 &= (\mathbf{r}_1 - \mathbf{r}_2 + \mathbf{r}_3 - \mathbf{r}_4)/2 \\
\mathbf{t}_2 &= (\mathbf{r}_1 - \mathbf{r}_3)/\sqrt{2} \\
\mathbf{t}_3 &= (\mathbf{r}_2 - \mathbf{r}_4)/\sqrt{2} \\
\mathbf{t}_4 &= (\mathbf{r}_1 + \mathbf{r}_2 + \mathbf{r}_3 + \mathbf{r}_4)/2
\end{aligned} \tag{4.82}$$

It follows from (4.80) and (4.81) that

$$q(\mathbf{v}) = \det \begin{pmatrix} d^2 v_2 & d^2 v_1 & 0 & \mathbf{t}_1.\mathbf{v} \\ d v_3 & d v_3 & d(v_1 + v_2) & \mathbf{t}_2.\mathbf{v} \\ -d v_3 & d v_3 & -d(v_1 - v_2) & \mathbf{t}_3.\mathbf{v} \\ -2v_1 & -2v_2 & 0 & \mathbf{t}_4.\mathbf{v} \end{pmatrix} + O(d^5) \tag{4.83}$$

On expanding the determinant on the right-hand side of (4.83) and rearranging some of the terms the following equations are obtained

$$\begin{aligned}
q(\mathbf{v}) = -4d^2(\mathbf{t}_1.\mathbf{v}) \det &\begin{pmatrix} 0 & v_3 & v_2 \\ -v_3 & v_3 & -v_1 + v_2 \\ -v_1 & -v_2 & 0 \end{pmatrix} \\
+2d^3(\mathbf{t}_2.\mathbf{v}) \det &\begin{pmatrix} v_2 & v_1 & 0 \\ -v_3 & v_3 & -v_1 + v_2 \\ -v_1 & -v_2 & 0 \end{pmatrix} \\
-2d^3(\mathbf{t}_3.\mathbf{v}) \det &\begin{pmatrix} v_2 & v_1 & 0 \\ v_3 & v_3 & v_1 + v_2 \\ -v_1 & -v_2 & 0 \end{pmatrix}
\end{aligned}$$

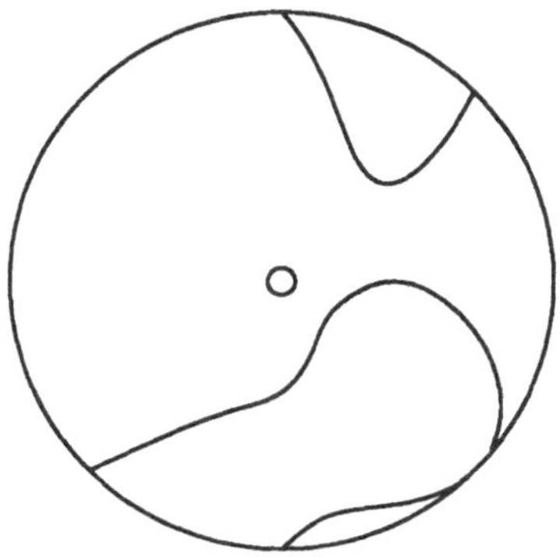

Fig. 4.4. A quartic curve q obtained from a square of base points

$$+2d^4(\mathbf{t}_4.\mathbf{v})\det\begin{pmatrix} v_2 & v_1 & 0 \\ 0 & v_3 & v_2 \\ -v_3 & v_3 & -v_1+v_2 \end{pmatrix} + O(d^5)$$

$$= -4d^2(\mathbf{t}_1.\mathbf{v})v_3(v_1^2+v_2^2)+2d^3(\mathbf{t}_2.\mathbf{v})(v_1-v_2)(v_1^2-v_2^2)+2d^3(\mathbf{t}_3.\mathbf{v})(v_1+v_2)(v_1^2-v_2^2)$$
$$- 4d^4 v_1v_2v_3(\mathbf{t}_4.\mathbf{v}) + O(d^5) \tag{4.84}$$

The quartic q does not depend on the angular velocity, thus it is assumed without loss of generality that the image velocity field arises from translational velocity alone. If the image velocity field is smooth then it follows from (4.82) that the vectors $\mathbf{t}_i$ have magnitudes

$$\|\mathbf{t}_1\| = O(d^2) \qquad \|\mathbf{t}_2\| = O(d)$$
$$\|\mathbf{t}_3\| = O(d) \qquad \|\mathbf{t}_4\| = O(1) \tag{4.85}$$

If (4.85) holds, in particular if the image velocity field is smooth, then the first four terms on the right-hand side of (4.84) are all $O(d^4)$ in magnitude. If any one of the equalities of (4.85) fails to hold, in that one of the $\|\mathbf{t}_i\|$ is too large, then the term on the right-hand side of (4.84) with coefficient $\mathbf{t}_i.\mathbf{v}$ becomes the leading order term of q. The leading order term then has a linear factor $\mathbf{t}_i.\mathbf{v}$, in agreement with the Corollary of Theorem 4.17.

In Fig. 4.4 an example is shown of a quartic of the form (4.84). The image velocity field is obtained from a moving rigid plane with normal $(1/\sqrt{2}, 0, 1/\sqrt{2})^\top$ moving with a translational velocity in the direction $(1,1,1)^\top$. The value of d is $d = 0.1$ radians. The disk in Fig. 4.4 is a projection of that part of the unit sphere with $v_3 \geq 0$. The point $\mathbf{q} = (0,0,1)^\top$ projects to the centre of the disk, and the projection is chosen such that the line $v_3 = 0$ projects to the rim of the disk.

Further information about the quartic (4.84) is obtained in the following three results.

Theorem 4.20. *The leading order terms of the quartic q of (4.84) have an isolated singular point, i.e. an acnode, at $\mathbf{q} = (0,0,1)^\top$.*

Proof. The leading order terms of (4.84) near $\mathbf{q}$ are those involving the highest power of v_3. These terms are

$$-4d^2 t_{13} v_3^2 (v_1^2 + v_2^2) - 4d^4 t_{43} v_1 v_2 v_3^2 \tag{4.86}$$

where the coefficients t_{ij} are defined by $t_{ij} = (\mathbf{t}_i)_j$. The expression (4.86) has a singular point at $\mathbf{q}$ because it is of degree two in v_1, v_2. It remains to find the type of the singular point. It is assumed that the true angular velocity associated with the image velocity vectors is zero. There is no loss of generality in making this assumption because q is independent of angular velocity. With this assumption, the vectors $\mathbf{r}_i$ on the right-hand side of (4.72) are given by

$$\mathbf{r}_i = \dot{\mathbf{q}}_i \times \mathbf{q}_i = (\hat{\mathbf{v}} \times \mathbf{q}_i) K_i \qquad (1 \le i \le 4) \tag{4.87}$$

where $\hat{\mathbf{v}}$ is the true translational velocity of the surface giving rise to the four image velocity vectors. It follows from (4.87) that

$$(\mathbf{r}_i)_3 = (\hat{v}_1 y_i - \hat{v}_2 x_i) K_i \tag{4.88}$$

Equation (4.88) is applied to the definition of the $\mathbf{t}_i$ in (4.82) and the coordinates given in (4.77) are used to substitute for x_i, y_i to obtain

$$\begin{aligned} t_{13} &= \frac{1}{2}(\mathbf{r}_1 - \mathbf{r}_2 + \mathbf{r}_3 - \mathbf{r}_4)_3 \\ &= \frac{d\hat{v}_1}{2\sqrt{2}}(K_1 - K_2 - K_3 + K_4) - \frac{d\hat{v}_2}{2\sqrt{2}}(K_1 + K_2 - K_3 - K_4) \end{aligned} \tag{4.89}$$

$$\begin{aligned} t_{43} &= (\mathbf{r}_1 + \mathbf{r}_2 + \mathbf{r}_3 + \mathbf{r}_4)_3 \\ &= \frac{d\hat{v}_1}{2\sqrt{2}}(K_1 + K_2 - K_3 - K_4) - \frac{d\hat{v}_2}{2\sqrt{2}}(K_1 - K_2 - K_3 + K_4) \end{aligned} \tag{4.90}$$

It follows from (4.89) and (4.90) that, in general, t_{13} and t_{43} are both of the same order of magnitude. If the inverse distance is expanded as a Taylor series in x and y, then t_{13} and t_{43} depend on the terms linear in x and y. The second term of (4.86) is thus much smaller than the first term. It follows that the leading order terms of q at $\mathbf{q}$ are

$$-4d^2 t_{13} v_3^2 (v_1^2 + v_2^2) \tag{4.91}$$

The quartic defined by (4.91) has an isolated singular point at $\mathbf{q} = (0, 0, 1)^\top$. □

Theorem 4.21. *If the image velocity field is smooth then the positions of the zeros of the quartic of (4.84) have increased stability near to the line $v_3 = 0$, in comparison with the stability of the positions of the zeros of the quartic at other values of $\mathbf{v}$.*

Proof. It is assumed without loss of generality that the angular velocity associated with the image velocity field is zero. The leading order terms of (4.84) near to $v_3 = 0$ are

$$2d^3(v_1^2 - v_2^2)[(\mathbf{t}_2.\mathbf{v})(v_1 - v_2) + (\mathbf{t}_3.\mathbf{v})(v_1 + v_2)] \tag{4.92}$$

A smooth image velocity field possesses a good linear approximation, thus there exist functions a, b, c of $\mathbf{v}$ such that

$$\mathbf{r}_i.\mathbf{v} = a + bx_i + cy_i + O(d^2) \qquad (1 \le i \le 4) \qquad (4.93)$$

The increased stability of the positions of the zeros of $q(\mathbf{v})$ near to the line $v_3 = 0$ arises because the linear approximation to $\mathbf{r}_i.\mathbf{v}$ given in (4.93) is sufficient to determine (4.92) to leading order. In detail, it follows from (4.77) and the definition (4.82) of the $\mathbf{t}_i$ that

$$\begin{aligned} \mathbf{t}_2.\mathbf{v} &= (\mathbf{r}_1 - \mathbf{r}_3).\mathbf{v}/\sqrt{2} \\ &= b(x_1 - x_3)/\sqrt{2} + c(y_1 - y_3)/\sqrt{2} + O(d^2) \\ &= d(b + c) + O(d^2) \end{aligned} \qquad (4.94)$$

$$\begin{aligned} \mathbf{t}_3.\mathbf{v} &= (\mathbf{r}_2 - \mathbf{r}_4).\mathbf{v}/\sqrt{2} \\ &= b(x_2 - x_4)/\sqrt{2} + c(y_2 - y_4)/\sqrt{2} + O(d^2) \\ &= -d(b - c) + O(d^2) \end{aligned} \qquad (4.95)$$

It follows from (4.94) and (4.95) that the leading order terms of (4.92) are

$$4d^2(v_2^2 - v_1^2)(bv_2 - cv_1) \qquad (4.96)$$

To complete the proof it is shown that the $O(d^2)$ term of (4.93) makes a contribution to (4.92) that is small in comparison with the leading order term (4.96). To do this, it is necessary to show that in general $bv_2 - cv_1 \neq O(d)$. It follows from the hypothesis of a smooth image velocity field that there exist constants k_1, k_2, k_3 such that the inverse distances K_i have a linear approximation of the form

$$K_i = k_1 + k_2 x_i + k_3 y_i + O(d^2) \qquad (1 \le i \le 4) \qquad (4.97)$$

The expansion of $\mathbf{r}_i.\mathbf{v}$ and the application of (4.97) yields

$$\begin{aligned} \mathbf{r}_i.\mathbf{v} &= (\dot{\mathbf{q}}_i \times \mathbf{q}_i).\mathbf{v} \\ &= [(\hat{\mathbf{v}} \times \mathbf{q}_i).\mathbf{v}]K_i \\ &= [(\mathbf{v} \times \hat{\mathbf{v}})_1 x_i + (\mathbf{v} \times \hat{\mathbf{v}})_2 y_i + (\mathbf{v}\mathbf{v})_3 z_i]K_i \qquad (1 \le i \le 4) \end{aligned} \qquad (4.98)$$

The comparison of (4.93) and (4.98) yields

$$\begin{aligned} b &= (\mathbf{v} \times \hat{\mathbf{v}})_1 k_1 + (\mathbf{v} \times \hat{\mathbf{v}})_3 k_2 \\ c &= (\mathbf{v} \times \hat{\mathbf{v}})_2 k_1 + (\mathbf{v} \times \hat{\mathbf{v}})_3 k_3 \end{aligned} \qquad (4.99)$$

from which it follows that

$$bv_2 - cv_1 = k_1[(\mathbf{v} \times \hat{\mathbf{v}})_1 v_2 - (\mathbf{v} \times \hat{\mathbf{v}})_2 v_1] + (\mathbf{v} \times \hat{\mathbf{v}})_3(k_2 v_2 - k_3 v_1) \qquad (4.100)$$

It follows from (4.100) that in general $bv_2 - cv_1 \neq O(d)$, as required. □

In Theorem 4.21 only the positions of the zeros of q are considered. The coefficients of the terms involving v_3 will, in general, change markedly in response to perturbations $\delta\dot{\mathbf{q}}_i$ of size $O(d^2)$ in the image velocity vectors $\dot{\mathbf{q}}_i$, but these perturbations have little effect on the zeros of $q(\mathbf{v})$ near to the line $v_3 = 0$. A consequence of Theorem 4.19 is that the improved accuracy of Theorem 4.21 is impossible if $|v_3| \neq O(d)$.

The quartic q has four intersections with the line $v_3 = 0$. In general only two of these intersections are possible values of $\hat{\mathbf{v}}$. The other two intersections are close to $(1,1,0)^\top$ and $(1,-1,0)^\top$, regardless of the value of $\hat{\mathbf{v}}$. It follows that $\hat{v}_1$ and $\hat{v}_2$ satisfy, to a first approximation, a quadratic constraint near to $v_3 = 0$. If, by chance, $\hat{\mathbf{v}}$ is close to $(1,1,0)^\top$ or to $(1,-1,0)^\top$, then q is near tangential to $v_3 = 0$. These observations are summarised in the following proposition.

Proposition 4.22. *Let $\dot{\mathbf{q}}_i$ for $1 \leq i \leq 4$ be a smooth image velocity field arising from rigid motion, and let the base points $\mathbf{q}_i$ be at the vertices of the square defined by (4.77). The true translational velocity, $\hat{\mathbf{v}}$, of the rigid surface lies on the line $v_3 = 0$ only if*

$$bv_2 - cv_1 = 0$$

holds for some $\mathbf{v}$ with $v_3 = 0$, where b, c are the linear forms in $\mathbf{v}$ defined by (4.99)

Proof. It follows from (4.96) that the leading order terms of $q(\mathbf{v})$ near to $v_3 = 0$ are

$$4d^2(v_2^2 - v_1^2)(bv_2 - cv_1) \tag{4.101}$$

The result follows immediately because the factor $v_2^2 - v_1^2$ of (4.101) is independent of the true translational velocity, and the remaining factor $bv_2 - cv_1$ is of degree two in $\mathbf{v}$. □

4.3.5 Symmetric Arrangements of Base Points: The Tripod

The leading order terms of q are obtained for a second highly symmetrical arrangement of base points, the tripod. It turns out that the quartics obtained from the tripod are very different in appearance from the quartics obtained from the square of base points.

Let d be a small quantity. The tripod of base points is defined by

$$\begin{aligned}
\mathbf{q}_1 &= (d, 0, \sqrt{1-d^2})^\top \\
\mathbf{q}_2 &= (-d/2, \sqrt{3}d/2, \sqrt{1-d^2})^\top \\
\mathbf{q}_3 &= (-d/2, -\sqrt{3}d/2, \sqrt{1-d^2})^\top \\
\mathbf{q}_4 &= (0, 0, 1)^\top
\end{aligned}$$

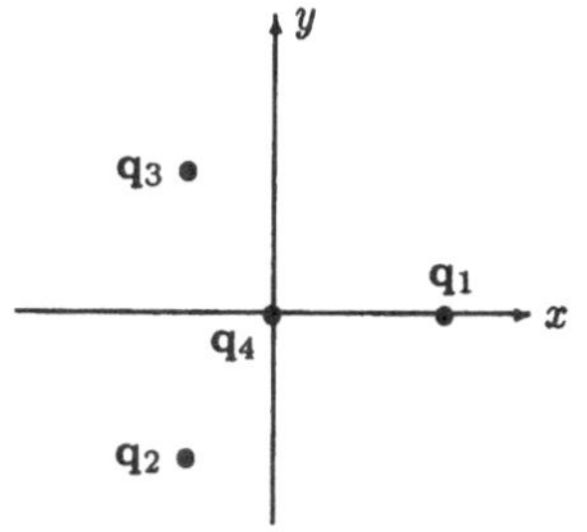

Fig. 4.5. The tripod of base points

as illustrated in Fig. 4.5. The orthogonal matrix U defined by

$$U = \begin{pmatrix} \sqrt{2}/\sqrt{3} & -1/\sqrt{6} & -1/\sqrt{6} & 0 \\ 0 & 1/\sqrt{2} & -1/\sqrt{2} & 0 \\ 1/\sqrt{3} & 1/\sqrt{3} & 1/\sqrt{3} & 0 \\ 0 & 0 & 0 & 1 \end{pmatrix}$$

is used to simplify the calculation of the leading order terms of q. The terminology and general form of the calculation are similar to those employed for the square of base points in Sect. 4.3.4. It follows from the definition of q that $q(\mathbf{v}) = \det(UA|U\mathbf{l})$. The expansion of UA yields

$$UA = \begin{pmatrix} \sqrt{3}dv_3/\sqrt{2} & 0 & \sqrt{3}dv_1/\sqrt{2} \\ 0 & \sqrt{3}dv_3/\sqrt{2} & \sqrt{3}dv_2/\sqrt{2} \\ -\sqrt{3}(1-d^2/2)v_1 & -\sqrt{3}(1-d^2/2)v_2 & -\sqrt{3}d^2v_3 \\ -v_1 & -v_2 & 0 \end{pmatrix} + O(d^3) \quad (4.102)$$

The expansion of $U\mathbf{l}$ yields

$$U\mathbf{l} = \frac{1}{\sqrt{6}}\left((2\mathbf{r}_1 - \mathbf{r}_2 - \mathbf{r}_3).\mathbf{v}, \sqrt{3}(\mathbf{r}_2 - \mathbf{r}_3).\mathbf{v}, \sqrt{2}(\mathbf{r}_1 + \mathbf{r}_2 + \mathbf{r}_3).\mathbf{v}, \mathbf{r}_4.\mathbf{v}\right)^\top \quad (4.103)$$

The vectors $\mathbf{s}_i$ for $1 \le i \le 4$ are defined by

$$U\mathbf{l} = (\mathbf{s}_1.\mathbf{v}, \mathbf{s}_2.\mathbf{v}, \mathbf{s}_3.\mathbf{v} + \sqrt{3}(1-d^2/2)\mathbf{r}_4.\mathbf{v}, \mathbf{s}_4.\mathbf{v})^\top$$

It follows that

$$\begin{aligned} \mathbf{s}_1 &= (2\mathbf{r}_1 - \mathbf{r}_2 - \mathbf{r}_3)/\sqrt{6} \\ \mathbf{s}_2 &= (\mathbf{r}_2 - \mathbf{r}_3)/\sqrt{2} \\ \mathbf{s}_3 &= (\mathbf{r}_1 + \mathbf{r}_2 + \mathbf{r}_3)/\sqrt{3} - \sqrt{3}(1-d^2/2)\mathbf{r}_4 \\ \mathbf{s}_4 &= \mathbf{r}_4 \end{aligned} \quad (4.104)$$

It follows from (4.72), (4.102) and (4.103) that

$$q(\mathbf{v}) = \det \begin{pmatrix} \sqrt{3}dv_3/\sqrt{2} & 0 & \sqrt{3}dv_1/\sqrt{2} & \mathbf{s}_1.\mathbf{v} \\ 0 & \sqrt{3}dv_3/\sqrt{2} & \sqrt{3}dv_2/\sqrt{2} & \mathbf{s}_2.\mathbf{v} \\ 0 & 0 & -\sqrt{3}d^2v_3 & \mathbf{s}_3.\mathbf{v} \\ -v_1 & -v_2 & 0 & \mathbf{s}_4.\mathbf{v} \end{pmatrix} + O(d^5) \quad (4.105)$$

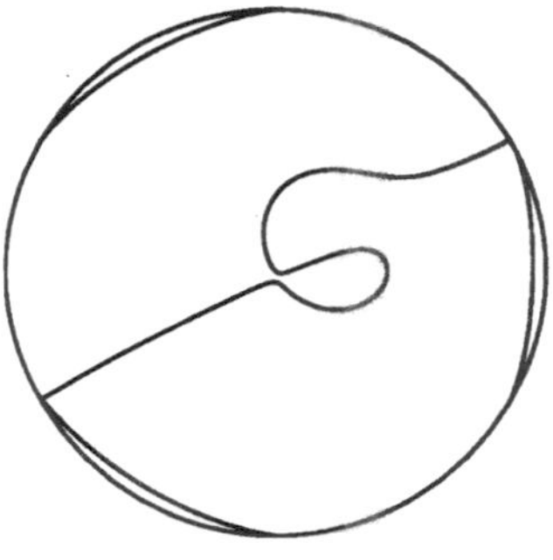

Fig. 4.6. A quartic curve q obtained from a tripod of base points

The expansion of the determinant in (4.105) and the rearrangement of the terms yields

$$q(\mathbf{v}) = \frac{\sqrt{3}}{2} v_3 (\mathbf{s}_1.\mathbf{v}) \det \begin{pmatrix} dv_3 & dv_2 \\ 0 & -d^2 v_3 \end{pmatrix} + \frac{\sqrt{3}}{\sqrt{2}} d^3 v_3 (\mathbf{s}_2.\mathbf{v}) \det \begin{pmatrix} v_3 & 0 \\ -v_1 & -v_2 \end{pmatrix}$$

$$-\frac{3}{2}(\mathbf{s}_3.\mathbf{v}) \det \begin{pmatrix} dv_3 & 0 & dv_1 \\ 0 & dv_3 & dv_2 \\ -v_1 & -v_2 & 0 \end{pmatrix} + \frac{3\sqrt{3}}{2} d^2 v_3 (\mathbf{s}_4.\mathbf{v}) \det \begin{pmatrix} -dv_3 & 0 \\ 0 & -dv_3 \end{pmatrix} + O(d^5)$$

from which it follows that

$$\begin{aligned} q(\mathbf{v}) &= -\frac{\sqrt{3}}{2} d^3 (\mathbf{s}_1.\mathbf{v}) v_1 v_3^2 - \frac{3}{\sqrt{2}} d^3 (\mathbf{s}_2.\mathbf{v}) v_2 v_3^2 - \frac{3}{2} d^2 (\mathbf{s}_3.\mathbf{v}) v_3 (v_1^2 + v_2^2) \\ &\quad + \frac{3\sqrt{3}}{2} d^4 (\mathbf{s}_4.\mathbf{v}) v_3^3 + O(d^5) \\ &= -\frac{\sqrt{3}}{2} d^2 v_3 [d(\mathbf{s}_1.\mathbf{v}) v_1 v_3 + \sqrt{6} d (\mathbf{s}_2.\mathbf{v}) v_2 v_3 + \sqrt{3} (\mathbf{s}_3.\mathbf{v})(v_1^2 + v_2^2) \\ &\quad - 3d^2 (\mathbf{s}_4.\mathbf{v}) v_3^2] + O(d^5) \qquad (4.106) \end{aligned}$$

In Fig. 4.6 the quartic of (4.106) is shown for the same moving plane that produced the quartic plane curve of Fig. 4.4. The leading order terms of q split into the linear factor v_3 and a cubic factor. The quartic satisfies $q(0,0,1) = 0$ exactly, because $(0,0,1)^\top$ is one of the base points. In contrast with the quartic (4.84) obtained from the square of base points, the values of q do not show increased stability near to the line $v_3 = 0$. However, the leading order terms of q do have a singular point at $\mathbf{q}$.

Proposition 4.23. *The cubic factor of the leading order terms of the quartic q of (4.106) has a singular point at $\mathbf{q} = (0,0,1)^\top$.*

Proof. It is assumed without loss of generality that the image velocity field arises from translational velocity alone. The quantities s_{ij} are defined by $s_{ij} = (\mathbf{s}_i)_j$.

On the right-hand side of (4.106) the terms of the cubic factor linear in v_1 and v_2 are

$$dv_3^2[s_{13}v_1 - \sqrt{6}\, s_{23}v_2 + 3d(\mathbf{s}_4.\mathbf{v})] \tag{4.107}$$

Let c_{31} be the coefficient of $v_3v_1^2$ in the cubic factor. To prove the result it is shown that the coefficients of $v_3^2v_1$ and $v_3^2v_2$ in (4.107) are each $O(d^2)$, whilst c_{31} is $O(d)$, but $c_{31} \neq O(d^2)$. The vectors $\mathbf{r}_i = \dot{\mathbf{q}}_i \times \mathbf{q}_i$ in the definition (4.72) of $q(\mathbf{v})$ are given by

$$\mathbf{r}_i = (\hat{\mathbf{v}} \times \mathbf{q}_i)K_i \tag{4.108}$$

It follows that $r_{i3} = O(d)$, thus $s_{i3} = O(d)$. Thus the coefficients of $v_3^2v_1$ and $v_3^2v_2$ in (4.107) are each $O(d^2)$.

In the case of c_{31} (4.106) yields

$$\begin{aligned} c_{31} &= d\, s_{11} + \sqrt{3}\,[s_{33} - \sqrt{3}(1 - d^2/2)s_{43}] \\ &= d\, s_{11} + \sqrt{3}\, s_{33} + O(d^2) \end{aligned} \tag{4.109}$$

Equations (4.104) and (4.108) yield expressions for s_{11} and s_{33},

$$\begin{aligned} \sqrt{6}\, s_{11} &= 2r_{11} - r_{21} - r_{31} \\ &= 2\hat{v}_2K_1 - (\hat{v}_2\sqrt{3}\, d\hat{v}_3/2)K_2 - (\hat{v}_2 + \sqrt{3}\, d\hat{v}_3/2)K_3 + O(d^2) \\ &= \hat{v}_2(2K_1 - K_2 - K_3) + \sqrt{3}\, d\hat{v}_3(K_2 - K_3)/2 + O(d^2) \\ \sqrt{3}\, s_{33} &= -d\hat{v}_2K_1 + \sqrt{3}\, d\hat{v}_1(K_2 - K_3)/2 + d\hat{v}_2(K_2 + K_3)/2 \end{aligned} \tag{4.110}$$

Equations (4.109) and (4.110) yield

$$c_{31} = -d\hat{v}_2K_1 + \sqrt{3}d\hat{v}_1(K_2 - K_3)/2 + O(d^2)$$

It follows that $c_{31} = O(d)$ and $c_{31} \neq O(d^2)$. □

4.4 The Linear System of Quartics

In Sect. 4.3.4 and Sect. 4.3.5 the quartic constraints on the translational velocity are obtained from quadruples of image velocity vectors with highly symmetric arrangements of base points. If the positions of the base points are arbitrary then the equation for the quartic is very complicated and it seems difficult to describe its general appearance. In this section a different approach to the analysis of the quartic constraint is taken. Instead of examining just one quartic, the whole family of quartics obtained from a particular fixed choice of base points is considered. The different quartics in the family arise from different choices for the image velocity vectors.

The quartic constraints q obtained from a fixed set of base points form a vector space of curves, because they depend linearly on the image velocity vectors. Vector spaces of curves are standard constructions in algebraic geometry; they are known as linear systems (Semple & Roth 1949). Let L be the linear system of quartic plane curves obtained for a fixed choice of base points. Then L is a subspace of the 15 (Euclidean) dimensional space of all quartic plane curves. The space L is, by definition, independent of any choice of the image velocity vectors. It depends only on the base points of the image velocity vectors.

It is shown that for almost all choices of base points, L has vector space dimension five. The quartics corresponding to the points of L are, in general, non-singular and each quartic determines the associated image velocity field up to a component due to angular velocity alone. A base point of a linear system is a point contained in all the curves in the linear system. The space L is characterised by ten base points, in that there are ten values of $\mathbf{v}$ through which all the quartics q in L pass and, conversely, any quartic passing through these ten points is in L. The base points of a linear system should not be confused with the base points of the image velocity vectors.

4.4.1 The Dimension of the Linear System

Let $\mathbf{q}_i$, $1 \leq i \leq 4$, be a fixed set of base points on the unit sphere. The linear system L is obtained as the image of a linear transformation T from the space of image velocity fields with four velocity vectors based at the $\mathbf{q}_i$ to the space of quartics. A general quartic plane curve is defined by a polynomial equation with 15 coefficients. The usual convention in projective geometry is to identify any two equations that differ by a non-zero scale factor. In this section the identification is not made. The space of all quartics is parameterised by the points of $\mathbb{R}^{15}$. Each vector $\mathbf{r}_i = \dot{\mathbf{q}}_i \times \mathbf{q}_i$ lies in the tangent space to the unit sphere at $\mathbf{q}_i$. Each of these tangent spaces is a copy of $\mathbb{R}^2$. Let $\langle \mathbf{r} \rangle$ be the point in the product $(\mathbb{R}^2)^4$ of the four tangent spaces defined by

$$\langle \mathbf{r} \rangle = (\mathbf{r}_1, \mathbf{r}_2, \mathbf{r}_3, \mathbf{r}_4)^\top$$

The linear transformation $T : (\mathbb{R}^2)^4 \to \mathbb{R}^{15}$ is defined by

$$T(\langle \mathbf{r} \rangle) = q \tag{4.111}$$

where the definition of the quartic q is taken from (4.72),

$$q(\mathbf{v}) = \begin{pmatrix} (\mathbf{v}.\mathbf{q}_1)\mathbf{q}_1^\top - \mathbf{v}^\top \\ (\mathbf{v}.\mathbf{q}_2)\mathbf{q}_2^\top - \mathbf{v}^\top \\ (\mathbf{v}.\mathbf{q}_3)\mathbf{q}_3^\top - \mathbf{v}^\top \\ (\mathbf{v}.\mathbf{q}_4)\mathbf{q}_4^\top - \mathbf{v}^\top \end{pmatrix} \tag{4.112}$$

It follows from (4.112) that q is a linear function of $\langle \mathbf{r} \rangle$, thus T is linear, as claimed. It follows that L is a linear system because it is the image of T in $\mathbb{R}^{15}$. Let $\dim(L)$ to be the dimension of L as a vector space. The following two results establish that $\dim(L) = 5$ for almost all choices of base points.

Proposition 4.24. *Let L be the linear system of quartic curves associated with a quadruple of base points. The dimension of L satisfies*

$$\dim(L) \leq 5$$

Proof. The result holds if all four base points coincide, because then (4.112) yields $\dim(L) = 0$. Let two base points, $\mathbf{q}_3$ and $\mathbf{q}_4$ say, be distinct and let J be the linear subspace of $(\mathbb{R}^2)^4$ containing exactly those $\langle \mathbf{r} \rangle$ with $\mathbf{r}_4 = 0$ and with $\mathbf{r}_3$ restricted such that it is normal to the plane defined by $\mathbf{q}_3$ and $\mathbf{q}_4$. The subspace J is spanned by the vectors in the tangent planes to the unit sphere at $\mathbf{q}_1$, $\mathbf{q}_2$, together with the vector $\mathbf{q}_3 \times \mathbf{q}_4$ in the tangent plane at $\mathbf{q}_3$; thus J has dimension five. If $\langle \mathbf{r} \rangle$ results from an arbitrary image velocity field then it can be reduced to an element of J by subtracting from it an image velocity field arising from angular velocity alone. Proposition 4.14 shows that this reduction leaves $q(\mathbf{v})$ unchanged, thus $T(J) = L$. As the transformation T is linear it follows that $\dim(L) \le \dim(J) = 5$. □

Theorem 4.25. *Let L be the linear system of quartics associated with a quadruple of base points $\mathbf{q}_i$. Then, for almost all choices of the $\mathbf{q}_i$ it is the case that* $\dim(L) = 5$.

Proof. It is shown that $\dim(L) = 5$ for the following choice of base points:

$$\begin{aligned} \mathbf{q}_1 &= (1,0,0)^\top \\ \mathbf{q}_2 &= (0,1,0)^\top \\ \mathbf{q}_3 &= (0,0,1)^\top \\ \mathbf{q}_4 &= (1/\sqrt{3}, 1/\sqrt{3}, 1/\sqrt{3})^\top \end{aligned} \tag{4.113}$$

An image velocity field due to angular velocity alone is subtracted from the given image velocity field to ensure that $\mathbf{r}_4 = 0$, and that $\mathbf{r}_3$ is normal to the plane spanned by $\mathbf{q}_3$ and $\mathbf{q}_4$. The vector $\mathbf{r}_3$ is thus parallel to $\mathbf{q}_3 \times \mathbf{q}_4 = (1/\sqrt{3})(-1,1,0)^\top$. Each quartic in L is unaffected by this change in the angular velocity.

On substituting the values of $\mathbf{q}_i$ given by (4.113) into (4.112) it follows that

$$q(\mathbf{v}) = \frac{1}{3}\det\begin{pmatrix} 0 & -v_2 & -v_3 & r_{12}v_2 + r_{13}v_3 \\ -v_1 & 0 & -v_3 & r_{21}v_1 + r_{23}v_3 \\ -v_1 & -v_2 & 0 & r_{31}(v_1 - v_2) \\ v_2 + v_3 - 2v_1 & v_1 + v_3 - 2v_2 & v_1 + v_2 - 2v_3 & 0 \end{pmatrix} \tag{4.114}$$

where the r_{ij} are defined by $r_{ij} = (\mathbf{r}_i)_j$. The expansion of the determinant in (4.114) yields

$$q(\mathbf{v}) = p_{12}(\mathbf{v})r_{12} + p_{13}(\mathbf{v})r_{13} + p_{21}(\mathbf{v})r_{21} + p_{23}(\mathbf{v})r_{23} + p_{31}(\mathbf{v})r_{31} \tag{4.115}$$

where the p_{ij} are quartic polynomials in $\mathbf{v}$ defined by

$$\begin{aligned} p_{12} &: v_3^2 v_2(v_2 - v_1) + v_3 v_2(v_2^2 + 2v_2 v_1 - v_1^2) - v_2^2 v_1(v_2 + v_1) \\ p_{13} &: v_3^3(v_2 - v_1) + v_3^2(v_2^2 + 2v_2 v_1 - v_1^2) - v_3 v_2 v_1(v_2 + v_1) \end{aligned}$$

$$\begin{aligned}
p_{21} &: v_3^2 v_1(v_1 - v_2) + v_3 v_1(v_1^2 - v_2^2 + 2v_2 v_1) - v_2 v_1^2(v_2 + v_1) \\
p_{23} &: v_3^3(v_1 - v_2) + v_3^2(v_1^2 - v_2^2 + 2v_2 v_1) - v_3 v_2 v_1(v_2 + v_1) \\
p_{31} &: (v_2 - v_1)[v_3^2(v_2 + v_1) + v_3(v_2^2 - 2v_2 v_1 + v_1^2) - v_2 v_1(v_2 + v_1)] \quad (4.116)
\end{aligned}$$

It is shown that the p_{ij} are linearly independent. Let $a_{12}, a_{13}, a_{21}, a_{23}, a_{31}$ be scalars independent of $\mathbf{v}$ such that for all values of $\mathbf{v}$

$$a_{12}p_{12}(\mathbf{v}) + a_{13}p_{13}(\mathbf{v}) + a_{21}p(\mathbf{v}) + a_{23}p_{23}(\mathbf{v}) + a_{31}p_{31}(\mathbf{v}) = 0 \quad (4.117)$$

The terms in (4.117) involving the monomial v_3^3 are identically zero, thus $a_{13} - a_{23} = 0$. The monomial $v_2^3 v_1$ occurs only in p_{12} thus $a_{12} = 0$. Similarly, $v_2 v_1^3$ occurs only in p_{21}, thus $a_{21} = 0$. Equation (4.117) reduces to

$$a_{13}[p_{13}(\mathbf{v}) + p_{23}(\mathbf{v})] + a_{31}p(\mathbf{v}) = 0 \quad (4.118)$$

The quartics p_{13} and p_{23} are both divisible by v_3, but p_{31} is not divisible by v_3. It thus follows from (4.118) that $a_{31} = 0$, $a_{13} = 0$. Thus all the coefficients a_{ij} are zero. It follows that the p_{ij} are linearly independent and that in consequence L has dimension five for the choice (4.113) of base points.

This single example is enough to establish the theorem. For a general choice of base points an expression for q of the form (4.115) is obtained, where the p_{ij} are quartics in $\mathbf{v}$ with coefficients depending on the base points $\mathbf{q}_i$. The condition that the p_{ij} are linearly dependent is a polynomial constraint on the coefficients of the p_{ij} and hence a polynomial constraint

$$c(\mathbf{q}_1, \mathbf{q}_2, \mathbf{q}_3, \mathbf{q}_4) = 0 \quad (4.119)$$

on the coordinates of the base points. If the base points (4.113) are chosen then the constraint (4.119) is *not* satisfied. It follows that (4.119) is non-trivial, i.e. $c(\mathbf{q}_1, \mathbf{q}_2, \mathbf{q}_3, \mathbf{q}_4) \neq 0$ for almost all choices of base points. It follows that $\dim(L) \geq 5$. The result $\dim(L) = 5$ follows on applying Proposition 4.24. □

Theorem 4.25 is used to obtain a converse to Proposition 4.14.

Theorem 4.26. *For almost all choices of base points, q determines $\langle\dot{\mathbf{q}}\rangle$ up to an image velocity field due to angular velocity alone.*

Proof. Let $\langle\dot{\mathbf{q}}'\rangle$, $\langle\dot{\mathbf{q}}\rangle$ be two image velocity fields on the same four base points such that for all $\mathbf{v}$

$$q(\langle\dot{\mathbf{q}}'\rangle, \mathbf{v}) = q(\langle\dot{\mathbf{q}}\rangle, \mathbf{v}) \quad (4.120)$$

Let $\mathbf{r}_i'$ and $\mathbf{r}_i$ be defined by $\mathbf{r}_i' = \dot{\mathbf{q}}_i' \times \mathbf{q}_i'$, $\mathbf{r}_i = \dot{\mathbf{q}} \times \mathbf{q}_i$. The transformation T defined by (4.111) is linear, thus it follows from (4.120) that $T(\langle\mathbf{r}'\rangle - \langle\mathbf{r}\rangle) = 0$. It is clear that $\langle\mathbf{r}'\rangle - \langle\mathbf{r}\rangle = \langle\mathbf{r}' - \mathbf{r}\rangle$. It follows that $\langle\mathbf{r}' - \mathbf{r}\rangle$ is contained in the null space of T.

It follows from Theorem 4.25 that $\dim(L) = 5$ for almost all choices of base points, thus the null space of T has dimension $3 = 8 - 5$. The points $\langle\mathbf{r}''\rangle$ of $(\mathbb{R}^2)^4$ arising from image velocity fields due to angular velocity alone form a three-dimensional subspace of the null space of T, thus the set of such $\langle \mathbf{r}'' \rangle$ comprise the entire null space of T. It follows that $\langle \mathbf{r}' - \mathbf{r} \rangle$ arises from angular velocity alone. There exists an angular velocity $\mathbf{w}$ such that

$$\mathbf{r}_i' - \mathbf{r}_i = -(\mathbf{w} \times \mathbf{q}_i) \times \mathbf{q}_i \qquad (1 \le i \le 4) \tag{4.121}$$

The vector product of (4.121) with $\mathbf{q}_i$ yields

$$\dot{\mathbf{q}}_i' = \dot{\mathbf{q}}_i + \mathbf{w} \times \mathbf{q}_i \qquad (1 \le i \le 4)\square$$

4.4.2 Singular Points of Quartics

In Sect. 4.3.4 and Sect. 4.3.5 explicit expressions are obtained for the leading order terms of q when the base points form a square or a tripod. These expressions possess singular points. This raises the question of whether q possesses a singular point for a general choice of base points. It is shown in the next theorem that q is in general non-singular.

Theorem 4.27. *For a general fixed choice of base points q is without a singular point for almost all choices of the image velocity vectors.*

Proof. The quartic q is expanded in the form (4.115), where the p_{ij} are quartic polynomials with coefficients depending on the base points $\mathbf{q}_i$. It follows from Bertini's Theorem that for a fixed choice of base points, q has a singular point for a general, possibly complex, choice of image velocity vectors if and only if the p_{ij} have a common singular point. The condition that the p_{ij} have a common singular point is an algebraic constraint on the coefficients of the p_{ij}, and hence an algebraic constraint on the coordinates of the base points.

It is shown that the algebraic constraint on the coordinates of the base points is non-trivial. To do this it is sufficient to show that the p_{ij} do not have a common singular point when the base points are given by (4.113). Note first that $(0,0,1)^\top$ is not a singular point of p_{13} or p_{23} and $(1,1,1)^\top$ is not a singular point of p_{21}, because $\partial p_{21}/\partial v_3 \neq 0$ at $\mathbf{v} = (1,1,1)^\top$. It follows from (4.116) that

$$v_1 p_{12}(\mathbf{v}) + v_2 p_{21}(\mathbf{v}) = 2v_1^2 v_2^2 (2v_3 - v_1 - v_2) \tag{4.122}$$

It follows from (4.122) that if $\mathbf{v}$ is a common singular point of the p_{ij} then

$$v_1 = 0 \quad \text{or} \quad v_2 = 0 \quad \text{or} \quad v_3 = \frac{1}{2}(v_1 + v_2) \tag{4.123}$$

The gradient of p_{31} is calculated,

$$\begin{aligned}
\frac{\partial p_{31}}{\partial v_1} &= v_3^2 + 2v_3(v_1 - v_2) - v_2^2 - 2v_1 v_2 \\
\frac{\partial p_{31}}{\partial v_2} &= v_3^2 + 2v_3(v_2 - v_1) - 2v_1 v_2 - v_1^2 \\
\frac{\partial p_{31}}{\partial v_3} &= 2v_3(v_2 + v_1) + (v_2 - v_1)^2
\end{aligned} \tag{4.124}$$

The first option $v_1 = 0$ of (4.123) is chosen. The substitution of 0 for v_1 in the equations of (4.125) yields

$$\nabla p_{31}(0, v_2, v_3) = (v_3^2 - 2v_3v_2 - v_2^2, v_3^2 + 2v_3v_2, 2v_3v_2 + v_2^2)^\top \qquad (4.125)$$

It follows from (4.125) that $\nabla p_{31}(0, v_2, v_3) \neq 0$ for all v_2, v_3 provided either $v_2 \neq 0$ or $v_3 \neq 0$. The first option of (4.123) thus does not yield a common singular point of the p_{ij}. A similar argument rules out the second option of (4.123), $v_2 = 0$. The final option is $v_3 = (1/2)(v_1 + v_2)$. In this case

$$\partial p_{31}(\mathbf{v})/\partial v_3 = 2(v_1^2 + v_2^2) \qquad (4.126)$$

The right-hand side of (4.126) is zero if and only if $v_1 = \pm i v_2$. The substitution of iv_2 for v_1 in the first equation of (4.125) yields

$$\partial p_{31}(\mathbf{v})/\partial v_1 = (3 \pm 3i/2)v_2^3 \neq 0$$

The p_{ij} thus do not possess a common singular point.

It has been established that for a general but fixed choice of base points with real coordinates, q does not have a singular point for a general choice of image velocity vectors with complex coordinates. The condition which ensures that q has a singular point is a polynomial f in the components of the velocity vectors. The coefficients of f are real because they are obtained from the coordinates of the base points. The polynomial constraint f is non-trivial for a general complex choice of image velocity vectors thus f is non-trivial for a general real choice of image velocity vectors. □

4.4.3 Base Points of the Linear System

The base points of a linear system of curves are those points through which every curve of the linear system passes. In the next three theorems it is shown that the linear system L of quartics has exactly ten base points. The ten base points determine L in the sense that any quartic passing through the ten points is in L. The first step is to obtain an alternative description of the base points. It is recalled from (4.112) that q is defined by

$$q(\mathbf{v}) = \det(A|\mathbf{l}) \qquad (4.127)$$

where

$$A = \begin{pmatrix} (\mathbf{v}.\mathbf{q}_1)\mathbf{q}_1^\top - \mathbf{v}^\top \\ (\mathbf{v}.\mathbf{q}_2)\mathbf{q}_2^\top - \mathbf{v}^\top \\ (\mathbf{v}.\mathbf{q}_3)\mathbf{q}_3^\top - \mathbf{v}^\top \\ (\mathbf{v}.\mathbf{q}_4)\mathbf{q}_4^\top - \mathbf{v}^\top \end{pmatrix} \qquad (4.128)$$

and

$$\mathbf{l} = ((\dot{\mathbf{q}}_1 \times \mathbf{q}_1).\mathbf{v}, (\dot{\mathbf{q}}_2 \times \mathbf{q}_2).\mathbf{v}, (\dot{\mathbf{q}}_3 \times \mathbf{q}_3).\mathbf{v}, (\dot{\mathbf{q}}_4 \times \mathbf{q}_4).\mathbf{v})^\top$$

Theorem 4.28. *The base points of L comprise the four base points, $\mathbf{q}_i$, of the image velocity vectors and the points at which A has rank two or less.*

Proof. It follows from the equation (4.127) for q and the definitions of A and $\mathbf{l}$ that the $\mathbf{q}_i$ are base points of L. If A has rank two or less at a point $\mathbf{v}$ then the matrix $(A|\mathbf{l})$ has rank three or less for all choices of the image velocity vectors, $\dot{\mathbf{q}}_i$. It follows that for all choices of the $\dot{\mathbf{q}}_i$,

$$q(\mathbf{v}) = \det(A\,|\,\mathbf{l}) = 0$$

Hence $\mathbf{v}$ is a base point of L.

To prove the converse, it suffices to show that any base point $\mathbf{b}$ of L at which A has rank three is a base point of one of the image velocity vectors. By hypothesis, $q(\langle\dot{\mathbf{q}}\rangle, \mathbf{b}) = 0$ for all choices of the $\dot{\mathbf{q}}_i$. As A has rank three, three of the rows of $A(\mathbf{b})$ are linearly independent. It is assumed without loss of generality that these rows are the first three. Then the coefficient of $\mathbf{r}_4.\mathbf{b}$ in the expansion of q obtained from (4.127) is non-zero. However, $q(\mathbf{b}) = 0$ for all choices of $\mathbf{r}_4$. It follows that for all choices of $\dot{\mathbf{q}}_4$,

$$\mathbf{r}_4.\mathbf{b} = (\dot{\mathbf{q}}_4 \times \mathbf{q}_4).\mathbf{b} = 0 \tag{4.129}$$

Equation (4.129) holds for all $\dot{\mathbf{q}}_4$ only if $\mathbf{b} = \mathbf{q}_4$. □

The aim now is to prove that there are, in general, exactly six values of $\mathbf{v}$ at which A has rank two or less. The next theorem is a preliminary.

Theorem 4.29. *Let $\mathbf{q}_i$, $1 \le i \le 4$, be a set of base points and let f_{123}, f_{124} be cubic homogeneous polynomials in $\mathbf{v}$ defined as in Sect. 4.3.2 by*

$$f_{123}(\mathbf{v}) = \det\begin{pmatrix} (\mathbf{v}.\mathbf{q}_1)\mathbf{q}_1^\top - \mathbf{v}^\top \\ (\mathbf{v}.\mathbf{q}_2)\mathbf{q}_2^\top - \mathbf{v}^\top \\ (\mathbf{v}.\mathbf{q}_3)\mathbf{q}_3^\top - \mathbf{v}^\top \end{pmatrix} \tag{4.130}$$

$$f_{124}(\mathbf{v}) = \det\begin{pmatrix} (\mathbf{v}.\mathbf{q}_1)\mathbf{q}_1^\top - \mathbf{v}^\top \\ (\mathbf{v}.\mathbf{q}_2)\mathbf{q}_2^\top - \mathbf{v}^\top \\ (\mathbf{v}.\mathbf{q}_4)\mathbf{q}_4^\top - \mathbf{v}^\top \end{pmatrix} \tag{4.131}$$

Then f_{123} and f_{124} intersect in general at nine distinct points, three of which are $\mathbf{q}_1$, $\mathbf{q}_2$, $\mathbf{q}_1 \times \mathbf{q}_2$.

Proof. It follows immediately from (4.130) and (4.131) that f_{123} and f_{124} intersect at $\mathbf{q}_1$, $\mathbf{q}_2$ and $\mathbf{q}_1 \times \mathbf{q}_2$. The condition, that f_{123} and f_{124} have strictly less than nine intersections is a polynomial constraint on the coefficients of f_{123} and f_{124}. Thus to show that f_{123} and f_{124} have nine distinct intersections for almost all quadruples $\mathbf{q}_i$ it is enough to find just *one* quadruple of base points for which f_{123} and f_{124} have the required property. Let four base points be chosen as follows:

$$\begin{aligned} \mathbf{q}_1 &= (1/\sqrt{2}, 0, 1/\sqrt{2})^\top \\ \mathbf{q}_2 &= (-1/\sqrt{2}, 0, \sqrt{2})^\top \\ \mathbf{q}_3 &= (0, 0, 1)^\top \\ \mathbf{q}_4 &= (q_1, q_2, q_3)^\top \end{aligned} \tag{4.132}$$

The substitution of the values of the $\mathbf{q}_i$ given by (4.132) into (4.130) yields

$$f_{123}(\mathbf{v}) = -\frac{1}{2}\sqrt{1-d^2}\, v_2(v_1 + iv_3)(v_1 - iv_3) \tag{4.133}$$

The cubic f_{123} thus splits into three lines. The point $(0,1,0)^\top$ is the intersection of two of these lines, thus it is a singular point of f_{123}. The substitution of the values of the $\mathbf{q}_i$ given by (4.132) into (4.131) yields

$$f_{124}(\mathbf{v}) = \det \begin{pmatrix} (v_3 - v_1)/2 & -v_2 & (v_1 - v_3)/2 \\ -(v_1 + v_3)/2 & -v_2 & -(v_1 + v_3)/2 \\ (\mathbf{v}.\mathbf{q}_4)q_1 - v_1 & (\mathbf{v}.\mathbf{q}_4)q_2 - v_2 & (\mathbf{v}.\mathbf{q}_4)q_3 - v_3 \end{pmatrix} \tag{4.134}$$

The intersections of f_{124} with f_{123} are readily obtained from (4.133) and (4.134). The cubic f_{124} intersects the line $v_2 = 0$ at the three points

$$(1,0,1)^\top \qquad (-1,0,1)^\top \qquad (-q_3, 0, q_1)^\top$$

Let r, s be defined by

$$r = \frac{q_3 - iq_1}{1 - q_2} \qquad s = \frac{q_3 - iq_1}{1 + q_2} \tag{4.135}$$

The cubic f_{124} intersects the component $v_1 + iv_3$ of f_{123} at the three points

$$(0,1,0)^\top \qquad (-i, r, 1)\top \qquad (-i, s, -1)^\top \tag{4.136}$$

and finally, f_{124} intersects the component $v_1 - iv_3$ of f_{123} at the three points

$$(0,1,0)^\top \qquad (i, \overline{r}, 1)^\top \qquad (i, \overline{s}, -1)^\top \tag{4.137}$$

It follows from (4.135), (4.136) and (4.137) that f_{123} and f_{124} intersect at eight *distinct* points. The intersection at $(0,1,0)^\top$ is counted with multiplicity two, to make the total of $9 = 3 \times 3$ intersections required by Bézout's Theorem.

Two polynomials f_{123} and f_{124} with nine distinct intersections are obtained by perturbing $\mathbf{q}_3$. Let δ be a small positive real number, and let $\mathbf{q}_3$ be perturbed to

$$(0, \delta, \sqrt{1-\delta^2})^\top$$

The cubic f_{124} is unchanged by this perturbation. The perturbed version of f_{123} still passes through $(0,1,0)^\top$, but now f_{123} has a unique tangent line at $(0,1,0)^\top$ equal to

$$(0,0,-\delta)^\top + O(\delta^2)$$

The tangent of f_{124} at $(0,1,0)^\top$ is $q_2(q_1, 0, -q_3)^\top$. If $\delta \neq 0$, then f_{123} and f_{124} meet at $(0,1,0)^\top$ with distinct tangents. It follows that f_{123} and f_{124} have a transverse intersection at $(0,1,0)^\top$. Furthermore, the seven intersections of f_{123} and f_3 with multiplicity one obtained at $\delta = 0$ remain distinct provided δ is sufficiently small. It follows that f_{123} and f_{124} intersect transversely at eight distinct points for all sufficiently small values of δ. The ninth intersection of f_{124} and f_{123} is then also transverse and distinct from the other eight intersections. Thus a pair of polynomials, f_{123} and f_{124}, with nine distinct intersections is obtained, as required. □

It is shown in Sect. 5.4.1 that the polynomials f_{123}, f_{124} define cubic plane curves that are, in general, non-singular.

Theorem 4.30. *For almost all choices of base points L has exactly ten base points.*

Proof. In view of Theorem 4.28 it suffices to show that the matrix A defined by (4.128) has rank two or less at exactly six points and that none of these six points are base points of the image velocity vectors. The two cubic polynomials f_{123}, f_{124} defined by (4.130) and (4.131) are determinants of triples of rows of A. It is shown in Theorem 4.29 that f_{123}, f_{124} meet at six points $\mathbf{u}_i$, which are distinct from $\mathbf{q}_1$, $\mathbf{q}_2$, $\mathbf{q}_1 \times \mathbf{q}_2$. At each point $\mathbf{u}_i$ the two vectors

$$(\mathbf{v}.\mathbf{q}_1)\mathbf{q}_1 - \mathbf{v} \qquad \text{and} \qquad (\mathbf{v}.\mathbf{q}_2)\mathbf{q}_2 - \mathbf{v} \tag{4.138}$$

are linearly independent. It follows from (4.130) and (4.131) that $(\mathbf{v}.\mathbf{q}_3)\mathbf{q}_3 - \mathbf{v}$ and $(\mathbf{v}.\mathbf{q}_4)\mathbf{q}_4 - \mathbf{v}$ are each linear combinations of the vectors of (4.138). Thus the two vectors of (4.138) span the row space of A. It follows that A has rank two or less at each of the points $\mathbf{u}_i$. A routine calculation, which is omitted, shows that A has rank three at the four base points, $\mathbf{q}_i$, and at $\mathbf{q}_1 \times \mathbf{q}_2$. □

Theorem 4.31. *For almost all choices of base points of the image velocity vectors, any quartic passing through the ten base points of the linear system L is in L.*

Proof. Let q be any quartic passing through the ten base points of L. Let q' be a quartic in L such that $q'(\mathbf{q}_1 \times \mathbf{q}_2) = 1$. Let q'' be defined by

$$q''(\mathbf{v}) = q(\mathbf{v}) - q(\mathbf{q}_1 \times \mathbf{q}_2)q'(\mathbf{v})$$

If q'' is in L then so is q, because L is a vector space of curves. Furthermore, q'' passes through all ten base points of L and $q''(\mathbf{q}_1 \times \mathbf{q}_2) = 0$. It is thus assumed, without loss of generality, that $q(\mathbf{q}_1 \times \mathbf{q}_2) = 0$. The quartic q then passes through all the nine distinct intersections of the cubics f_{123} and f_{124} defined by (4.130) and (4.131). It follows from Noether's Theorem that there exist two linear forms in $\mathbf{v}$, namely $\mathbf{a}$ and $\mathbf{b}$, such that

$$q(\mathbf{v}) = a(\mathbf{v})f_{123}(\mathbf{v}) + b(\mathbf{v})f_{124}(\mathbf{v}) \tag{4.139}$$

It follows from the definition of q that

$$q(\mathbf{q}_3) = q(\mathbf{q}_4) = 0,$$

thus, in general,

$$a(\mathbf{q}_4) = b(\mathbf{q}_3) = 0$$

Thus, there exist vectors $\dot{\mathbf{q}}_4$ and $\dot{\mathbf{q}}_3$ such that

$$\begin{aligned} a(\mathbf{v}) &= (\dot{\mathbf{q}}_4 \times \mathbf{q}_4).\mathbf{v} \\ b(\mathbf{v}) &= -(\dot{\mathbf{q}}_3 \times \mathbf{q}_3).\mathbf{v} \end{aligned} \tag{4.140}$$

The result follows from (4.139) and (4.140). □

Corollary. If q is a quartic with real coefficients passing through the ten base points of the linear system then there exists a set of image velocity vectors $\dot{\mathbf{q}}_i$ with real coefficients such that q is given by (4.127).

4.5 The Derivatives of the Image Velocity Field

If a large number of image velocity vectors are available then the derivatives of the image velocity field with respect to position on the projection surface can be estimated. The derivatives encode information about the velocity of the camera and about the shape of the surface. The camera velocity can be reconstructed from the derivatives to second order evaluated at a point on the projection surface (Waxman *et al.* 1987). The difficulty with this approach to reconstruction is that it is hard to obtain accurate estimates of the second order derivatives.

In Sect. 4.5.1 expressions are obtained for the derivatives to second order of an image velocity field. There are in all twelve derivatives, six for each component of the image velocity field. In order to simplify the evaluation of the derivatives the velocity field is projected onto a plane rather than a sphere. In Sect. 4.5.2 it is shown that the derivatives yield three homogeneous polynomial constraints of degree three on the translational velocity of the rigid surface. In Sect. 4.5.3 it is shown that the time to contact of the camera with an obstacle can be obtained from the derivatives of the image velocity field to first order.

4.5.1 The Derivatives to Second Order

The expressions for the derivatives of the image velocity field on a projection plane are obtained. Coordinates x, y, z are chosen in $\mathbb{R}^3$ such that the optical centre is at the origin and such that the projection plane is given by $z = 1$. A point $\mathbf{x} = (x, y, z)^\top$ of $\mathbb{R}^3$ projects to the point $(x/z, y/z, 1)^\top$. Let X, Y be the coordinates on the projection plane, chosen such that

$$X = \frac{x}{z} \qquad\qquad Y = \frac{y}{z}$$

The derivatives of X and Y with respect to time are

$$\begin{aligned} \dot{X} &= \frac{\dot{x}}{z} - \frac{x\,\dot{z}}{z^2} \\ \dot{Y} &= \frac{\dot{y}}{z} - \frac{y\,\dot{z}}{z^2} \end{aligned} \tag{4.141}$$

Let $\mathbf{x}$ be a point on a rigid surface moving with translational velocity $\mathbf{v}$ and angular velocity $\mathbf{w}$, taken about an axis through the origin. The velocity $\dot{\mathbf{x}}$ of $\mathbf{x}$ is given by

$$\dot{\mathbf{x}} = \mathbf{v} + \mathbf{w} \times \mathbf{x} \tag{4.142}$$

It follows from (4.142) that the components $\dot{x}$, $\dot{y}$, $\dot{z}$ of $\dot{\mathbf{x}}$ are

$$\begin{aligned} \dot{x} &= v_1 + w_2 z - w_3 y \\ \dot{y} &= v_2 + w_3 x - w_1 z \\ \dot{z} &= v_3 + w_1 y - w_2 x \end{aligned} \tag{4.143}$$

Let K' be the inverse distance defined by $K' = 1/z$. The superscript $'$ is used to distinguish $K' = 1/z$ from the inverse distance $K = 1/\|\mathbf{x}\|$ defined in Sect. 4.1. It follows from (4.143) and the definition of K' that

$$\begin{aligned} z^{-1}\dot{x} &= v_1 K' + w_2 - w_3 Y \\ z^{-1}\dot{y} &= v_2 K' + w_3 X - w_1 \\ z^{-1}\dot{z} &= v_3 K' + w_1 Y - w_2 X \end{aligned} \tag{4.144}$$

On substituting the values of $z^{-1}\dot{x}$, $z^{-1}\dot{y}$, $z^{-1}\dot{z}$ given by (4.144) into (4.141) the following expressions for $\dot{X}$, $\dot{Y}$ are obtained,

$$\begin{aligned} \dot{X} &= (v_1 - X\,v_3)K' - w_3 Y + w_2(1 + X^2) - w_1 X\,Y \\ \dot{Y} &= (v_2 - Y\,v_3)K' + w_3 X - w_1(1 + Y^2) + w_2 X\,Y \end{aligned} \tag{4.145}$$

The equations (4.145) are equivalent to the equation (4.6) describing the image velocity field obtained by projection onto the unit sphere centred at the origin. On eliminating the inverse distance K' from (4.145) it follows that

$$\begin{aligned} \dot{X}(v_2 - Y\,v_3) - \dot{Y}(v_1 - X\,v_3) = w_1 v_1 + w_2 v_2 - (w_1 v_3 + w_3 v_1)X - (w_2 v_3 + w_3 v_2)Y + \\ (w_2 v_2 + w_3 v_3)X^2 - (w_1 v_2 + w_2 v_1)X\,Y + (w_1 v_1 + w_3 v_3)Y^2 \end{aligned} \tag{4.146}$$

Equation (4.145) yields expressions for the derivatives of the image velocity field to second order evaluated at $(0,0)^\top$. Let $k_1, \ldots, k_6$ be the six coefficients of the Taylor series of the inverse distance K', expanded as a function of X and Y.

$$K' = k_1 + k_2 X + k_3 Y + k_4 X^2 + k_5 X\,Y + k_6 Y^2 + R_{K'} \tag{4.147}$$

It is assumed that the absolute value of the remainder term $R_{K'}$ is third order in X, Y. This assumption ensures that the derivatives of the surface exist to second order at the point $(0,0,z)^\top$.

The Taylor coefficients of $\dot{X}$ to second order are denoted by $f_1, \ldots, f_6$ and the Taylor coefficients of $\dot{Y}$ to second order are denoted by $g_1, \ldots, g_6$. The Taylor expansions of $\dot{X}$, $\dot{Y}$ as functions of X, Y are

$$\begin{aligned} \dot{X} &= f_1 + f_2 X + f_3 Y + f_4 X^2 + f_5 X\,Y + f_6 Y^2 + R_X \\ \dot{Y} &= g_1 + g_2 X + g_3 Y + g_4 X^2 + g_5 X\,Y + g_6 Y^2 + R_Y \end{aligned} \tag{4.148}$$

where the remainder terms R_X, R_Y are each third order in X, Y. It follows from (4.145), (4.147) and (4.148) that the f_i and the g_i are given by

$$\begin{aligned}
f_1 &= v_1k_1 + w_2 \\
f_2 &= v_1k_2 - v_3k_1 \\
f_3 &= v_1k_3 - w_3 \\
f_4 &= v_1k_4 - v_3k_2 + w_2 \\
f_5 &= v_1k_5 - v_3k_3 - w_1 \\
f_6 &= v_1k_6 \\
g_1 &= v_2k_1 - w_1 \\
g_2 &= v_2k_2 + w_3 \\
g_3 &= v_2k_3 - v_3k_1 \\
g_4 &= v_2k_4 \\
g_5 &= v_2k_5 - v_3k_2 + w_2 \\
g_6 &= v_2k_6 - v_3k_3 - w_1
\end{aligned} \tag{4.149}$$

The velocity of the camera can be recovered from the full set of derivatives to second order by solving an equation of degree three for one component of the translational velocity. The coefficients of the equation are functions of the derivatives. One such equation is obtained by Longuet-Higgins & Prazdny (1980). The derivatives to first order are not sufficient to determine the camera velocity, but they do impose bounds on w_3 and on the time to contact of the camera with the surface giving rise to the image velocity field. These bounds are obtained by Subbarao (1990).

The image velocity field $(\dot{X}, \dot{Y})^\top$ produces a first order deformation of the image in the region surrounding $(0, 0)$. In a time Δt an image point with coordinates $(X, Y)^\top$ moves to the point with coordinates

$$(X, Y)^\top + \Delta t(\dot{X}, \dot{Y})^\top + O(\Delta t^2)$$

The deformation is thus given by

$$\begin{pmatrix} X \\ Y \end{pmatrix} \mapsto \begin{pmatrix} X \\ Y \end{pmatrix} + \Delta t \begin{pmatrix} f_1 \\ g_1 \end{pmatrix} + \Delta t \begin{pmatrix} f_2 & f_3 \\ g_2 & g_3 \end{pmatrix} \begin{pmatrix} X \\ Y \end{pmatrix} + \Delta t R_2 + O(\Delta t^2) \tag{4.150}$$

where R_2 is second order in X, Y.

The first order approximation to the deformation on the right-hand side of (4.150) can be written as a composition of five simpler deformations, each of which has a geometric interpretation. The simpler deformations T, S_i, $1 \le i \le 4$ are defined by

$$\begin{aligned}
T\begin{pmatrix} X \\ Y \end{pmatrix} &= \begin{pmatrix} X \\ Y \end{pmatrix} + \Delta t \begin{pmatrix} f_1 \\ g_1 \end{pmatrix} && \text{uniform shift} \\
S_1 &= I + \frac{1}{2}\Delta t(f_2 + g_3)\begin{pmatrix} 1 & 0 \\ 0 & 1 \end{pmatrix} && \text{dilatation}
\end{aligned}$$

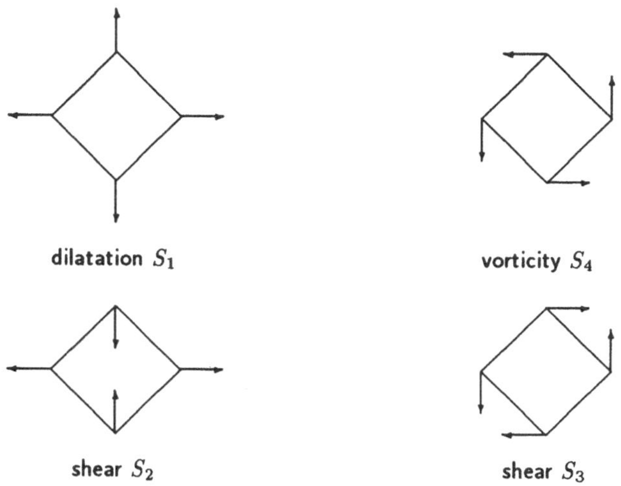

Fig. 4.7. Components of the deformation produced by the image velocities

$$
\begin{aligned}
S_2 &= I + \frac{1}{2}\Delta t(f_2 - f_3)\begin{pmatrix} 1 & 0 \\ 0 & -1 \end{pmatrix} && \text{shear} \\
S_3 &= I + \frac{1}{2}\Delta t(f_3 + g_2)\begin{pmatrix} 0 & 1 \\ 1 & 0 \end{pmatrix} && \text{shear} \\
S_4 &= I + \frac{1}{2}\Delta t(f_3 - g_2)\begin{pmatrix} 0 & 1 \\ -1 & 0 \end{pmatrix} && \text{vorticity} \qquad (4.151)
\end{aligned}
$$

The matrix I appearing in (4.151) is the 2×2 identity matrix. It follows from (4.150) and (4.151) that the first order deformation produced by the image velocity field is

$$\begin{pmatrix} X \\ Y \end{pmatrix} \mapsto TS_1S_2S_3S_4\begin{pmatrix} X \\ Y \end{pmatrix} + \Delta t R_2 + O(\Delta t^2)$$

The dilatation and the two components of shear are all independent of the angular velocity $\mathbf{w}$ of the rigid surface giving rise to the image velocity field. The vorticity depends on the component w_3 of $\mathbf{w}$ parallel to the z axis, but it is independent of w_1 and w_2. The dilatation and the vorticity are unchanged if the X and Y coordinate axes are rotated. The two shear components transform together under a rotation. The qualitative appearance of the deformations S_i is shown in Fig. 4.7, taken from Longuet-Higgins & Prazdny (1980). The decomposition of the first order image deformation into dilatation, shear and vorticity is described by Koenderink & Van Dorn (1976) and Subbarao (1990).

4.5.2 Polynomial Constraints on the Translational Velocity

The equations (4.149) for the derivatives of the image velocity field have formed the basis of numerous algorithms for estimating the camera velocity. A typical algorithm is described by Waxman *et al.* (1987). The translational velocity $\mathbf{v}$ is obtained by solving a cubic polynomial equation. In this subsection a method is given for eliminating $\mathbf{w}$ and the Taylor coefficients k_i from the equations (4.149). After the elimination there remain three polynomial equations in the translational velocity $\mathbf{v}$.

The k_i are eliminated immediately by beginning with equation (4.146) rather than with the two equations of (4.145). The Taylor expansions (4.148) of $\dot{X}$, $\dot{Y}$ are substituted into (4.146) and the coefficients of $1, X, Y, X^2, XY, Y^2$ on the left-hand side of (4.146) are equated to the corresponding coefficients on the right-hand side of (4.146) to yield

$$\begin{aligned}
w_1v_1 + w_2v_2 &= -g_1v_1 + f_1v_2 \\
w_1v_3 + w_3v_1 &= g_2v_1 - f_2v_2 - g_1v_3 \\
w_2v_3 + w_3v_2 &= g_3v_1 - f_3v_2 + f_1v_3 \\
w_2v_2 + w_3v_3 &= -g_4v_1 + f_4v_2 + g_2v_3 \\
w_1v_2 + w_2v_1 &= g_5v_1 - f_5v_2 + (f_2 - g_3)v_3 \\
w_1v_1 + w_3v_3 &= -g_6v_1 + f_6v_2 - f_3v_3
\end{aligned} \tag{4.152}$$

The vector $\mathbf{w}$ is eliminated from (4.152). The first three equations of (4.152) are written in matrix form as $\Xi\mathbf{w} = \mathbf{a}$ and the remaining three equations are written in matrix form as $\Upsilon\mathbf{w} = \mathbf{b}$, where the matrices Ξ, Υ are defined by

$$\Xi = \begin{pmatrix} v_1 & v_2 & 0 \\ v_3 & 0 & v_1 \\ 0 & v_3 & v_2 \end{pmatrix} \qquad \Upsilon = \begin{pmatrix} 0 & v_2 & v_3 \\ v_2 & v_1 & 0 \\ v_1 & 0 & v_3 \end{pmatrix}$$

and the vectors $\mathbf{a}$, $\mathbf{b}$ are defined by

$$\mathbf{a} = \begin{pmatrix} -g_1v_1 + f_1v_2 \\ g_2v_1 - f_2v_2 - g_1v_3 \\ g_3v_1 - f_3v_2 + f_1v_3 \end{pmatrix} \qquad \mathbf{b} = \begin{pmatrix} -g_4v_1 + f_4v_2 + g_2v_3 \\ g_5v_1 - f_5v_2 + (f_2 - g_3)v_3 \\ -g_6v_1 + f_6v_2 - f_3v_3 \end{pmatrix}$$

It follows from the equations $\Xi\mathbf{w} = \mathbf{a}$ and $\Upsilon\mathbf{w} = \mathbf{b}$ that

$$\det(\Xi)\mathbf{w} = \text{Adjoint}(\Xi)\mathbf{a} \qquad \det(\Upsilon)\mathbf{w} = \text{Adjoint}(\Upsilon)\mathbf{b} \tag{4.153}$$

A direct calculation yields

$$\det(\Xi) = \det(\Upsilon) = -v_3(v_1^2 + v_2^2)$$

The subtraction of the second equation of (4.153) from the first yields

$$\text{Adjoint}(\Xi)\mathbf{a} - \text{Adjoint}(\Upsilon)\mathbf{b} = 0 \tag{4.154}$$

Any vector $\mathbf{v}$ that satisfies (4.152) also satisfies (4.154). The converse holds provided Ξ and Υ are invertible, that is, provided $v_3(v_1^2 + v_2^2) \neq 0$. The algebraic manipulations leading from (4.152) to (4.154) have introduced the spurious solution $\mathbf{v} = (0,0,1)^\top$. The value $(0,0,1)^\top$ for $\mathbf{v}$ satisfies (4.154), but it does not, in general, lead to a solution of the equations (4.152).

The adjoints of Ξ and Υ are

$$\text{Adjoint}(\Xi) = \begin{pmatrix} -v_1v_3 & -v_2^2 & v_1v_2 \\ -v_2v_3 & v_1v_2 & -v_1^2 \\ v_3^2 & -v_1v_3 & -v_2v_3 \end{pmatrix}$$

$$\text{Adjoint}(\Upsilon) = \begin{pmatrix} v_1v_3 & -v_2v_3 & -v_1v_3 \\ -v_2v_3 & -v_1v_3 & v_2v_3 \\ -v_1^2 & v_1v_2 & -v_2^2 \end{pmatrix}$$

The three components of the vector equation (4.154) are expressed in the form $p_1 = 0$, $p_2 = 0$, $p_3 = 0$, where the p_i are the following homogeneous cubic polynomials in $\mathbf{v}$:

$$p_1(\mathbf{v}) = v_3^2[(f_2 - g_3)v_2 - (f_3 + g_2)v_1] +$$

$$v_3[(g_1 - f_5)v_2^2 + (f_6 - f_4 + g_5)v_2v_1 + (g_1 + g_4 - g_6)v_1^2] + v_2^2(f_2v_2 - g_2v_1) - v_2v_1(f_3v_2 - g_3v_1)$$

$$p_2(\mathbf{v}) = v_3^2[(g_2 + f_3)v_2 + (f_2 - g_3)v_1] +$$

$$v_3[(f_4 - f_1 - f_6)v_2^2 + (g_6 - f_5 - g_4)v_2v_1 + (g_5 - f_1)v_1^2] - v_2v_1(f_2v_2 - g_2v_1) + v_1^2(f_3v_2 - g_3v_1)$$

$$p_3(\mathbf{v}) = v_2^3 f_6 + v_2^2 v_1(f_5 - g_6) + v_2v_1^2(f_4 - g_5) - v_1^3 g_4$$

Each condition $p_i = 0$, $i = 1,2,3$ can be interpreted as a constraint which must hold if a value of the corresponding component w_i of $\mathbf{w}$ is to be found consistent with the spatial derivatives of the image velocity field to second order. The equation $p_3 = 0$ is given by Longuet-Higgins & Prazdny (1980). It reduces to a cubic polynomial equation in the ratio v_1/v_2. If at least one coefficient of p_3 is non-zero then it has at most three distinct roots. In the general case only one of the roots yields a velocity $\mathbf{v}$ that is compatible with the two equations $p_1 = 0$ and $p_2 = 0$.

A fourth polynomial, p_4, is obtained from p_1 and p_2,

$$p_4(\mathbf{v}) = v_3^{-1}(p_1(\mathbf{v}) + v_2 p_2(\mathbf{v}))$$

The polynomial p_4 defines a cubic plane curve with a node. The advantage of p_4 over p_1 and p_2 is that it has an algebraic parameterisation by $\mathbb{P}^1$. This parameterisation can be used to search for the common zeros of p_1 and p_2. The full expression for p_4 is

$$p_4(\mathbf{v}) = v_3[2(f_2 - g_3)v_2v_1 + (f_3 + g_2)(v_2^2 - v_1^2)] + v_2^3(f_4 - f_1 - f_6)$$

$$+ v_2^2v_1(g_1 + g_6 - 2f_5 - g_4) + v_2v_1^2(2g_5 - f_4 + f_6 - f_1) + v_1^3(g_1 + g_4 - g_6) \quad (4.155)$$

The common zeros of p_1 and p_2 lie on p_4, with the possible exception of any zeros for which $v_3 = 0$. It is shown below that there are no special difficulties associated with the line $v_3 = 0$ in that if $\hat{\mathbf{v}} = (\hat{v}_1, \hat{v}_2, 0)^\top$ is a translational velocity compatible with the derivatives of the image velocity field to second order then $p_4(\hat{\mathbf{v}}) = 0$.

Proposition 4.32. *The homogeneous polynomial p_4 has a singular point at $(0,0,1)^\top$. If either $f_3 + g_2 \neq 0$ or $f_2 - g_3 \neq 0$, then the singular point is a simple node and the two tangents at the node are at right angles.*

Proof. The polynomial p_4 has a singular point at $(0,0,1)^\top$ because there are no terms on the right-hand side of (4.155) of degree one or less in v_1 or v_2. In order to find the nature of the singular point spherical polar coordinates are used,

$$\begin{aligned} v_1 &= \sin(\theta)\cos(\phi) \\ v_2 &= \sin(\theta)\sin(\phi) \\ v_3 &= \cos(\theta) \end{aligned}$$

On substituting for $\mathbf{v}$ on the right-hand side of (4.155) it follows that

$$\begin{aligned} p_4(\mathbf{v}) &= \sin^2(\theta)[2(f_2 - g_3)\cos(\phi)\sin(\phi) + (f_3 + g_2)(\sin^2(\phi) - \cos^2(\phi))] \\ &\quad + O(\theta^3) \end{aligned} \tag{4.156}$$

The hypotheses ensure that the coefficient of $\sin^2(\theta)$ on the right-hand side of (4.156) is not identically zero. The nature of the singular point is determined by the term of p_4 which is second order in θ. The tangents to p_4 at $(0,0,1)^\top$ correspond to those values of ϕ for which the coefficient of $\sin^2(\theta)$ in (4.156) is equal to zero,

$$2(f_2 - g_3)\cos(\phi)\sin(\phi) + (f_3 + g_2)[\sin^2(\phi) - \cos^2(\phi)] = 0$$

It follows that the values of ϕ yielding the two tangent lines satisfy

$$\tan(2\phi) = \frac{f_3 + g_2}{f_2 - g_3} \tag{4.157}$$

Equation (4.157) for ϕ has two solutions, differing by 90° as required. □

Corollary. The cubic plane curve p_4 is a rational curve parameterised by the pencil of lines centred at the node of p_4.

It is shown that if $\hat{\mathbf{v}} = (\hat{v}_1, \hat{v}_2, 0)^\top$ is a translational velocity of a rigid surface giving rise to the image velocity field then $p_4(\hat{\mathbf{v}}) = 0$. It follows from (4.155) that

$$\begin{aligned} &p_4(\hat{\mathbf{v}}) = p_4(\hat{v}_1, \hat{v}_2, 0) \\ &= \hat{v}_2^3(f_4 - f_1 - f_6) + \hat{v}_2^2\hat{v}_1(g_1 + g_6 - 2f_5 - g_4) + \hat{v}_2\hat{v}_1^2(2g_5 - f_4 + f_6 - f_1) + \hat{v}_1^3(g_1 + g_4 - g_6) \end{aligned} \tag{4.158}$$

The coefficients on the right-hand side of (4.158) are evaluated using (4.149), to yield

$$\begin{aligned} f_4 - f_1 - f_6 &= \hat{v}_1(k_4 - k_1 - k_6) \\ g_1 + g_6 - 2f_5 - g_4 &= \hat{v}_2(k_1 + k_6 - k_4) - 2\hat{v}_1 k_5 \\ 2g_5 - f_4 + f_6 - f_1 &= 2\hat{v}_2 k_5 - \hat{v}_1(k_4 - k_6 + k_1) \\ g_1 + g_4 - g_6 &= v_2(k_1 + k_4 - k_6) \end{aligned} \tag{4.159}$$

It follows from (4.158) and (4.159) that

$$\begin{aligned} p_4(\hat{\mathbf{v}}) &= \hat{v}_2^3\hat{v}_1(k_4 - k_1 - k_6) + \hat{v}_2^3 v_1(k_1 + k_6 - k_4) - \hat{v}_2^2\hat{v}_1^3(k_4 - k_6 + k_1) \\ &\quad + \hat{v}_2\hat{v}_1^3(k_1 + k_4 - k_6) \\ &= 0 \end{aligned}$$

4.5.3 Time to Contact

If the camera is heading towards an obstacle then the time to contact can be estimated from the image velocity field, even though the distance to the obstacle and the magnitude of the velocity of the camera cannot be estimated separately. The distance to the obstacle and the magnitude of the velocity are both determined up to a single unknown scale factor. This scale factor cancels out when the ratio τ of the distance to the velocity is taken. The ratio τ is the estimate of the time to contact. If the camera is heading straight for an obstacle with a constant translational velocity then τ is exactly the time that elapses before collision.

Let r be the distance to the obstacle and let v_3 be the component of the translational velocity of the obstacle away from the camera. Then τ is given by the equation $\tau = -r/v_3$. The minus sign is inserted to ensure that the time to contact is positive if the gap between the vehicle and the obstacle is decreasing. The value of τ can be estimated from the changes in area produced by the image velocities. Let A be the area of a region S in the image arising from the projection of an identifiable region on the rigid surface giving rise to the image velocity field. The area A of S is a function of time t. The fractional change in A during a short interval of time Δt is, by definition,

$$\frac{A(t + \Delta t) - A(t)}{A(t)} \tag{4.160}$$

Let $\dot{A}$ be the derivative of A with respect to time. Then the fractional change in A is

$$\frac{A(t + \Delta t) - A(t)}{A(t)} = \frac{\dot{A}}{A}\Delta t + O(\Delta t^2) \tag{4.161}$$

The first order deformation of S produced by the image velocity field is described by the uniform shift T and the elementary deformations S_i, $1 \leq i \leq 4$, defined in (4.151). The only one of T, S_i to produce a change in area is the dilatation S_1. It follows from the definition of S_1 that

$$A(t+\Delta t) = \det(S_1)A + \Delta t\, R_3 + O(\Delta t^2)$$

where R_3 is third order in X, Y. The determinant of S_1 is obtained from (4.151),

$$\det(S_1) = 1 + (f_2 + g_3)\Delta t + O(\Delta t^2)$$

It follows that

$$\begin{aligned} \frac{A(t+\Delta t) - A(t)}{A(t)} &= \frac{(\det(S_1) - 1)A(t) + \Delta t\, R_3}{A(t)} + O(\Delta t^2) \\ &= (f_2 + g_3)\Delta t + \Delta t A^{-1} R_3 + O(\Delta t^2) \end{aligned} \qquad (4.162)$$

It follows from (4.161) and (4.162) that

$$\frac{\dot{A}}{A} = f_2 + g_3 + A^{-1}R_3 + O(\Delta t) \qquad (4.163)$$

An expression for the time to contact is obtained by using the equations of (4.149) to substitute for f_2 and g_3 in (4.163). The substitution yields

$$\frac{\dot{A}}{A} = -2v_3k_1 + v_1k_2 + v_2k_3 + A^{-1}R_3 + O(\Delta t)$$

The Taylor coefficients k_i are defined by (4.147). The coefficient k_1 is the inverse distance to the rigid surface. If the camera is heading towards the surface then v_1 and v_2 are small, but v_3 is large. Under these conditions

$$\begin{aligned} \frac{\dot{A}}{A} &\sim -2v_3k_1 \\ &\sim 2\tau \end{aligned}$$

If the area A is increasing then the time to contact is positive. If the area is decreasing then the time to contact is negative. In this case the object is receding from the camera. Further details can be found in Maybank (1987a).

References

Grzywacz N.M. & Hildreth E.C. 1987 Incremental rigidity scheme for recovering structure from motion: position-based versus velocity-based methods. *J. Optical Soc. America*, Series A **4**, 503-518.

Hardy G.H. 1967 *A Course of Pure Mathematics*. Cambridge: Cambridge University Press.

Horn B.K.P. 1987 Motion fields are hardly ever ambiguous. *International J. Computer Vision* **1**, 263-278.

Jerian C. & Jain R. 1990 Polynomial methods for structure from motion. *IEEE Trans. Pattern Analysis and Machine Intelligence* **12**, 1150-1166.

Koenderink J.J. & Van Dorn A.J. 1976 Local structure of movement parallax of the plane. *J. Optical Soc. America*, **66**, 717-723.

Longuet-Higgins H.C. 1984 The visual ambiguity of a moving plane. *Proc. Royal Soc. London, Series B* **223**, 165-175.

Longuet-Higgins H.C. & Prazdny K. 1980 The interpretation of a moving retinal image. *Proc. Royal Soc. Lond, Series B* **208**, 385-397.

Maybank S.J. 1985 The angular velocity associated with the optical flow field due to a rigid moving body. *Proc. Royal Soc. London, Series A* **401**, 317-326.

Maybank S.J. 1987a Apparent area of a rigid moving body. *Image and Vision Computing* **15**, 111-113.

Maybank S.J. 1987b *A theoretical study of optical flow.* PhD thesis, University of London, Birkbeck College.

Maybank S.J. 1990 Rigid velocities compatible with five image velocity vectors. *Image and Vision Computing* **8**, 18-23.

Murray D.W. & Buxton B.F. 1990 *Experiments in the Machine Interpretation of Visual Motion.* Cambridge, Massachusetts: The MIT Press.

Semple J.G. & Roth R. 1949 *Introduction to Algebraic Geometry.* Oxford: Clarendon Press, reprinted 1985.

Sokolnikoff I.S. & Redheffer R.M. 1966 *Mathematics of Physics and Modern Engineering.* USA: McGraw-Hill.

Subbarao M. 1990 Bounds on time-to-collision and rotational component from first-order derivatives of image flow. *Computer Vision, Graphics, and Image Processing* **50**, 329-341.

Waxman A.M., Kamgar-Parsi B. & Subbarao M. 1987 Closed form solutions to image flow equations for 3D structure and motion. *International J. Computer Vision* **1**, 239-258.

5 Reconstruction from Minimal Data

It is shown in Sect. 2.2.1 that reconstruction from image correspondences is algebraic in that it can be expressed as the problem of finding the common zeros of a set of polynomial equations. This fact has profound consequences, especially if the number n of image correspondences is small. The algebraic nature of reconstruction is particularly apparent in the minimal case when n is just sufficient to ensure that only a finite number of reconstructions are possible. In reconstruction with known camera calibration five pairs of corresponding points are sufficient to ensure at most a finite number of reconstructions. The minimal number for reconstruction up to a collineation is seven. Reconstruction from image velocities is also an algebraic problem and the minimal number of image velocity vectors is five. The term 'minimal data' is used to refer either to point correspondences or to image velocities, depending on the context. The algebraic properties of reconstruction are important in applications because they have to be taken into account by any algorithm for reconstruction which makes full use of the available information.

It follows immediately from the algebraic nature of reconstruction that in the minimal case the number N of reconstructions is constant for almost all choices of the data. The total N includes both real and complex reconstructions. The complex reconstructions are not physically realisable, but they are an inescapable feature of the algebra. If for some particular choice of minimal data a total of N' solutions are obtained such that $N' \neq N$ then either $N' > N$ and there are infinitely many solutions or $N' < N$ and at least one of the N' solutions is a repeated solution. The number N is defined to be the degree of reconstruction.

In reconstruction from image correspondences and reconstruction from image velocities the degree N is ten. In reconstruction up to a collineation the degree N is three. A degree of ten is high, indicating that reconstruction is difficult regardless of the algorithm employed. The degree three for reconstruction up to a collineation suggests a more tractable algebraic problem.

This chapter is divided into four sections. In Sects. 5.1 and 5.2 two contrasting methods for reconstruction are described. Each method uses the minimum number, five, of image correspondences. One method is due to Kruppa (1913) and the other is due to Demazure (1988). Kruppa's method is based on projective geometry and the epipolar transformation. In contrast, Demazure uses the algebraic variety of essential matrices in $\mathbf{P}^8$. It turns out that the degree of reconstruction from image correspondences is equal to the degree of the variety of essential matrices. In Sect. 5.3 a method due to Sturm (1869) for reconstruction up to a collineation is described. Sturm uses projective geometric methods to show that there are exactly three pairs of epipoles compatible with seven image correspondences. In Sect. 5.4 reconstruction from image velocities is described. Two proofs are given of the fact that there are, in general, exactly ten rigid velocities compatible with five image velocity vectors. The first proof is based on the quartic polynomial constraint on the translational velocity obtained in Sect. 4.3 and the second proof is based on the theory of critical surfaces.

5.1 Kruppa's Method

Kruppa (1913) uses the properties of the epipolar transformation to recover the epipoles compatible with five image correspondences. The epipolar transformation is described in Sect. 2.3.1. Let $\mathbf{p}$ be the epipole in the first image and let $\mathbf{p}'$ be the epipole in the second image. In the first part of Kruppa's method two polynomial constraints are obtained on the coordinates of $\mathbf{p}$ and $\mathbf{p}'$. These constraints arise from the fact that the epipolar transformation is a homography, preserving the cross ratio between epipolar lines. Two further constraints on $\mathbf{p}$ and $\mathbf{p}'$ are obtained from the camera calibration. There are thus four constraints on the four unknown variables specifying the positions of $\mathbf{p}$ and $\mathbf{p}'$. From these four constraints two sextic (degree six) plane curves are obtained, such that the epipoles $\mathbf{p}$ compatible with the image correspondences arise from the intersections of these two curves. Not every intersection yields an epipole. Certain of the intersections produce spurious solutions which are introduced by the algebraic manipulations required to obtain the sextic curves.

In the description of Kruppa's method references to the second image are indicated by adding a $'$ as a superscript. For example, $\mathbf{p}$, $\mathbf{p}'$ are the epipoles in the first and second images respectively. It is recalled from Sect. 1.4.2 that $\overline{\wedge}$ denotes a homographic correspondence.

5.1.1 The Homography

Let five image correspondences $\mathbf{q}_i \leftrightarrow \mathbf{q}_i'$, $1 \le i \le 5$, be given in general position and let projective coordinates be chosen in the two image planes such that the coordinates of $\mathbf{q}_i$, $\mathbf{q}_i'$ for $1 \le i \le 4$ are given by

$$
\begin{aligned}
\mathbf{q}_1 &= (1,0,0)^\top & \mathbf{q}_1' &= (1,0,0)^\top \\
\mathbf{q}_2 &= (0,1,0)^\top & \mathbf{q}_2' &= (0,1,0)^\top \\
\mathbf{q}_3 &= (0,0,1)^\top & \mathbf{q}_3' &= (0,0,1)^\top \\
\mathbf{q}_4 &= (1,1,1)^\top & \mathbf{q}_4' &= (1,1,1)^\top
\end{aligned} \tag{5.1}
$$

The coordinates of $\mathbf{q}_5$ and $\mathbf{q}'_5$ are

$$\mathbf{q}_5 = (q_1, q_2, q_3)^\top \qquad \mathbf{q}'_5 = (q'_1, q'_2, q'_3)^\top$$

Let the q_{ij} be defined by $q_{ij} = (\mathbf{q}_i)_j$ and let l_i, l'_i be the epipolar lines defined by

$$l_i = \langle \mathbf{q}_i, \mathbf{p} \rangle \qquad l'_i = \langle \mathbf{p}', \mathbf{q}'_i \rangle \qquad (1 \le i \le 5) \tag{5.2}$$

In the next theorem two constraints on $\mathbf{p}$, $\mathbf{p}'$ are obtained using the fact that the epipolar transformation preserves cross ratios.

Theorem 5.1. *The lines l_i, l'_i defined by (5.2) are in homographic correspondence, $l_i \barwedge l'_i$, if and only if*

$$\begin{aligned} \frac{p_2(p_1 - p_3)}{p_1(p_2 - p_3)} &= \frac{p'_2(p'_1 - p'_3)}{p'_1(p'_2 - p'_3)} \\ \frac{p_2(p_1 q_3 - p_3 q_1)}{p_1(p_2 q_3 - p_3 q_2)} &= \frac{p'_2(p'_1 q'_3 - p'_3 q'_1)}{p'_1(p'_2 q'_3 - p'_3 q'_2)} \end{aligned} \tag{5.3}$$

Proof. A point $\mathbf{x}$ is on the line l_i if and only if $(\mathbf{p} \times \mathbf{q}_i).\mathbf{x} = 0$. The pencil of lines through $\mathbf{p}$ is parameterised by taking the intersection of each line with the fixed line $\langle \mathbf{q}_1, \mathbf{q}_2 \rangle$. The line l_i intersects $\langle \mathbf{q}_1, \mathbf{q}_2 \rangle$ at the point $\mathbf{x}_i$ defined by

$$\begin{aligned} \mathbf{x}_i &= (\mathbf{p} \times \mathbf{q}_i) \times (\mathbf{q}_1 \times \mathbf{q}_2) \\ &= [(\mathbf{p} \times \mathbf{q}_i).\mathbf{q}_2]\mathbf{q}_1 - [(\mathbf{p} \times \mathbf{q}_i).\mathbf{q}_1]\mathbf{q}_2 \end{aligned} \tag{5.4}$$

The coefficients of $\mathbf{q}_1$ and $\mathbf{q}_2$ on the right-hand side of (5.4) furnish projective coordinates for l_i. In effect, l_i is represented by the following point of $\mathbb{P}^1$,

$$((\mathbf{p} \times \mathbf{q}_i).\mathbf{q}_2, -(\mathbf{p} \times \mathbf{q}_i).\mathbf{q}_1)^\top$$

The details of the choice of parameterisation of the epipolar lines do not affect the final equations, because the equations are based on cross ratios. The cross ratio of four lines is independent of the choice of parameterisation. However, a good choice of parameterisation simplifies the calculations leading to the equations.

Let the inhomogeneous coordinate θ_i of l_i be defined by

$$\theta_i = -\frac{(\mathbf{p} \times \mathbf{q}_i).\mathbf{q}_1}{(\mathbf{p} \times \mathbf{q}_i).\mathbf{q}_2} \tag{5.5}$$

It follows from (5.1) and (5.5) that

$$\theta_1 = 0 \qquad \theta_2 = \infty \qquad \theta_3 = \frac{p_2}{p_1} \tag{5.6}$$

Let τ be the cross ratio of the four lines through $\mathbf{p}$ with inhomogeneous coordinates θ_i, $1 \le i \le 4$. It follows from the definition of the cross ratio that

$$\begin{aligned}\tau &= \left(\frac{\theta_1 - \theta_3}{\theta_2 - \theta_3}\right) \Big/ \left(\frac{\theta_1 - \theta_4}{\theta_2 - \theta_4}\right) \\ &= \theta_3/\theta_4 \\ &= p_2/(p_1\theta_4) \end{aligned} \tag{5.7}$$

It follows from the choice of coordinates (5.1) and (5.2) that

$$\theta_4 = \frac{p_3 - p_2}{p_3 - p_1} \tag{5.8}$$

Equations (5.7) and (5.8) yield

$$\tau = \frac{p_2}{p_1}\left(\frac{p_3 - p_1}{p_3 - p_2}\right) \tag{5.9}$$

Similarly, the cross ratio τ' of the lines l_i', $1 \le i \le 4$, is given by

$$\tau' = \frac{\theta_3'}{\theta_4'} = \frac{p_2'(p_1' - p_3')}{p_1'(p_2' - p_3')} \tag{5.10}$$

The first equation of (5.3) is obtained from (5.9) and (5.10) by setting $\tau = \tau'$. The second equation of (5.3) is obtained in a similar way by equating the cross ratio of the lines l_1, l_2, l_3, l_5 to the cross ratio of the lines l_1', l_2', l_3', l_5'.

It remains to show that if (5.3) holds then $l_i \barwedge l_i'$. Let A be the matrix defining the unique homographic correspondence between the points of $\langle \mathbf{q}_1, \mathbf{q}_2 \rangle$ and the points of $\langle \mathbf{q}_1', \mathbf{q}_2' \rangle$ such that

$$A\begin{pmatrix}1\\ \theta_i\end{pmatrix} = \begin{pmatrix}1\\ \theta_i'\end{pmatrix} \qquad (1 \le i \le 3) \tag{5.11}$$

It follows from (5.6) and (5.11) that

$$A = \begin{pmatrix} p_1'p_2 & 0 \\ 0 & p_2'p_1 \end{pmatrix} \tag{5.12}$$

Equations (5.12), (5.8) and the first equation of (5.3) yield

$$\begin{aligned} A\begin{pmatrix}1\\ \theta_4\end{pmatrix} &= \begin{pmatrix} p_1'p_2 \\ p_2'p_1\theta_4\end{pmatrix} \\ &= \begin{pmatrix} p_1'p_2(p_3 - p_1) \\ p_2'p_1(p_3 - p_2)\end{pmatrix} \\ &= \begin{pmatrix} p_1' - p_3' \\ p_2' - p_3'\end{pmatrix} \\ &= \begin{pmatrix}1\\ \theta_4'\end{pmatrix} \end{aligned}$$

Similarly, (5.12) and the second equation of (5.3) yield

$$A\begin{pmatrix}1\\ \theta_5\end{pmatrix}=\begin{pmatrix}1\\ \theta_5'\end{pmatrix}$$

The linear transformation A extends to a collineation from the first image to the second image, also denoted by A, with the property $A(l_i) = l_i'$. It follows that $l_i \barwedge l_i'$. □

The two equations of (5.3) are simplified by applying the the quadratic transformations Φ, Φ' defined by

$$\begin{aligned}\mathbf{u} &= \Phi(\mathbf{p}) &&= (p_2p_3, p_3p_1, p_1p_2)^{\mathsf{T}}\\ \mathbf{u}' &= \Phi'(\mathbf{p}') &&= (p_2'p_3', p_3'p_1', p_1'p_2')^{\mathsf{T}}\end{aligned} \tag{5.13}$$

On applying Φ and Φ' to (5.3) and cancelling u_1u_2 and $u_1'u_2'$ from the numerator and denominator of the left and right hand sides respectively the following two equations are obtained:

$$\frac{u_3-u_1}{u_3-u_2}=\frac{u_3'-u_1'}{u_3'-u_2'} \tag{5.14}$$

$$\frac{u_3q_3-u_1q_1}{u_3q_3-u_2q_2}=\frac{u_3'q_3'-u_1'q_1'}{u_3'q_3'-u_2'q_2'} \tag{5.15}$$

Theorem 5.2. *Let* $\mathbf{u}$, $\mathbf{u}'$ *be points of* $\mathbb{P}^2$ *such that the coordinates of* $\mathbf{u}$, $\mathbf{u}'$ *satisfy (5.14) and (5.15). Let* Σ *be the map defined by* $\Sigma(\mathbf{u}) = \mathbf{u}'$. *Then* Σ *is a quadratic transformation.*

Proof. Define the vectors $\mathbf{a}$, $\mathbf{b}$, $\mathbf{c}$, $\mathbf{d}$ by

$$\mathbf{a}=(-1,0,1)^{\mathsf{T}} \qquad \mathbf{c}=(-q_1,0,q_3)^{\mathsf{T}}$$

$$\mathbf{b}=(0,-1,1)^{\mathsf{T}} \qquad \mathbf{d}=(0,-q_2,q_3)^{\mathsf{T}}$$

The equations (5.14) and (5.15) have the form

$$\frac{\mathbf{a}.\mathbf{u}}{\mathbf{b}.\mathbf{u}}=\frac{\mathbf{a}'.\mathbf{u}'}{\mathbf{b}'.\mathbf{u}'} \qquad \frac{\mathbf{c}.\mathbf{u}}{\mathbf{d}.\mathbf{u}}=\frac{\mathbf{c}'.\mathbf{u}'}{\mathbf{d}'.\mathbf{u}'} \tag{5.16}$$

On multiplying out the denominators of the equations of (5.16) and rearranging terms it follows that

$$\begin{aligned}[(\mathbf{b}.\mathbf{u})\mathbf{a}'-(\mathbf{a}.\mathbf{u})\mathbf{b}'].\mathbf{u}' &= 0\\ [(\mathbf{d}.\mathbf{u})\mathbf{c}'-(\mathbf{c}.\mathbf{u})\mathbf{d}'].\mathbf{u}' &= 0\end{aligned} \tag{5.17}$$

Equation (5.17) yields

$$\Sigma(\mathbf{u})=\mathbf{u}'=[(\mathbf{b}.\mathbf{u})\mathbf{a}'-(\mathbf{a}.\mathbf{u})\mathbf{b}']\times[(\mathbf{d}.\mathbf{u})\mathbf{c}'-(\mathbf{c}.\mathbf{u})\mathbf{d}'] \tag{5.18}$$

In (5.18) the components of $\mathbf{u}'$ are obtained as homogeneous quadratic forms in the components of $\mathbf{u}$. The symmetry between $\mathbf{u}$ and $\mathbf{u}'$ ensures that the components of $\mathbf{u}$ are obtainable as homogeneous quadratic forms in the components of $\mathbf{u}'$. It follows that Σ is quadratic and invertible. Thus Σ satisfies the definition of a quadratic transformation. □

5.1.2 Constraints Arising from the Camera Calibration

Two further constraints on $\mathbf{p}$, $\mathbf{p}'$ are obtained using the information contained in the camera calibration. These two constraints, together with (5.14) and (5.15), are sufficient to ensure that there are only a finite number of pairs of epipoles $\mathbf{p}$, $\mathbf{p}'$.

The camera calibration determines the image ω of the absolute conic Ω, and conversely, specifying the image ω of Ω is equivalent to specifying the camera calibration. The details are given in Sect. 3.1.1. The tangents drawn from $\mathbf{p}$ to ω correspond under the epipolar transformation to the tangents drawn from $\mathbf{p}'$ to ω'. This correspondence of the tangent epipolar lines is not a special property of Ω. It holds for a general curve c in $\mathbb{P}^3$. In detail, let s, s' be the curves obtained by projecting c into the first and second images respectively. A line l in the first image through $\mathbf{p}$ tangent to s is the projection of a unique epipolar plane Π. The plane Π is tangent to c, because l is tangent to s. The line l' obtained by projecting Π into the second image is tangent to s' and $l \barwedge l'$ because both lines are projections of the same epipolar plane Π.

Let D be the matrix of ω in the coordinate system chosen such that the image points $\mathbf{q}_i$, $1 \le i \le 4$, have the coordinates given in (5.1). A point $\mathbf{x}$ is on ω if and only if $\mathbf{x}^\top A\mathbf{x} = 0$. Coefficients δ_{ij}, δ_i are defined by the following equation to agree with the notation of Kruppa (1913):

$$\text{Adjoint}(D) = \begin{pmatrix} -\delta_{23} & \delta_3 & \delta_2 \\ \delta_3 & -\delta_{13} & \delta_1 \\ \delta_2 & \delta_1 & -\delta_{12} \end{pmatrix}$$

The matrix, Adjoint(D), defines the dual conic to ω. A line $\mathbf{l}$ with the equation $\mathbf{l}.\mathbf{x} = 0$ is tangent to ω if and only if $\mathbf{l}^\top\text{Adjoint}(D)\mathbf{l} = 0$. As in Sect. 5.1.1, the epipolar lines are parameterised by their points of intersection with the line $x_3 = 0$. The line joining $\mathbf{p}$ to the point $\mathbf{x} = (x_1, x_2, 0)^\top$ is tangent to ω, as illustrated in Fig. 5.1, if and only if the line $\langle\mathbf{p}, \mathbf{x}\rangle$ is on the dual conic to ω. The vector of coefficients of $\langle\mathbf{p}, \mathbf{x}\rangle$ is $\mathbf{p} \times \mathbf{x}$. The condition that $\langle\mathbf{p}, \mathbf{x}\rangle$ is tangent to ω is thus

$$(\mathbf{p} \times \mathbf{x})^\top\text{Adjoint}(A)(\mathbf{p} \times \mathbf{x}) = 0 \tag{5.19}$$

The expansion of (5.19) yields

$$A_{11}x_1^2 + 2A_{12}x_1x_2 + A_{22}x_2^2 = 0 \tag{5.20}$$

where the A_{ij} are the following quadratic functions of $\mathbf{p}$:

$$\begin{aligned} A_{11} &= \delta_{12}p_2^2 + \delta_{31}p_3^2 + 2\delta_1 p_2 p_3 \\ A_{22} &= \delta_{12}p_1^2 + \delta_{23}p_3^2 + 2\delta_2 p_1 p_3 \\ A_{12} &= \delta_3 p_3^2 - \delta_{12}p_1p_2 - \delta_1 p_1 p_3 - \delta_2 p_2 p_3 \end{aligned} \tag{5.21}$$

An equation

$$A'_{11}x_1'^2 + 2A'_{12}x'_1x'_2 + A'_{22}x_2'^2 = 0 \tag{5.22}$$

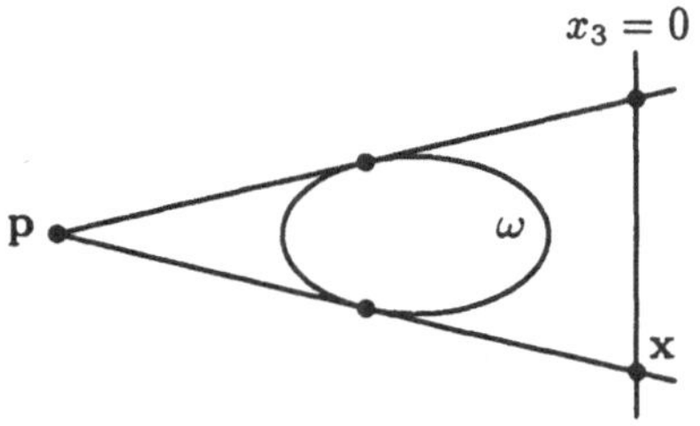

Fig. 5.1. The tangents to ω drawn from **p**

similar in form to (5.20) is obtained from the second image.

Theorem 5.3. *The points* $\mathbf{u} = \Phi(\mathbf{p})$, $\mathbf{u}' = \Phi'(\mathbf{p}')$ *satisfy the following two homogeneous polynomial equations:*

$$\frac{\delta_{12}u_3^2 + \delta_{13}u_2^2 + 2\delta_1 u_3 u_2}{\delta_3 u_1 u_2 - \delta_{12}u_3^2 - \delta_1 u_2 u_3 - \delta_2 u_1 u_3} = \frac{\delta'_{12}u_3'^2 + \delta'_{13}u_2'^2 + 2\delta'_1 u'_3 u'_2}{\delta'_3 u'_1 u'_2 - \delta'_{12}u_3'^2 - \delta'_1 u'_2 u'_3 - \delta'_2 u'_1 u'_3} \tag{5.23}$$

$$\frac{\delta_{12}u_3^2 + \delta_{23}u_1^2 + 2\delta_2 u_1 u_3}{\delta_3 u_1 u_2 - \delta_{12}u_3^2 - \delta_1 u_2 u_3 - \delta_2 u_1 u_3} = \frac{\delta'_{12}u_3'^2 + \delta'_{23}u_1'^2 + 2\delta'_2 u'_1 u'_3}{\delta'_3 u'_1 u'_2 - \delta'_{12}u_3'^2 - \delta'_1 u'_2 u'_3 - \delta'_2 u'_1 u'_3} \tag{5.24}$$

Proof. Let ω be the image of the absolute conic, and let $\mathbf{x} = (x_1, x_2, 0)^\top$ be the intersection of a line through **p** tangent to ω with the line $x_3 = 0$. The components x_1, x_2 of **x** satisfy the quadratic equation (5.20)

$$A_{11}x_1^2 + 2A_{12}x_1x_2 + A_{22}x_2^2 = 0$$

It follows from equations (5.11) and (5.12), obtained in the proof of Theorem 5.1, that

$$\frac{x_1}{x_2} = \left(\frac{p_1 p'_2}{p'_1 p_2}\right)\frac{x'_1}{x'_2} \tag{5.25}$$

Let ρ be defined by

$$\rho = p_1 p'_2 / p'_1 p_2 \tag{5.26}$$

It follows from (5.22) and (5.25) that

$$A'_{11}\rho^2 x_1^2 + 2A'_{12}\rho x_1 x_2 + A'_{22}x_2^2 = 0 \tag{5.27}$$

Equations (5.20) and (5.27) are quadratic in x_1/x_2. The solutions of each equation yield the points at which the line $x_3 = 0$ intersects the lines through **p** tangent to ω. The solutions of the two equations are thus identical. It follows that the coefficients of the two equations differ by a single scalar multiple. On taking ratios of the coefficients Kruppa's equations are obtained,

$$\frac{\rho A'_{11}}{A'_{12}} = \frac{A_{11}}{A_{12}} \qquad \text{and} \qquad \frac{A'_{22}}{\rho A'_{12}} = \frac{A_{22}}{A_{12}} \tag{5.28}$$

It follows from (5.26) and (5.28) that

$$\frac{p_1 A_{11}}{p_2 A_{12}} = \frac{p'_1 A'_{11}}{p'_2 A'_{12}} \tag{5.29}$$

$$\frac{p_2 A_{22}}{p_1 A_{12}} = \frac{p'_2 A'_{22}}{p'_1 A'_{12}} \tag{5.30}$$

Equation (5.21) and the analogous equation for the A'_{ij} are used to substitute for the A_{ij} and the A'_{ij} in (5.29) and (5.30). The following two equations are obtained:

$$\frac{p_1}{p_2}\left(\frac{\delta_{12}p_2^2 + \delta_{31}p_3^2 + 2\delta_1 p_2 p_3}{\delta_3 p_3^2 - \delta_{12}p_1 p_2 - \delta_1 p_1 p_3 - \delta_2 p_2 p_3}\right) = \frac{p'_1}{p'_2}\left(\frac{\delta'_{12}p'^2_2 + \delta'_{31}p'^2_3 + 2\delta'_1 p'_2 p'_3}{\delta'_3 p'^2_3 - \delta'_{12}p'_1 p'_2 - \delta'_1 p'_1 p'_3 - \delta'_2 p'_2 p'_3}\right)$$

$$\frac{p_2}{p_1}\left(\frac{\delta_{12}p_1^2 + \delta_{23}p_3^2 + 2\delta_2 p_1 p_3}{\delta_3 p_3^2 - \delta_{12}p_1 p_2 - \delta_1 p_1 p_3 - \delta_2 p_2 p_3}\right) = \frac{p'_2}{p'_1}\left(\frac{\delta'_{12}p'^2_1 + \delta'_{23}p'^2_3 + 2\delta'_2 p'_1 p'_3}{\delta'_3 p'^2_3 - \delta'_{12}p'_1 p'_2 - \delta'_1 p'_1 p'_3 - \delta'_2 p'_2 p'_3}\right) \tag{5.31}$$

The quadratic transformations Φ, Φ' defined by (5.13) and (5.31) are used to transform the components p_i, p'_i in (5.31). Equations (5.23) and (5.24) are obtained after cancelling common factors from the numerator and the denominator of each side. □

5.1.3 The Two Sextics

Four equations, (5.14), (5.15), (5.23) and (5.24), in the coordinates of $\mathbf{u}$ and $\mathbf{u}'$ have been obtained. There are four unknowns, two arising from $\mathbf{u}$ and two arising from $\mathbf{u}'$. The number of equations is equal to the number of unknowns, thus there are at most a finite number of solutions. The components of $\mathbf{u}'$ are eliminated from the four equations to leave two equations in the components of $\mathbf{u}$.

Theorem 5.4. *The point* $\mathbf{u}$ *obtained from the epipole* $\mathbf{p}$ *by applying the quadratic transformation* Φ *defined by (5.13) satisfies two sextic polynomial constraints.*

Proof. The quadratic transformation Σ of (5.18) is used to substitute for $\mathbf{u}'$ in (5.23) and (5.24). The substitution yields two equations in $\mathbf{u}$,

$$\frac{h_1(\mathbf{u})}{k(\mathbf{u})} = \frac{g_1(\mathbf{u})}{l(\mathbf{u})} \quad \text{and} \quad \frac{h_2(\mathbf{u})}{k(\mathbf{u})} = \frac{g_2(\mathbf{u})}{l(\mathbf{u})} \tag{5.32}$$

The polynomials h_1 and h_2 are the numerators on the left-hand sides of (5.23) and (5.24). The polynomial k is the common denominator of the left-hand sides of (5.23) and (5.24). The polynomials g_1, g_2, l are obtained by substituting $\Sigma(\mathbf{u})$ for $\mathbf{u}'$ on the right-hand sides of (5.23) and (5.24). The polynomials h_1, h_2, k

are of degree two in $\mathbf{u}$, whilst g_1, g_2, l are of degree four in $\mathbf{u}$. The equations of (5.32) yield two sextic plane curves f_1 and f_2 defined by

$$\begin{aligned} f_1 &= h_1(\mathbf{u})l(\mathbf{u}) - g_1(\mathbf{u})k(\mathbf{u}) \\ f_2 &= h_2(\mathbf{u})l(\mathbf{u}) - g_2(\mathbf{u})k(\mathbf{u}) \end{aligned} \tag{5.33}$$

The point $\mathbf{u}$ is on f_1 and f_2 as required. □

Let f_1, f_2 be the curves defined by (5.33). The two curves have a total of 36 intersections, counted with the correct multiplicities. The epipoles $\mathbf{p}$ compatible with the five image correspondences of (5.1) yield points $\Phi(\mathbf{p})$ at which f_1 and f_2 intersect, but not every intersection of f_1 and f_2 is obtained in this way. There are additional intersections arising from the algebraic transformations by which f_1 and f_2 are obtained. These additional intersections do not yield epipoles compatible with the five image correspondences. The associated solutions are said to be spurious. In the following theorem it is shown that 26 of the intersections of f_1 and f_2 yield spurious solutions.

Theorem 5.5. *The sextic curves f_1 and f_2 obtained in Theorem 5.4 have 26 intersections which do not yield epipoles compatible with the original five image correspondences.*

Proof. The intersections yielding spurious solutions include

- the points $\mathbf{u}$ on f_1 and f_2 with at least one component u_i equal to zero;
- the fundamental points of the quadratic transformation Σ defined by (5.18);
- the points of intersection of the denominators k and l in (5.32).

The intersections of f_1 and f_2 for each of the three cases are counted. The details of the algebra are postponed until after the theorem.

The point $(1,0,0)^\top$ is an order three point of f_1 and $(0,1,0)^\top$ is an order one point of f_1. Similarly, $(1,0,0)^\top$ is an order one point of f_2 and $(0,1,0)^\top$ is an order three point of f_2. The point $(0,0,1)^\top$ is a point of tangency of f_1 and f_2. This yields a total of $3 \times 1 + 1 \times 3 + 2 = 8$ spurious solutions.

The three fundamental points of Σ are each of order two on f_1 and f_2. This yields a total of $2 \times 2 + 2 \times 2 + 2 \times 2 = 12$ spurious solutions.

The curves k and l have eight intersections, because k is of degree two and l is of degree four. These eight intersections include $(1,0,0)^\top$ and $(0,1,0)^\top$ which have already been counted in the first of the three cases. This yields a total of $6 = 8 - 2$ spurious solutions.

On adding the intersections from each of the three cases a total of $8+12+6 = 26$ is obtained. □

Corollary. There are at most ten distinct epipoles compatible with five image correspondences in general position. The corollary follows because f_1 and f_2 have 36 intersections, counted with the correct multiplicities. There remain ten intersections after subtracting the intersections yielding spurious solutions listed in the proof of Theorem 5.5.

The details of the algebra underlying Theorem 5.5 are given for the first two cases. The curves f_1 and f_2 are examined in the neighbourhood of each intersection. A straightforward calculation shows that the quadratic transformation Σ defined by (5.18) takes the values

$$\begin{aligned} \Sigma(1,0,0)^\top &= (1,0,0)^\top \\ \Sigma(0,1,0)^\top &= (0,1,0)^\top \\ \Sigma(0,0,1)^\top &= (0,0,1)^\top \end{aligned} \tag{5.34}$$

It follows from (5.34) that in the neighbourhood of $\mathbf{u} = (1,0,0)^\top$ the point $\mathbf{u}' = \Sigma(\mathbf{u})$ is given by

$$\mathbf{u}' = (u_1^2 + r_1, r_2, r_3)^\top \tag{5.35}$$

where r_1, r_2, r_3 are quadratic polynomials in $\mathbf{u}$, such that the coefficient of u_1^2 in each of r_1, r_2 and r_3 is zero. On using (5.35) to substitute for $\mathbf{u}'$ in the first equation of (5.33) it is found that h_1 is of degree one in u_1, k is of degree one in u_1, g_1 is of degree one in u_1 and l is of degree two in u_1. It follows that the highest power of u_1 occurring in f_1 is u_1^3. Thus $(1,0,0)^\top$ is a singular point of order three on f_1. The order of $(0,1,0)^\top$ on f_1 and the orders of $(1,0,0)^\top$ and $(0,1,0)^\top$ on f_2 are obtained similarly.

It is shown that f_1 and f_2 are tangential at $(0,0,1)^\top$. It follows from (5.34) that

$$\mathbf{u}' = (s_1, s_2, s_3 + u_3^2)^\top \tag{5.36}$$

where s_1, s_2, s_3 are quadratic polynomials in $\mathbf{u}$, such that the coefficient of u_3^2 in s_1, s_2 and s_3 is equal to zero. On using (5.36) to substitute for certain of the components of $\mathbf{u}'$ in the equation of (5.33) for f_1 the following equation is obtained.

$$\begin{aligned} f_1(\mathbf{u}) &= h_1(\mathbf{u})l(\mathbf{u}) - g_1(\mathbf{u})k(\mathbf{u}) \\ &= (\delta_{12}u_3^2 + 2\delta_1 u_3 u_2)(-\delta'_{12}u_3'^2 - \delta'_1 u'_2 u'_3 - \delta'_2 u'_1 u'_3) \\ &\quad -(\delta'_{12}u_3'^2 + 2\delta'_1 u'_3 u'_2)(-\delta_{12}u_3^2 - \delta_1 u_2 u_3 - \delta_2 u_1 u_3) + O(u_1^2, u_1u_2, u_2^2) \\ &= u_3u'_3(\delta_{12}\delta'_2 u_3 u'_1 + \delta_1\delta'_{12}u_2u'_3 - \delta'_{12}\delta_2 u_1 u'_3 - \delta'_1\delta_{12}u_3u'_2) + O(u_1^2, u_1u_2, u_2^2) \end{aligned} \tag{5.37}$$

It follows from (5.34) and (5.37) that $f_1(0,0,1) = 0$. Near to $(0,0,1)^\top$ the curve f_1 is given by

$$f_1(\mathbf{u}) = u_3^4(\delta_{12}\delta'_2 s_1 + \delta_1\delta'_{12}u_2u_3 - \delta'_{12}\delta_2 u_1u_3 - \delta_{12}\delta'_1 s_2) + O(u_1^2, u_1u_2, u_2^2) \tag{5.38}$$

The coefficient of u_3^5 in f_1 is obtained from (5.38). A similar calculation shows that $f_2(0,\ 0,1) = 0$, and that the coefficients of u_3^5 in f_1 and f_2 are equal. It follows that f_1 and f_2 have a common tangent line at $(0,0,1)^\top$ as required.

In the second case listed in Theorem 5.5 the fundamental points of Σ are of order at least two on both f_1 and f_2 because the right-hand sides of (5.23) and (5.24) are ratios of quadratic polynomials in $\mathbf{u}'$.

The Corollary to Theorem 5.5 establishes that there are at most ten pairs of epipoles compatible with five general image correspondences. The possibility is not ruled out that some of the remaining ten intersections of f_1 and f_2 yield spurious solutions. One way of checking the ten intersections is to find a single example in which each intersection yields a reconstruction. In the next subsection a completely different approach to reconstruction is described. In this second approach it is relatively straightforward to show that there are in general at least ten reconstructions compatible with five image correspondences.

5.2 Demazure's Method

Demazure's method is based on the properties of the variety of essential matrices in $\mathbb{P}^8$. The defining equations for the variety are given in (2.89). Each image correspondence $\mathbf{q} \leftrightarrow \mathbf{q}'$ defines a hyperplane in $\mathbb{P}^8$, namely the set of solutions E to the homogeneous linear equation $\mathbf{q}'^{\top} E\mathbf{q} = 0$. The reconstruction problem for a set of image correspondences $\mathbf{q}_i \leftrightarrow \mathbf{q}_i'$, $1 \le i \le n$, is solved by finding a point E on the variety of essential matrices such that $\mathbf{q}_i'^{\top} E\mathbf{q}_i = 0$ for $1 \le i \le n$. Demazure (1988) uses methods from algebraic geometry to show that there are, in general, at least ten reconstructions compatible with five image correspondences. The result that there are exactly ten reconstructions can then be obtained either by a detailed examination of the algebra, or by applying the Corollary to Theorem 5.5.

The algebraic variety of essential matrices is defined as follows. Let $\mathbb{P}^8(\mathbb{C})$ be the projective space formed by taking the nine dimensional space of 3×3 matrices with complex entries, removing the zero matrix, and then identifying any two matrices that differ by a non-zero scalar multiple. Two matrices E_1 and E_2 are identified if and only if $E_1 = \lambda E_2$, for some $\lambda \neq 0$. The symbol E denotes a 3×3 matrix or a point of $\mathbb{P}^8$, depending on context. No difficulties arise from this ambiguity because all the equations involving E are homogeneous in the coordinates of E. It is recalled from Sect. 2.2 that a 3×3 matrix is an essential matrix if and only if it is the product of an orthogonal matrix and a skew symmetric matrix. Let $\mathcal{M}$ to be the *closure* of the set of essential matrices in $\mathbb{P}^8$. It is established in Sect. 5.2.1 below that $\mathcal{M}$ is the algebraic variety defined by the nine equations of degree three obtained from the coefficients of (2.89). The variety $\mathcal{M}$ is called the variety of essential matrices even though it contains some rank one matrices which are not essential matrices. The essential matrices with complex entries are included in $\mathcal{M}$ even though they do not yield physically realisable reconstructions. They are a necessary part of the algebra.

Each image correspondence $\mathbf{q} \leftrightarrow \mathbf{q}'$ is compatible only with the essential matrices E that satisfy

$$\mathbf{q}^{\top} E\mathbf{q}' = 0 \tag{5.39}$$

Equation (5.39) defines a hyperplane H_q in $\mathbb{P}^8$. The essential matrices that satisfy (5.39) are precisely those matrices contained in $H_q \cap \mathcal{M}$. The image correspondences $\mathbf{q}_i \leftrightarrow \mathbf{q}_i'$ for $1 \le i \le 5$ give rise to a three dimensional linear subspace L of $\mathbb{P}^8$ defined by

$$L = H_{q_1} \cap H_{q_2} \cap H_{q_3} \cap H_{q_4} \cap H_{q_5} \tag{5.40}$$

The essential matrices compatible with the five image correspondences are contained in $L \cap \mathcal{M}$.

The properties of the variety of essential matrices are discussed in Sect. 5.2.1 and Sect. 5.2.2. It is shown in Sect. 5.2.3 that a general three dimensional linear subspace L of $\mathbb{P}^8$ has the form (5.40) for some choice of five image correspondences. The subspace L does not in general meet those points of $\mathcal{M}$ which are not essential matrices. It follows from these results that the number N of reconstructions compatible with five image correspondences is equal to the number of points in $L \cap \mathcal{M}$, where L is an arbitrary three dimensional subspace of $\mathbb{P}^8$. The number N is finite and it is equal to the degree of $\mathcal{M}$ as an algebraic variety. The degree of an algebraic variety is defined in Hartshorne (1977). It is shown in Sect. 5.2.4 that $\mathcal{M}$ has degree ten.

5.2.1 The Variety of Essential Matrices

It is recalled from Sect. 2.2.1 that the essential matrices satisfy

$$EE^\top E = \frac{1}{2}\mathrm{tr}(EE^\top)E \tag{5.41}$$

The set of solutions to (5.41) in $\mathbb{P}^8$ is a closed algebraic variety $\mathcal{N}$ which contains the essential matrices. The set $\mathcal{M}$ is defined to be the closure of the set of essential matrices, thus $\mathcal{M} \subseteq \mathcal{N}$. Further, it is shown in Sect. 2.2.1 that any matrix in $\mathcal{N}$ that is not of rank one is in $\mathcal{M}$. In order to show that $\mathcal{N} = \mathcal{M}$ it remains only to demonstrate that the rank one matrices contained in $\mathcal{N}$ are also in $\mathcal{M}$.

Theorem 5.6. *Let $E = \mathbf{u} \otimes \mathbf{v}$ be a rank one matrix. Then E satisfies (5.41) if and only if $\mathbf{u}.\mathbf{u} = 0$ or $\mathbf{v}.\mathbf{v} = 0$. Any rank one matrix satisfying (5.41) is the limit of a sequence of essential matrices.*

Proof. The product $EE^\top E$ is evaluated,

$$\begin{aligned} EE^\top E &= (\mathbf{u}\otimes\mathbf{v})(\mathbf{u}\otimes\mathbf{v})(\mathbf{u}\otimes\mathbf{v}) \\ &= (\mathbf{v}.\mathbf{v})(\mathbf{u}.\mathbf{u})(\mathbf{u}\otimes\mathbf{v}) \\ &= \mathrm{tr}(EE^\top)(\mathbf{u}\otimes\mathbf{v}) \\ &= \mathrm{tr}(EE^\top)E \end{aligned} \tag{5.42}$$

It follows from (5.42) that E satisfies (5.41) if and only if

$$\mathrm{tr}(EE^\top) = (\mathbf{v}.\mathbf{v})(\mathbf{u}.\mathbf{u}) = 0$$

Thus E satisfies (5.41) if and only if $\mathbf{u}.\mathbf{u} = 0$ or $\mathbf{v}.\mathbf{v} = 0$.

It is shown that a rank one matrix $E = \mathbf{u} \otimes \mathbf{v}$ satisfying (5.41) is the limit of a sequence of essential matrices. The case $\mathbf{u}.\mathbf{u} = \mathbf{v}.\mathbf{v} = 0$ is considered first. There exists a sequence of vectors $\mathbf{v}_n$ such that $\mathbf{v}_n.\mathbf{v}_n \neq 0$ and such that $\mathbf{v}_n \to \mathbf{v}$. If for each n, $\mathbf{u} \otimes \mathbf{v}_n$ is the limit of a sequence of essential matrices, then $\mathbf{u} \otimes \mathbf{v}$ is also the limit of a sequence of essential matrices. It is thus sufficient to consider the case $\mathbf{u}.\mathbf{u} = 0$, $\mathbf{v}.\mathbf{v} \neq 0$. (If $\mathbf{u}.\mathbf{u} \neq 0$, $\mathbf{v}.\mathbf{v} = 0$, then E is replaced by $E^{\top}$.) On pre- and post-multiplying E by orthogonal matrices and scaling E as necessary, a reduction is made to the case $\mathbf{u} = (0, 1, i)^{\top}$, $\mathbf{v} = (1, 0, 0)^{\top}$. The matrix E is given by

$$E = \mathbf{u} \otimes \mathbf{v} = \begin{pmatrix} 0 & 0 & 0 \\ 1 & 0 & 0 \\ i & 0 & 0 \end{pmatrix}$$

Let $n > 0$ be an integer and let E_n be the matrix defined by

$$E_n = \begin{pmatrix} 0 & 1/n & i/n \\ 1 & 0 & 0 \\ i & 0 & 0 \end{pmatrix} \tag{5.43}$$

Let θ be the angle defined by $\cos(\theta) + i\sin(\theta) = -1/n$. It follows from the equation

$$\begin{pmatrix} 0 & 1/n & i/n \\ 1 & 0 & 0 \\ i & 0 & 0 \end{pmatrix} = \begin{pmatrix} 0 & -1 & -i \\ 1 & 0 & 0 \\ i & 0 & 0 \end{pmatrix} \begin{pmatrix} 1 & 0 & 0 \\ 0 & \cos(\theta) & -\sin(\theta) \\ 0 & \sin(\theta) & \cos(\theta) \end{pmatrix}$$

that each E_n is an essential matrix. The limit of (5.43) yields $\lim_{n \to \infty} E_n = E$.□

Corollary. The set $\mathcal{M}$, defined as the closure of the set of essential matrices, is an algebraic variety because it is the set of zeros of the polynomial equations (2.89).

5.2.2 Properties of the Variety of Essential Matrices

Three fundamental properties of the algebraic variety $\mathcal{M}$ of essential matrices are obtained. It is shown firstly that $\mathcal{M}$ is irreducible, secondly that $\mathcal{M}$ has dimension five and thirdly that the singular points of $\mathcal{M}$ are contained in the set of rank one matrices.

In proving the irreducibility of $\mathcal{M}$ it is more convenient to work in the affine space $\mathbf{C}^8$ rather than in the projective space $\mathbf{P}^8$. This is equivalent to 'dehomogenising' (5.41). It is enough to set one of the entries of E, for example E_{33}, equal to one. The plane at infinity $H_\infty = \mathbf{P}^8 \setminus \mathbf{C}^8$ is then $E_{33} = 0$. Let $\mathcal{M}'$ be the affine variety $\mathcal{M}' = \mathcal{M} \setminus \mathcal{M} \cap H_\infty$. The projective variety $\mathcal{M}$ is the closure of $\mathcal{M}'$ in $\mathbf{P}^8$. The varieties $\mathcal{M}$ and $\mathcal{M}'$ have the same dimension. The variety $\mathcal{M}$ is irreducible if and only if $\mathcal{M}'$ is irreducible.

It is convenient to introduce some concepts from algebraic geometry. The ring of polynomials in n variables $x_1, x_2, \ldots, x_n$ with coefficients in $\mathbf{C}$ is denoted either by $\mathbf{C}[x_1, \dots, x_n]$ or more briefly by $\mathbf{C}[\mathbf{x}]$. An affine algebraic variety $\mathcal{A}$ is the set of common zeros in $\mathbf{C}^n$ of a finite number of polynomials f_i in $\mathbf{C}[\mathbf{x}]$ for $1 \le i \le m$. The affine algebraic varieties are closely associated with certain subsets of $\mathbf{C}[\mathbf{x}]$ known as ideals. The ideal $\mathcal{I}(\langle f_i \rangle)$ in $\mathbf{C}[\mathbf{x}]$ generated by the polynomials f_i for $1 \le i \le m$, is defined to be the set of all linear combinations

$$\sum_{i=1}^{m} g_i f_i$$

where the g_i are arbitrary elements of $\mathbf{C}[\mathbf{x}]$. The ideal $\mathcal{I}(\langle f_i \rangle)$ consists of polynomial functions which vanish on $\mathcal{A}$. However, $\mathcal{I}(\langle f_i \rangle)$ need not contain all the polynomial functions vanishing on $\mathcal{A}$. It follows from the definition of $\mathcal{I}(\langle f_i \rangle)$ that if f is any element of $\mathcal{I}(\langle f_i \rangle)$ and if g is any element of $\mathbf{C}[\mathbf{x}]$ then gf is in $\mathcal{I}(\langle f_i \rangle)$. An ideal $\mathcal{I}$ is a prime ideal if whenever a product $g_1 g_2$ of polynomials is in $\mathcal{I}$ then at least one of g_1, g_2 is in $\mathcal{I}$. Let $\mathcal{I}$ be an ideal and let $\sqrt{\mathcal{I}}$ be the subset of $\mathbf{C}[\mathbf{x}]$ defined such that f is in $\sqrt{\mathcal{I}}$ if and only if some power f^n of f is in $\mathcal{I}$. The set $\sqrt{\mathcal{I}}$ is an ideal called the radical ideal of $\mathcal{I}$.

Let $\mathcal{A}$ be an affine algebraic variety, and let $\mathcal{I}(\mathcal{A})$ be the set of polynomials in $\mathbf{C}[\mathbf{x}]$ that vanish on $\mathcal{A}$. Then $\mathcal{I}(\mathcal{A})$ is a radical ideal called the ideal of $\mathcal{A}$. Let f_i for $1 \le i \le m$ be a set of polynomials such that $\mathcal{A}$ is the set of common zeros of the f_i. It follows from the definition of $\mathcal{I}(\mathcal{A})$ that $\mathcal{I}(\langle f_i \rangle) \subset \mathcal{I}(\mathcal{A})$. In fact $\mathcal{I}(\mathcal{A})$ is the radical ideal of $\mathcal{I}(\langle f_i \rangle)$. The correspondence $\mathcal{A} \mapsto \mathcal{I}(\mathcal{A})$ is one to one and onto from the set of affine algebraic varieties to the set of radical ideals. The correspondence reverses inclusions in that $\mathcal{A} \subseteq \mathcal{B}$ if and only if $\mathcal{I}(\mathcal{A}) \supseteq \mathcal{I}(\mathcal{B})$.

An irreducible algebraic variety is one that cannot be expressed as the union of two or more simpler algebraic varieties. More formally, an algebraic variety $\mathcal{N}$ is defined to be irreducible if there do not exist algebraic varieties $\mathcal{N}_1$, $\mathcal{N}_2$ such that $\mathcal{N} = \mathcal{N}_1 \cup \mathcal{N}_2$, $\mathcal{N} \neq \mathcal{N}_1$, $\mathcal{N} \neq \mathcal{N}_2$. An algebraic variety $\mathcal{N}$ is irreducible if whenever f, g are any two homogeneous polynomials in $\mathbf{C}[\mathbf{x}]$ then $f(\mathcal{N})g(\mathcal{N}) = 0$ if and only if either $f(\mathcal{N}) = 0$ or $g(\mathcal{N}) = 0$. To see this, note first that if $\mathcal{N}$ has a decomposition $\mathcal{N} = \mathcal{N}_1 \cup \mathcal{N}_2$ into two distinct non-empty algebraic varieties $\mathcal{N}_1$, $\mathcal{N}_2$ then there exist polynomials f, g such that $f(\mathcal{N}_1) = 0$, $f(\mathcal{N}_2) \neq 0$ and $g(\mathcal{N}_1) \neq 0$, $g(\mathcal{N}_2) = 0$. It follows that $f(\mathcal{N})g(\mathcal{N}) = 0$, but $f(\mathcal{N}) \neq 0$, $g(\mathcal{N}) \neq 0$. Conversely, if there exist polynomials f, g such that $f(\mathcal{N})g(\mathcal{N}) = 0$, $f(\mathcal{N}) \neq 0$, $g(\mathcal{N}) \neq 0$ then a decomposition of $\mathcal{N}$ is obtained by defining $\mathcal{U}$, $\mathcal{V}$ to be the sets of zeros of f and g respectively in $\mathbf{P}^8$ and then setting $\mathcal{N}_1 = \mathcal{U} \cap \mathcal{N}$, $\mathcal{N}_2 = \mathcal{V} \cap \mathcal{N}$. The algebraic variety $\mathcal{N}$ is irreducible if and only if the ideal $\mathcal{I}(\mathcal{N})$ of $\mathbf{C}[\mathbf{x}]$ is a prime ideal.

A projective algebraic variety $\mathcal{N}$ in $\mathbf{P}^n$ is the set of common zeros of a finite number m of homogeneous polynomials f_i in n variables $x_1, \dots x_n$. If H is any hyperplane in $\mathbf{P}^n$ then $\mathcal{N} \setminus H$ is an affine algebraic variety. Certain of the properties of $\mathcal{N}$ can be deduced from similar properties of $\mathcal{N} \setminus H$, provided H is not contained in $\mathcal{N}$. For example, $\mathcal{N}$ is irreducible if and only if $\mathcal{N} \setminus H$ is irreducible, and if $\mathbf{p}$ is a singular point of $\mathcal{N} \setminus H$ then $\mathbf{p}$ is also a singular point of $\mathcal{N}$.

The notation $SO(3,\mathbf{C})$ is used for the group of complex orthogonal matrices with determinant +1.

Proposition 5.7. *The variety $\mathcal{M}$ is irreducible.*

Proof. It is enough to show that the affine variety $\mathcal{M}'$ of essential matrices is irreducible. Let f, g be homogeneous polynomials such that $f(\mathcal{M}')g(\mathcal{M}') = 0$ and let P be any point of $\mathcal{M}'$ such that $g(P) \neq 0$. It is shown that $f(\mathcal{M}') = 0$. There exists an open set U in $\mathbf{P}^8$ containing P such that g is non-zero at every point of U. It follows that $f(U \cap \mathcal{M}') = 0$. The set $U \cap \mathcal{M}'$ is open and non-empty in $\mathcal{M}'$, thus it is assumed without loss of generality that P is an essential matrix rather than a matrix of rank one. Let $P = RT_a$ and let h be the function from $\mathbf{C}^3$ to $\mathbf{C}$ defined at each point $\mathbf{b}$ of $\mathbf{C}^3$ by $h(\mathbf{b}) = f(RT_b)$. Then h is a polynomial which is identically zero at all points in an open neighbourhood of $\mathbf{a}$ in $\mathbf{C}^3$. It follows that $h(\mathbf{b}) = 0$ for all points $\mathbf{b}$ in $\mathbf{C}^3$. On allowing P to vary within $U \cap \mathcal{M}'$ it follows that $f(RT_b) = 0$ for all $\mathbf{b}$ and for all R in a non-empty open set O of $SO(3,\mathbf{C})$.

Let ST_b be an arbitrary essential matrix in $\mathcal{M}'$. It has been shown that $f(UT_b) = 0$ for all orthogonal matrices U in O. Let W be a fixed matrix in O and let H be the antisymmetric matrix such that $\exp(H) = SW^{\top}$. Let $k : \mathbf{C} \mapsto \mathbf{C}$ be the analytic function defined by

$$k(t) = f(\exp(Ht)WT_b)$$

If t is sufficiently close to zero then $\exp(Ht)W$ is in O. It follows that $k(t) = 0$ for all t in an open neighbourhood of 0. The function k is analytic thus $k(t) = 0$ for all t, in particular,

$$0 = k(1) = f(\exp(H)RT_b) = f(ST_b)$$

It follows that $f(E) = 0$ for all essential matrices E. The algebraic variety $\mathcal{M}'$ is contained in the closure of the set of essential matrices, thus $f(\mathcal{M}') = 0$. The algebraic variety $\mathcal{M}'$ is thus irreducible. □

The dimension of $\mathcal{M}$ is equal to the number of independent parameters required to parameterise a small open set of $\mathcal{M}$, or equivalently it is equal to the dimension of the tangent space at a general point of $\mathcal{M}$. Intuitively, $\mathcal{M}$ has dimension five because each essential matrix $E = RT_a$ depends on five parameters, three to describe the rotation R and two to describe the direction of the translation $\mathbf{a}$. It is only necessary to check the details.

Proposition 5.8. *The dimension of $\mathcal{M}$ is five.*

Proof. The algebraic variety $SO(3,\mathbf{C}) \times \mathbf{P}^2$ has dimension five. There is a differentiable (in fact, algebraic) map f from $SO(3,\mathbf{C}) \times \mathbf{P}^2$ onto an open dense subset of $\mathcal{M}$ defined by

$$f(R, \mathbf{a}) = RT_a$$

If E is given and if $E = RT_a$ then the direction of $\mathbf{a}$ is fixed and R can take only two values, which differ by a rotation of 180° about the axis $\mathbf{a}$. The proof of this fact as given in Proposition 2.10 assumes that $\|\mathbf{a}\| \neq 0$. The result is easily extended to cover the case $\|\mathbf{a}\| = 0$. It follows that f is locally one to one, i.e. the restriction of f to a small open neighbourhood of $(R, \mathbf{a})$ in $SO(3, \mathbf{C}) \times \mathbf{P}^2$ is one to one. It follows that the derivative of f is an isomorphism from the tangent space of $SO(3, \mathbf{C})$ at $(R, \mathbf{a})$ to the tangent space of $\mathcal{M}$ at RT_a. The dimension of $\mathcal{M}$ is equal to the dimension of the tangent space at a general point of $\mathcal{M}$. Hence the dimension of $\mathcal{M}$ is equal to the dimension of $SO(3, \mathbf{C}) \times \mathbf{P}^2$. □

The manifold $SO(3, \mathbf{C}) \times \mathbf{P}^2$ is a double cover of $\mathcal{M}$. If E is a general point of $\mathcal{M}$ then the set $f^{-1}(E)$ contains exactly two points.

A point $\mathbf{p}$ on an irreducible affine algebraic variety $\mathcal{N}$ in $\mathbf{C}^n$ is a singular point of $\mathcal{N}$ if the dimension of the tangent space of $\mathcal{N}$ at $\mathbf{p}$ is strictly greater than the dimension of $\mathcal{N}$. The tangent space of $\mathcal{N}$ at $\mathbf{p}$ is defined as follows. Let $\mathcal{N}$ be the set of common zeros of the polynomials f_i for $1 \leq i \leq m$. Let $\mathbf{\Delta}$ be a vector, and let t be a scalar parameter. The Taylor expansions at $\mathbf{p}$ of the f_i to first order in t yield

$$\begin{aligned} f_i(\mathbf{p} + t\mathbf{\Delta}) &= f_i(\mathbf{p}) + t\sum_{j=1}^{n} \left.\frac{\partial f_i}{\partial \mathbf{x}_j}\right|_{\mathbf{p}} \Delta_j + O(t^2) \\ &= t\sum_{j=1}^{n} \left.\frac{\partial f_i}{\partial \mathbf{x}_j}\right|_{\mathbf{p}} \Delta_j + O(t^2) \qquad (1 \leq i \leq m) \end{aligned}$$

The point $\mathbf{p} + t\mathbf{\Delta}$ is in the tangent space to $\mathcal{N}$ at $\mathbf{p}$ if and only if

$$f_i(\mathbf{p} + t\mathbf{\Delta}) = O(t^2) \qquad (1 \leq i \leq m)$$

It follows that $\mathbf{p} + t\mathbf{\Delta}$ is in the tangent space of $\mathcal{N}$ at $\mathbf{p}$ if and only if

$$\sum_{j=1}^{n} \left.\frac{\partial f_i}{\partial \mathbf{x}_j}\right|_{\mathbf{p}} \Delta_j = 0 \qquad (1 \leq i \leq m) \tag{5.44}$$

Let r be the dimension of $\mathcal{N}$. If the equations (5.44) fail to impose $n - r$ independent conditions on $\mathbf{\Delta}$ then the tangent space to $\mathcal{N}$ at $\mathbf{p}$ has dimension strictly greater than r. The point $\mathbf{p}$ is then by definition a singular point of $\mathcal{N}$. It is shown in Hartshorne (1977) that almost all points of an irreducible algebraic variety are non-singular. The singular points of the variety form a proper closed subset of measure zero.

Let $\mathcal{N}$ now be a projective variety in $\mathbf{P}^n$, let $\mathbf{p}$ be any point of $\mathcal{N}$ and let H be a hyperplane of $\mathbf{P}^n$ not contained in $\mathcal{N}$ and not containing $\mathbf{p}$. Then $\mathbf{p}$ is by definition a singular point of $\mathcal{N}$ if and only if $\mathbf{p}$ is a singular point of the affine algebraic variety $\mathcal{N} \setminus H$.

The tangent space of an affine algebraic variety need not be spanned by linear spaces which are tangent to the variety. For example the tangent space of the affine plane curve $x^3 - y^2 = 0$ at $(0,0)^\top$ is the whole plane $\mathbf{C}^2$. The point $(0,0)^\top$ is thus a singular point of the curve. There is only one line tangent to the curve at $(0,0)^\top$, namely, $t \mapsto (t,0)^\top$. The tangent line to the curve at $(0,0)^\top$ thus does not span the tangent space at $(0,0)^\top$.

The singular points of the variety of essential matrices are obtained.

Proposition 5.9. *A point E of $\mathcal{M}$ is a singular point of $\mathcal{M}$ if and only if $E = \mathbf{u} \otimes \mathbf{v}$, where $\mathbf{u}$, $\mathbf{v}$ are vectors such that $\mathbf{u}.\mathbf{u} = \mathbf{v}.\mathbf{v} = 0$.*

Proof. Suppose if possible that $\mathcal{M}$ has a singular point RT_a, where R is an orthogonal matrix and $\mathbf{a}$ is a vector such that $\mathbf{a}.\mathbf{a} \neq 0$. Let ST_b be any other point of $\mathcal{M}$ such that S is an orthogonal matrix and $\mathbf{b}.\mathbf{b} \neq 0$. Let U be an orthogonal matrix such that $\mathbf{b} = U\mathbf{a}$ and let ω be the collineation of $\mathbb{P}^8$ defined by $\omega(E) = SUR^\top EU^\top$. It follows from the definition of ω that $\omega(RT_a) = ST_b$. The variety $\mathcal{M}$ is invariant under ω, thus ST_b is a singular point of $\mathcal{M}$. It follows that almost all points of $\mathcal{M}$ are singular points. This is impossible because the singular points of $\mathcal{M}$ form a proper closed algebraic subset of $\mathcal{M}$. The variety $\mathcal{M}$ thus contains no singular points of the form RT_a, where $\mathbf{a}.\mathbf{a} \neq 0$.

Next suppose if possible that $\mathcal{M}$ has a singular point of the form RT_a, where $\mathbf{a}.\mathbf{a} = 0$. If one such point is a singular point then all points ST_b of $\mathcal{M}$, such that S is an orthogonal matrix and $\mathbf{b}.\mathbf{b} = 0$ are singular points of $\mathcal{M}$. It follows from the example given in the proof of Theorem 5.6 that there is a sequence R_iT_a, $i = 1, 2, \ldots$, of points of $\mathcal{M}$ converging to a point $\mathbf{u} \otimes \mathbf{a}$ of $\mathcal{M}$. The hypothesis that RT_a is a singular point of $\mathcal{M}$ ensures that every point R_iT_a, $i = 1, 2, \ldots$, is also a singular point of $\mathcal{M}$. It follows that $\mathbf{u} \otimes \mathbf{a}$ is a singular point of $\mathcal{M}$. It then follows that every point $\mathbf{u} \otimes \mathbf{v}$ of $\mathcal{M}$ such that $\mathbf{u}.\mathbf{u} \neq 0$, $\mathbf{v}.\mathbf{v} = 0$ is a singular point of $\mathcal{M}$. A contradiction is obtained. The equations defining $\mathcal{M}$, taken from (5.41), are

$$EE^\top E - \frac{1}{2}\mathrm{tr}(EE^\top)E = 0 \tag{5.45}$$

Let $\mathbf{\Delta}$ be a fixed vector. On substituting $\mathbf{u} \otimes \mathbf{v} + \mathbf{\Delta}$ for E on the left-hand side of (5.45) and taking the terms linear in $\mathbf{\Delta}$ the following expression is obtained:

$$(\mathbf{u} \otimes \mathbf{v})\mathbf{\Delta}^\top(\mathbf{u} \otimes \mathbf{v}) + \mathbf{\Delta}(\mathbf{v} \otimes \mathbf{v}) - \mathrm{tr}((\mathbf{u} \otimes \mathbf{v})\mathbf{\Delta}^\top)(\mathbf{u} \otimes \mathbf{v}) \tag{5.46}$$

A short calculation reduces (5.46) to

$$\mathbf{\Delta}(\mathbf{v} \otimes \mathbf{v}) \tag{5.47}$$

The expression (5.47) vanishes if and only if $\mathbf{\Delta}\mathbf{v} = 0$. It follows that Δ is subject to three independent linear constraints. The algebraic variety $\mathcal{M}$ is of dimension five and it is contained in a space of dimension $5 + 3 = 8$. The point $\mathbf{u} \otimes \mathbf{v}$ is thus not a singular point of $\mathcal{M}$. It follows that $\mathcal{M}$ has no singular points of the form RT_a, $\mathbf{a}.\mathbf{a} = 0$. A similar argument shows that $\mathbf{v} \otimes \mathbf{u}$ is not a singular point of $\mathcal{M}$.

The proof is completed by showing that every point $\mathbf{u} \otimes \mathbf{v}$ of $\mathcal{M}$ such that $\mathbf{u}.\mathbf{u} = \mathbf{v}.\mathbf{v} = 0$ is a singular point of $\mathcal{M}$. There exists an orthogonal matrix R

such that $R\mathbf{u} = \mathbf{v}$. Let ω be the collineation of $\mathbb{P}^8$ defined by $\omega(E) = RE$. It follows that $\omega(\mathbf{u} \otimes \mathbf{v}) = \mathbf{v} \otimes \mathbf{v}$. It is thus sufficient to prove that if $\mathbf{v}.\mathbf{v} = 0$ then $\mathbf{v} \otimes \mathbf{v}$ is a singular point of $\mathcal{M}$. On substituting $\mathbf{v} \otimes \mathbf{v} + \Delta$ for E in (5.45) and taking the terms linear in Δ the following expression is obtained:

$$(\mathbf{v} \otimes \mathbf{v})\mathbf{\Delta}^{\top}(\mathbf{v} \otimes \mathbf{v}) - \mathrm{tr}(\mathbf{\Delta}(\mathbf{v} \otimes \mathbf{v}))(\mathbf{v} \otimes \mathbf{v}) \qquad (5.48)$$

The expression (5.48) reduces to zero. It follows that the tangent space of $\mathcal{M}$ at each point $\mathbf{u} \otimes \mathbf{v}$ with $\mathbf{u}.\mathbf{u} = \mathbf{v}.\mathbf{v} = 0$ is the whole of $\mathbb{C}^8$. □

5.2.3 Linear Subspaces of $\mathbb{P}^8$

The variety $\mathcal{M}$ has dimension five, thus the intersection of $\mathcal{M}$ with a general linear subspace of L of $\mathbb{P}^8$ of dimension three contains only a finite number N of points. The number N is independent of the choice of L, provided L is sufficiently general. The degree of $\mathcal{M}$ is, by definition, equal to N (Hartshorne 1977). The aim in this subsection is to show that the degree of $\mathcal{M}$ is equal to the degree of reconstruction from image correspondences. The actual value of this degree is found in Sect. 5.2.4.

Each image correspondence $\mathbf{q} \leftrightarrow \mathbf{q}'$ defines a hyperplane H_q in $\mathbb{P}^8$ consisting of the points E such that $\mathbf{q}^{\top}E\mathbf{q}' = 0$. Let $\mathbf{q}_i \leftrightarrow \mathbf{q}_i'$, $1 \le i \le 5$, be a given set of image correspondences and let H_i be the set of points E of $\mathbb{P}^8$ such that $\mathbf{q}_i^{\top}E\mathbf{q}_i' = 0$. The linear subspace L of $\mathbb{P}^8$ is defined by

$$L = H_1 \cap H_2 \cap H_3 \cap H_4 \cap H_5 \qquad (5.49)$$

Proposition 5.10. *Let* $\mathbf{q}_i \leftrightarrow \mathbf{q}_i'$ *for* $1 \le i \le 5$, *be five image correspondences in general position. Then the linear subspace* L *of* $\mathbb{P}^8$ *defined by (5.49) has dimension three.*

Proof. Let L_i be the linear subspace of $\mathbb{P}^8$ defined by

$$L_i = H_1 \cap \ldots \cap H_i \qquad (1 \le i \le 5)$$

and let d_i be the dimension of L_i. If H_{i+1} contains L_i then $d_i = d_{i+1}$, and if H_{i+1} does not contain L_i then $d_i = d_{i+1} + 1$. Let E be any point of L_i. Then for almost all image correspondences $\mathbf{q} \leftrightarrow \mathbf{q}'$ it is the case that $\mathbf{q}^{\top}E\mathbf{q}' \neq 0$. The image correspondences are by hypothesis in general position, thus $\mathbf{q}_{i+1}^{\top}E\mathbf{q}_{i+1} \neq 0$. It follows that H_{i+1} does not contain L_i, hence $d_{i+1} + 1 = d_i$ for $1 \le i \le 4$. The result $d_5 = 3$ is obtained. □

The next step is to show that almost all subspaces of $\mathbb{P}^8$ of dimension three can be obtained as the intersection of five hyperplanes H_i, as defined by (5.49). The proof of this result uses the Grassmannian variety of subspaces of a projective space. The points of the Grassmannian variety $G(m,n)$ parameterise the m-dimensional linear subspaces of $\mathbb{P}^n$. It can be shown that $G(m,n)$ is an irreducible non-singular algebraic variety of dimension $(n-m)(m+1)$. Further information about $G(m,n)$ is given by Hartshorne (1977) and by Semple & Roth (1949).

Theorem 5.11. *Almost all linear subspaces L of $\mathbf{P}^8$ of dimension three can be obtained as the intersection of five hyperplanes H_i, as defined by (5.49).*

Proof. Let $G(3,8)$ be the Grassmannian of subspaces of dimension three in $\mathbf{P}^8$ and let $(\mathbf{P}^2 \times \mathbf{P}^2)^5$ be the space of quintuples of image correspondences. By definition, a point $\mathbf{x}$ of $(\mathbf{P}^2 \times \mathbf{P}^2)^5$ has the form

$$\mathbf{x} = (\mathbf{q}_1, \mathbf{q}_1', \ldots, \mathbf{q}_5, \mathbf{q}_5')$$

The image correspondences comprising $\mathbf{x}$ are $\mathbf{q}_i \leftrightarrow \mathbf{q}_i'$ for $1 \le i \le 5$. Let H_i, $1 \le i \le 5$, be the hyperplane of $\mathbf{P}^8$ consisting of those points E such that $\mathbf{q}_i^\top E \mathbf{q}_i' = 0$. Let f be the rational map from $(\mathbf{P}^2 \times \mathbf{P}^2)^5$ to $G(3,8)$ defined by

$$f(\mathbf{x}) = H_1 \cap \ldots \cap H_5$$

The map f is not defined on those points which do not yield elements of $G(3,8)$. For example, f is not defined if $\mathbf{q}_1 = \mathbf{q}_2$ and $\mathbf{q}_1' = \mathbf{q}_2'$.

It is sufficient to show that a general point of $G(3,8)$ is contained in the image of f. Let E_i, $1 \le i \le 4$, be four general points of $\mathbf{P}^8$ contained within a general linear subspace of $\mathbf{P}^8$ of dimension three. Let g_1, g_2 be the homogeneous cubic polynomials in $\mathbf{q}$ defined by

$$\begin{aligned} g_1(\mathbf{q}) &= (E_1\mathbf{q} \times E_2\mathbf{q}).E_3\mathbf{q} \\ g_2(\mathbf{q}) &= (E_1\mathbf{q} \times E_2\mathbf{q}).E_4\mathbf{q} \end{aligned} \tag{5.50}$$

The plane curves g_1 and g_2 intersect at nine points. Three of these intersections are the eigenvectors obtained by solving the following generalised eigenvalue equation for $\mathbf{q}$ and λ,

$$E_1\mathbf{q} = \lambda E_2\mathbf{q} \tag{5.51}$$

From the remaining six intersections of g_1 and g_2 five are selected to serve as image points $\mathbf{q}_i$. The corresponding points $\mathbf{q}_i'$ are defined by

$$\mathbf{q}_i' = E_1\mathbf{q}_i \times E_2\mathbf{q}_i \qquad (1 \le i \le 5) \tag{5.52}$$

The exclusion from the $\mathbf{q}_i$ of any solutions of (5.51) ensures that the right-hand side of each equation of (5.52) is a well defined point of $\mathbf{P}^2$. Let $\mathbf{p}$ be the point of $(\mathbf{P}^2 \times \mathbf{P}^2)^5$ defined by the $\mathbf{q}_i \leftrightarrow \mathbf{q}_i'$ for $1 \le i \le 5$. The construction of $\mathbf{p}$ from $\mathbf{x}$ yields a local inverse to the rational map $\mathbf{x} \mapsto f(\mathbf{x})$. A general point of $G(3,8)$ is thus contained in the image of f. □

The variety $\mathcal{M}$ of essential matrices has dimension five, thus the degree of $\mathcal{M}$ is by definition equal to the number of points in the intersection of $\mathcal{M}$ with a general linear subspace L of $\mathbf{P}^8$ of dimension $3 = 8-5$. Almost all such subspaces L are of the form $L = H_1 \cap \ldots \cap H_5$. The points of $\mathcal{M} \cap L$ yield the essential matrices compatible with the five image correspondences from which the H_i are obtained. It follows that the algebraic degree of $\mathcal{M}$ is equal to the degree of reconstruction from image correspondences.

5.2.4 Ten Distinct Intersections

It has been established in Sect. 5.2.3 that there are in general only a finite number of essential matrices compatible with five given image correspondences. Demazure (1988) gives an example of five image correspondences which are compatible with exactly ten essential matrices. The example is described in this subsection. An examination of the algebra shows that multiple solutions do not occur. It follows that there are, in general, exactly ten reconstructions from five image correspondences. From this fact and from the results obtained in Sect. 5.2.3 it is deduced that the degree of the variety of essential matrices is ten. In Demazure's example the ten essential matrices all have real entries.

The example begins with the definition of a set of image correspondences:

$$\begin{aligned} \mathbf{q}_1 &= \mathbf{q}_1' = (1,0,0)^\top \\ \mathbf{q}_2 &= \mathbf{q}_2' = (0,1,0)^\top \\ \mathbf{q}_3 &= \mathbf{q}_3' = (0,0,1)^\top \\ \mathbf{q}_4 &= \mathbf{q}_5' = \mathbf{d} \\ \mathbf{q}_5 &= \mathbf{q}_4' = \mathbf{e} \end{aligned} \tag{5.53}$$

The points $\mathbf{d}$, $\mathbf{e}$ are chosen arbitrarily in the image. The first step in the construction of the example is to find the essential matrices compatible with (5.53). Let $E = (e_{ij})$ be an essential matrix such that $\mathbf{q}_i^\top E \mathbf{q}_i' = 0$, $1 \le i \le 5$. It follows from the first three correspondences of (5.53) that $e_{11} = e_{22} = e_{33}$. It thus suffices to examine the set of essential matrices with all three entries on the leading diagonal equal to zero.

Theorem 5.12. *The set $\mathcal{S}$ of essential matrices $E = (e_{ij})$ such that $e_{11} = e_{22} = e_{33} = 0$ splits into seven components, $\mathcal{A}$, $\mathcal{A}\sigma_1$, $\mathcal{A}\sigma_2$, $\mathcal{A}\sigma_3$, $\mathcal{Q}_1$, $\mathcal{Q}_2$, $\mathcal{Q}_3$. The component $\mathcal{A}$ is the set of antisymmetric matrices, the σ_i are rotations of* 180° *about the coordinate axes and the $\mathcal{Q}_i$ are quadrics, i.e. algebraic varieties of degree two.*

Proof. The substitution of E into (5.45) yields

$$\begin{aligned} &\begin{pmatrix} e_{12}^2 + e_{13}^2 & e_{13}e_{23} & e_{32}e_{12} \\ e_{13}e_{23} & e_{21}^2 + e_{23}^2 & e_{21}e_{31} \\ e_{32}e_{12} & e_{21}e_{31} & e_{31}^2 + e_{32}^2 \end{pmatrix} \begin{pmatrix} 0 & e_{12} & e_{13} \\ e_{21} & 0 & e_{23} \\ e_{31} & e_{32} & 0 \end{pmatrix} \\ &= \frac{1}{2}\mathrm{tr}(EE^\top) \begin{pmatrix} 0 & e_{12} & e_{13} \\ e_{21} & 0 & e_{23} \\ e_{31} & e_{32} & 0 \end{pmatrix} \end{aligned} \tag{5.54}$$

Equating the diagonal elements on the left-hand side of (5.54) to zero yields

$$e_{13}e_{21}e_{23} + e_{32}e_{31}e_{12} = e_{32}e_{21}e_{31} + e_{13}e_{23}e_{12} = e_{13}e_{32}e_{12} + e_{23}e_{21}e_{31} = 0 \tag{5.55}$$

Suppose firstly that none of the e_{ij}, $i \neq j$, are zero. Then (5.55) yields

$$e_{21}^2 = e_{12}^2 \qquad e_{32}^2 = e_{23}^2 \qquad e_{13}^2 = e_{31}^2 \tag{5.56}$$

It follows from (5.56) that there exist $\epsilon_i = \pm 1$ for $i = 1, 2, 3$ such that

$$e_{21} = \epsilon_1 e_{12} \qquad e_{32} = \epsilon_2 e_{23} \qquad e_{13} = \epsilon_3 e_{31} \tag{5.57}$$

It follows from (5.55) and (5.57) that $\epsilon_1\epsilon_2\epsilon_3 = -1$. The four different choices for the ϵ_i yield the four components $\mathcal{A}$, $\mathcal{A}\sigma_1$, $\mathcal{A}\sigma_2$, $\mathcal{A}\sigma_3$ of $\mathcal{S}$ as follows:

$$\begin{array}{rlll}
\mathcal{A}: & e_{32} = -e_{23} & e_{13} = -e_{31} & e_{21} = -e_{12} \\
\mathcal{A}\sigma_1: & e_{32} = -e_{23} & e_{13} = e_{31} & e_{21} = e_{12} \\
\mathcal{A}\sigma_2: & e_{32} = e_{23} & e_{13} = -e_{31} & e_{21} = e_{12} \\
\mathcal{A}\sigma_3: & e_{32} = e_{23} & e_{13} = e_{31} & e_{21} = -e_{12}
\end{array}$$

The remaining three components $\mathcal{Q}_i$ of $\mathcal{S}$ are defined by

$$\mathcal{Q}_1 = \left\{ \begin{pmatrix} 0 & e_{12} & e_{13} \\ e_{21} & 0 & 0 \\ e_{31} & 0 & 0 \end{pmatrix} \middle| \; e_{21}^2 + e_{31}^2 = e_{12}^2 + e_{13}^2 \right\}$$

$$\mathcal{Q}_2 = \left\{ \begin{pmatrix} 0 & 0 & e_{13} \\ 0 & 0 & e_{23} \\ e_{31} & e_{32} & 0 \end{pmatrix} \middle| \; e_{32}^2 + e_{31}^2 = e_{13}^2 + e_{23}^2 \right\}$$

$$\mathcal{Q}_3 = \left\{ \begin{pmatrix} 0 & e_{12} & 0 \\ e_{21} & 0 & e_{23} \\ 0 & e_{32} & 0 \end{pmatrix} \middle| \; e_{12}^2 + e_{32}^2 = e_{21}^2 + e_{23}^2 \right\} \tag{5.58}$$

It follows from (5.54) and the definitions (5.58) that each $\mathcal{Q}_i$ is contained in $\mathcal{S}$.

The set $\mathcal{S}$ thus contains the seven components listed in the statement of the theorem. It remains to show that these seven components include all of $\mathcal{S}$. Let E be an arbitrary element of $\mathcal{S}$. If all the off-diagonal entries of E are non-zero then E is either in $\mathcal{A}$ or in one of the $\mathcal{A}\sigma_i$ for $i = 1, 2, 3$. If one of the entries e_{ij} of E is zero for $i \neq j$ then it follows from (5.55) that $e_{ji} = 0$. Then (5.54) is satisfied if and only if E is contained in one of $\mathcal{Q}_1$, $\mathcal{Q}_2$, $\mathcal{Q}_3$. □

Theorem 5.13. *If the points* $\mathbf{e}$, $\mathbf{d}$ *of (5.53) are in general position then each quadric* $\mathcal{Q}_i$ *defined by (5.58) contains exactly two essential matrices compatible with the image correspondences (5.53). The two essential matrices thus obtained have real entries.*

Proof. The proof is given for $\mathcal{Q}_1$. Let $E = (e_{ij})$ be an essential matrix contained in $\mathcal{Q}_1$. Then E is compatible with the image correspondences of (5.53) if and only if

$$\mathbf{e}^\top E\mathbf{d} = \mathbf{d}^\top E\mathbf{e} = 0 \tag{5.59}$$

It follows from (5.58) and the definition of $\mathcal{Q}_1$ in (5.59) that

$$\begin{aligned}
e_1(e_{12}d_2 - e_{13}d_3) - e_{21}e_2d_1 + e_{31}e_3d_1 &= 0 \\
d_1(e_{12}e_2 - e_{13}e_3) - e_{21}d_2e_1 + e_{31}d_3e_1 &= 0
\end{aligned} \tag{5.60}$$

The addition of the two equations in (5.60) yields

$$(e_1d_2 + e_2d_1)(e_{12} - e_{21}) + (e_3d_1 + e_1d_3)(e_{31} - e_{13}) = 0 \tag{5.61}$$

The subtraction of the second equation of (5.60) from the first yields

$$(e_1d_2 - e_2d_1)(e_{12} + e_{21}) - (e_1d_3 - e_3d_1)(e_{31} + e_{13}) = 0 \tag{5.62}$$

It follows from (5.61) and (5.62) that

$$(e_1^2d_2^2 - e_2^2d_1^2)(e_{12}^2 - e_{21}^2) = (e_1^2d_3^2 - e_3^2d_1^2)(e_{31}^2 - e_{13}^2) \tag{5.63}$$

It follows from the definition of $\mathcal{Q}_1$ in (5.58) that

$$e_{12}^2 - e_{21}^2 = e_{31}^2 - e_{13}^2 \tag{5.64}$$

If $e_{12}^2 - e_{21}^2 \neq 0$, and $e_{31}^2 - e_{13}^2 \neq 0$ then (5.63) and (5.64) together yield

$$e_1^2d_2^2 - e_2^2d_1^2 = e_1^2d_3^2 - e_3^2d_1^2 \tag{5.65}$$

Equation (5.65) contradicts the hypothesis that **d**, **e** are in general position. It follows that $e_{12}^2 - e_{21}^2 = 0$ and $e_{31}^2 - e_{13}^2 = 0$. The two equations reduce to $e_{12} = \pm e_{21}$ and $e_{31} = \pm e_{13}$. In the case $e_{12} = e_{21}$, (5.61) yields $e_{31} = e_{13}$. Equation (5.62) then yields

$$\frac{e_{21}}{e_{13}} = \frac{e_1d_3 - e_3d_1}{e_1d_2 - e_2d_1}$$

thus

$$E = \begin{pmatrix} 0 & e_1d_3 - e_3d_1 & e_1d_2 - e_2d_1 \\ e_1d_3 - e_3d_1 & 0 & 0 \\ e_1d_2 - e_2d_1 & 0 & 0 \end{pmatrix} \tag{5.66}$$

In the case $e_{12} = -e_{21}$, (5.62) yields $e_{31} = e_{13}$. Equation (5.61) then yields

$$\frac{e_{12}}{e_{13}} = -\frac{e_3d_1 + e_1d_3}{e_1d_2 + e_2d_1}$$

thus

$$E = \begin{pmatrix} 0 & -(e_3d_1 + e_1d_3) & e_1d_2 - e_2d_1 \\ e_3d_1 + e_1d_3 & 0 & 0 \\ e_1d_2 - e_2d_1 & 0 & 0 \end{pmatrix} \tag{5.67}$$

The two matrices (5.66) and (5.67) have real entries. □

Theorem 5.14. *There exists a set of image correspondences* $\mathbf{q}_i \leftrightarrow \mathbf{q}_i'$ *for* $1 \leq i \leq 5$*, compatible with exactly ten essential matrices. Each of the ten essential matrices has real entries.*

Proof. It follows from Theorems 5.12 and 5.13 that the five image correspondences of (5.53) are compatible with six essential matrices, two from each of

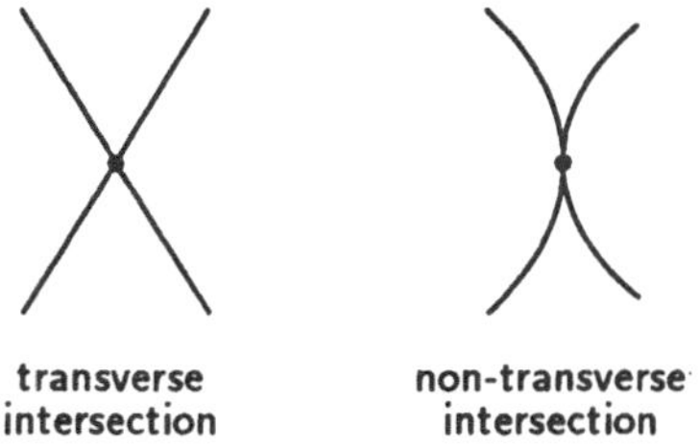

Fig. 5.2. Transverse and non-transverse intersections of two plane curves

the quadrics $\mathcal{Q}_i$, constructed as shown in the proof of Theorem 5.13. Each of these six matrices has real entries. The image correspondences of (5.53) are also compatible with three additional essential matrices with real entries obtained from the linear spaces $\mathcal{A}\sigma_i$. However, the image correspondences of (5.53) are compatible with *all* the matrices T_a in $\mathcal{A}$ which satisfy $\mathbf{d}^\top T_a \mathbf{e} = 0$. There is thus a one-parameter family of essential matrices compatible with (5.53).

A small perturbation is applied to the two image correspondences $\mathbf{q}_4 \leftrightarrow \mathbf{q}_4'$ and $\mathbf{q}_5 \leftrightarrow \mathbf{q}_5'$. If the perturbation is sufficiently small then the perturbed image correspondences are compatible with exactly nine essential matrices obtained from the $\mathcal{Q}_i$ and the $\mathcal{A}\sigma_i$. Each of these essential matrices has real entries. If the perturbation is at the same time sufficiently general then the degenerate intersection with $\mathcal{A}$ is resolved to yield one further essential matrix, also with real entries, namely T_a, where $\mathbf{a}$ is defined by

$$\mathbf{a} = (\mathbf{q}_4 \times \mathbf{q}_4') \times (\mathbf{q}_5 \times \mathbf{q}_5')$$

The required total of ten essential matrices with real entries is obtained. □

It follows from Theorem 5.14 that there are, in general, at least ten essential matrices compatible with five image correspondences. The analysis of Kruppa's method given in Sect. 5.1 shows that there are in general at most ten essential matrices compatible with five image correspondences. There are thus exactly ten essential matrices compatible with five image correspondences. This method of obtaining the degree of $\mathcal{M}$ is unsatisfactory because it mixes two different approaches to reconstruction. It is better to find a direct proof that an arbitrary linear subspace of $\mathbb{P}^8$ of dimension three intersects $\mathcal{M}$ at exactly ten points. This is done using the theory of transversality (Bruce & Giblin 1984). It is shown that $\mathcal{M}$ intersects a particular three dimensional linear subspace L of $\mathbb{P}^8$ transversally and that the intersection contains exactly ten points. The fact that the intersection is transverse ensures that each three dimensional subspace of $\mathbb{P}^8$ sufficiently close to L intersects $\mathcal{M}$ at ten points. This is enough to ensure that the degree of $\mathcal{M}$ is ten. The strategy of the proof is to show that a general three dimensional subspace containing only matrices with zeros on the leading diagonal intersects $\mathcal{M}$ transversally at ten points.

Two manifolds $\mathcal{N}_1$ and $\mathcal{N}_2$ in $\mathbb{C}^n$ are said to intersect transversally at a point $\mathbf{p}$ if the tangent spaces of $\mathcal{N}_1$ and $\mathcal{N}_2$ at $\mathbf{p}$ together span $\mathbb{C}^n$. The notation

$\mathcal{N}_1 \pitchfork \mathcal{N}_2$ means that the intersection of two manifolds $\mathcal{N}_1$ and $\mathcal{N}_2$ is transverse at each point of $\mathcal{N}_1 \cap \mathcal{N}_2$. The definition of $\mathcal{N}_1 \pitchfork \mathcal{N}_2$ includes the case in which $\mathcal{N}_1 \cap \mathcal{N}_2$ is empty. The simplest example is the intersection of two non-singular curves in the plane $\mathbf{C}^2$. If the curves are algebraic then a transverse intersection is the same thing as an intersection of multiplicity one. The tangent lines of each curve at the intersection point are unique and they are distinct, as illustrated in Fig. 5.2. If the intersection is not transverse then the curves are tangent to each other at the point of intersection. The two tangent lines then coincide.

Transverse intersections are defined here only for manifolds. The usual definition of a manifold excludes the possibility that the manifold can have a singular point. The theory of transversality is applied below to algebraic varieties which possess singular points, for example the variety of essential matrices. It is always possible to remove the singular points from the variety in order to leave a manifold to which the theory of transversality can be safely applied.

A key property of a transverse intersection of two manifolds is that it is stable under small perturbations of the manifolds. The converse is also true. An intersection of two manifolds that is not transverse is also not stable in that a small perturbation which changes the nature of the intersection can be found. For example if two plane curves are tangent at a point of intersection $\mathbf{p}$ then there exists a small perturbation which removes the tangency. If both curves are algebraic then there exists an algebraic perturbation that is small near to $\mathbf{p}$ and that removes the tangency. The algebraic perturbation may introduce new intersection points distant from $\mathbf{p}$.

Let U, V be two linear subspaces of $\mathbf{P}^n$ and let $U \oplus V$ be the smallest linear subspace of $\mathbf{P}^n$ containing U and V. The dimension of $U \oplus V$ is given by

$$\dim(U \oplus V) + \dim(U \cap V) = \dim(U) + \dim(V) \tag{5.68}$$

If $U \cap V$ is empty then (5.68) holds with $\dim(U \cap V) = -1$. The formula (5.68) applies to linear subspaces U, V in $\mathbf{C}^n$ provided $U \cap V$ is non-empty. Let $\mathcal{T}_1$ and $\mathcal{T}_2$ be the tangent spaces of the respective manifolds $\mathcal{N}_1$ and $\mathcal{N}_2$ at a point $\mathbf{p}$ in $N_1 \cap \mathcal{N}_2$. To prove that $\mathcal{N}_1$ and $\mathcal{N}_2$ intersect transversally at $\mathbf{p}$ it is sufficient to show that

$$\dim(\mathcal{T}_1) + \dim(\mathcal{T}_2) - \dim(\mathcal{T}_1 \cap \mathcal{T}_2) = n \tag{5.69}$$

The equations (5.68) and (5.69) together ensure that $\dim(\mathcal{T}_1 \oplus \mathcal{T}_2) = n$. It then follows that $\mathcal{T}_1 \oplus \mathcal{T}_2 = \mathbf{P}^n$, as required by the definition of a transverse intersection.

In the next theorem $\mathcal{D}$ is the five dimensional linear subspace of $\mathbf{P}^8$ formed from the non-zero 3×3 matrices with zeros on the leading diagonal.

Theorem 5.15. *Let H be a general hyperplane in $\mathbf{P}^8$ and let $\mathcal{M}$ be the variety of essential matrices. Let $\mathcal{D}'$, $\mathcal{M}'$ be the affine varieties defined by $\mathcal{D}' = \mathcal{D} \setminus \mathcal{D} \cap H$, $\mathcal{M}' = \mathcal{M} \setminus \mathcal{M} \cap H$, and let $\mathbf{p}$ be a general point of $\mathcal{D}' \cap \mathbf{M}'$. Then the intersection $\mathcal{D}' \cap \mathcal{M}'$ is transverse at $\mathbf{p}$.*

Proof. It is shown in Theorem 5.12 that $\mathcal{D} \cap \mathcal{M}$ splits into the seven components $\mathcal{A}$, $\mathcal{A}\sigma_i$, $\mathcal{Q}$ for $i = 1, 2, 3$. It follows that $\mathcal{D}' \cap \mathcal{M}'$ splits into seven components

$\mathcal{A}'$, $\mathcal{A}'_i$, $\mathcal{Q}'_i$ for $i = 1,2,3$. The spaces $\mathcal{D}'$, $\mathcal{M}'$ are algebraic varieties thus it suffices to find one point in each of the seven components of $\mathcal{D}' \cap \mathcal{M}'$ such that the intersection is transverse at that point.

Let $\mathcal{T}$ be the tangent space to $\mathcal{M}'$ at a general point T_a in $\mathcal{A}'$. The points in $\mathcal{T}$ are given by

$$T_a + T_w T_a + T_b$$

where $\mathbf{w}$ and $\mathbf{b}$ are variable three dimensional vectors. The diagonal entries of $T_a + T_w T_a + T_b$ are equal to the diagonal entries of $T_w T_a$. It follows that $T_a + T_w T_a + T_b$ is in $\mathcal{D}'$ if and only if

$$\begin{aligned} w_3 a_3 + w_2 a_2 &= 0 \\ w_1 a_1 + w_3 a_3 &= 0 \\ w_2 a_2 + w_1 a_1 &= 0 \end{aligned} \tag{5.70}$$

For a general choice of $\mathbf{a}$ (5.70) holds only if $\mathbf{w} = 0$. It follows that $\dim(\mathcal{T} \cap \mathcal{D}) = 2$. The application of (5.68) yields

$$\begin{aligned} \dim(\mathcal{T} \oplus \mathcal{D}') &= \dim(\mathcal{T}) + \dim(\mathcal{D}') - \dim(\mathcal{T} \cap \mathcal{D}') \\ &= 5 + 5 - 2 \\ &= 8 \end{aligned}$$

It follows that $\mathcal{T} \oplus \mathcal{D}' = \mathbf{C}^8$. The linear subspace $\mathcal{D}'$ thus intersects $\mathcal{M}'$ transversally at a general point of $\mathcal{A}'$. A similar argument shows that $\mathcal{D}'$ and $\mathcal{M}'$ intersect transversally at general points of $\mathcal{A}'\sigma_1$, $\mathcal{A}'\sigma_2$ and $\mathcal{A}'\sigma_3$.

The proof that $\mathcal{D}'$ meets $\mathcal{M}'$ transversally at a general point of each of the quadrics $\mathcal{Q}'_1$, $\mathcal{Q}'_2$, $\mathcal{Q}'_3$ is slightly more complicated. The details are given for $\mathcal{Q}'_1$. A general point E of $\mathcal{Q}'_1$ has the form

$$E = \begin{pmatrix} 0 & \cos(\theta) & \sin(\theta) \\ \cos(\phi) & 0 & 0 \\ \sin(\phi) & 0 & 0 \end{pmatrix} \tag{5.71}$$

Let $E = RT_a$, where R is an orthogonal matrix. The vector $\mathbf{a}$ and the matrix R are chosen such that

$$\begin{aligned} \mathbf{a} &= (0, -\sin(\theta), \cos(\theta))^{\mathsf{T}} \\ R &= \begin{pmatrix} 1 & 0 & 0 \\ 0 & -\cos(\phi - \theta) & \sin(\phi - \theta) \\ 0 & -\sin(\phi - \theta) & -\cos(\phi - \theta) \end{pmatrix} \end{aligned}$$

The tangent space $\mathcal{T}$ of $\mathcal{M}'$ at E is the set of points in $\mathbf{C}^8$ of the form

$$RT_a + RT_w T_a + RT_b$$

where $\mathbf{w}$, $\mathbf{b}$ are variable three dimensional vectors. Let c, s be defined by $c = \cos(\phi - \theta)$, $s = \sin(\phi - \theta)$. A point of $\mathcal{T}$ is in $\mathcal{D}'$ if and only if

$$\begin{aligned} -w_3 \cos(\theta) + w_2 \sin(\theta) &= 0 \\ (c\, w_3 + s\, w_2)\cos(\theta) - b_1 s &= 0 \\ (s\, w_3 - c\, w_2)\sin(\theta) - b_1 s &= 0 \end{aligned} \tag{5.72}$$

The equations (5.72) hold for a general choice of θ and ϕ if and only if

$$w_2 = w_3 = b_1 = 0$$

It follows that $\dim(\mathcal{T} \cap \mathcal{D}') = 2$ and that $\dim(\mathcal{T} \oplus \mathcal{D}') = 8$. The space $\mathcal{D}'$ thus intersects $\mathcal{M}'$ transversally at a general point of $\mathcal{Q}_1'$. The proofs that $\mathcal{D}'$ intersects $\mathcal{M}'$ transversally at a general point of $\mathcal{Q}_2'$ and $\mathcal{Q}_3'$ are similar. □

Theorem 5.16. *The degree of the variety $\mathcal{M}$ of essential matrices is ten.*

Proof. The notation of Theorem 5.15 is employed. It suffices to find a single three dimensional subspace L of $\mathbf{P}^8$ that intersects $\mathcal{M}$ transversally at ten points. The stability of transverse intersections under small perturbations ensures that any other three dimensional subspace of $\mathbf{P}^8$ sufficiently close to L also intersects $\mathcal{M}$ at ten points.

It is shown that a general subspace L of dimension three contained within the subspace $\mathcal{D}'$ defined in the proof of Theorem 5.15 intersects $\mathcal{M}'$ transversally. The general subspace L intersects $\mathcal{D}' \cap \mathcal{M}'$ at finitely many points and it does not contain any points at which the intersection $\mathcal{D}' \cap \mathcal{M}'$ is not transverse. Let $\mathbf{p}$ be a point at which L intersects $\mathcal{D}' \cap \mathcal{M}'$. Let $\mathcal{T}$ be the tangent space to $\mathcal{M}'$ at $\mathbf{p}$. It is shown in Theorem 5.15 that $\dim(\mathcal{T} \cap \mathcal{D}') = 2$. The dimension of $\mathcal{D}'$ is five, and L is a general three dimensional subspace of $\mathcal{D}'$, thus $L \oplus (\mathcal{T} \cap \mathcal{D}') = \mathcal{D}'$. It follows that

$$L \oplus \mathcal{T} = \mathcal{D}' \oplus \mathcal{T} = \mathbf{C}^8$$

Thus L intersects $\mathcal{M}$ transversally.

The subspace L has the form $\mathcal{D}' \cap H_4 \cap H_5$, where H_4, H_5 are hyperplanes in $\mathbf{C}^8$ arising from the image correspondences $\mathbf{q}_4 \leftrightarrow \mathbf{q}_4'$ and $\mathbf{q}_5 \leftrightarrow \mathbf{q}_5'$. It follows from Theorem 5.14 that $L \cap \mathcal{M}' = L \cap \mathcal{M}$ contains exactly ten points. Thus the degree of $\mathcal{M}$ is ten. □

The set $\mathcal{R} \subset (\mathbf{P}^2 \times \mathbf{P}^2)^5$ of sets of image correspondences $\mathbf{q}_i \leftrightarrow \mathbf{q}_i'$, $1 \leq i \leq 5$, compatible with exactly ten distinct essential matrices with real entries is open. This is because each of the ten essential matrices arises from a transverse intersection which is preserved under small perturbations of the points $\mathbf{q}_i$, $\mathbf{q}_i'$.

5.3 Reconstruction up to a Collineation

Sturm (1869) describes a geometrical method for obtaining the epipoles compatible with seven image correspondences. He assumes only that the images are formed by a linear projection. The full calibration of each camera is not required. The seven image correspondences are in general sufficient to ensure that there are only a finite number of compatible epipoles. It is shown in this section that there are in general exactly three possible sets of epipoles. The number of solutions obtained in the minimal case of reconstruction up to a collineation is thus significantly less than the ten solutions obtained in the case of reconstruction with known camera calibration. More image correspondences are required for reconstruction up to a collineation, but the underlying equations are simpler.

Sturm's method is described in Sect. 5.3.1. A more modern algebraic method for obtaining the three sets of epipoles compatible with seven image correspondences is described in Sect. 5.3.2.

5.3.1 Sturm's Method

Sturm's method uses the fact that the epipolar transformation preserves cross ratios. A point $\mathbf{p}$ in the first image and $\mathbf{p}'$ in the second image are possible epipoles compatible with the image correspondences $\mathbf{q}_i \leftrightarrow \mathbf{q}_i'$ for $1 \leq i \leq n$ if and only if

$$\langle \mathbf{p}, \mathbf{q}_i \rangle \barwedge \langle \mathbf{p}', \mathbf{q}_i' \rangle \qquad (1 \leq i \leq n) \tag{5.73}$$

As in Sect. 5.1, objects defined in the second image are given a superscript $'$.

The first step is to show that any six of the image correspondences constrain the epipole to lie on a cubic plane curve.

Theorem 5.17. *Let image correspondences $\mathbf{q}_i \leftrightarrow \mathbf{q}_i'$ for $1 \leq i \leq 6$, be given in general position. Then the epipoles $\mathbf{p}$, $\mathbf{p}'$ are each constrained to lie on a cubic plane curve.*

Proof. Coordinates are chosen in each image such that

$$\begin{aligned} \mathbf{q}_1 &= (1,0,0)^\top & \mathbf{q}_1' &= (1,0,0)^\top \\ \mathbf{q}_2 &= (0,1,0)^\top & \mathbf{q}_2' &= (0,1,0)^\top \\ \mathbf{q}_3 &= (0,0,1)^\top & \mathbf{q}_3' &= (0,0,1)^\top \\ \mathbf{q}_4 &= (1,1,1)^\top & \mathbf{q}_4' &= (1,1,1)^\top \end{aligned}$$

Let the components of the $\mathbf{q}_i$, $\mathbf{q}_i'$ be $q_{ij} = (\mathbf{q}_i)_j$ and $q_{ij}' = (\mathbf{q}_i')_j$. The equations (5.3) in Sect. 5.1.1 obtained in the first part of Kruppa's method are valid for reconstruction up to a collineation because they depend only on the fact that the epipolar transformation preserves the cross ratio. As in Sect. 5.1.1, quadratic transformations Φ, Φ' of the two images are defined by

$$\begin{aligned} \mathbf{u} &= \Phi(\mathbf{p}) = (p_2p_3, p_3p_1, p_1p_2)^\top \\ \mathbf{u}' &= \Phi'(\mathbf{p}') = (p_1'p_3', p_3'p_1', p_1'p_2')^\top \end{aligned}$$

Equation (5.73) yields three equations of the same form as (5.15),

$$\frac{u_3q_{i3} - u_1q_{i1}}{u_3q_{i3} - u_2q_{i2}} = \frac{u_3'q_{i3}' - u_1'q_{i1}'}{u_3'q_{i3}' - u_2'q_{i2}'} \qquad (i = 4,5,6) \tag{5.74}$$

On multiplying out the denominators of each of the equations in (5.74) three constraints on $\mathbf{u}'$ are obtained,

$$\mathbf{a}_i(\mathbf{u}).\mathbf{u}' = 0 \qquad (i = 4, 5, 6)$$

The vectors $\mathbf{a}_i$ are linear functions of $\mathbf{u}$ defined by

$$\mathbf{a}_i = \left(q'_{i1}(u_3 q_{i3} - u_2 q_{i2}), q'_{i2}(u_1 q_{i1} - u_3 q_{i3}), q'_{i3}(u_2 q_{i2} - u_1 q_{i1})\right)^\top \tag{5.75}$$

Let k be the cubic plane curve defined by

$$k(\mathbf{u}) = \mathbf{a}_4.(\mathbf{a}_5 \times \mathbf{a}_6) \tag{5.76}$$

The condition that the three equations of (5.74) have a solution for $\mathbf{u}'$ is $k(\mathbf{u}) = 0$. It follows from the definition (5.76) of k that the curve contains the three points $(1,0,0)^\top$, $(0,1,0)^\top$, $(0,0,1)^\top$. These are the fundamental points of the quadratic transformation Φ^{-1}. Let f be the curve defined in the image plane by

$$f(\mathbf{p}) = k(\Phi^{-1}(\mathbf{u})) \tag{5.77}$$

The curve f is of degree three because k contains each of the fundamental points of Φ^{-1}. The curve f is the required cubic constraint on the epipole $\mathbf{p}$. It contains the fundamental points of Φ, namely, $(1,0,0)^\top$, $(0,1,0)^\top$, $(0,0,1)^\top$. A similar argument leads to a cubic constraint on $\mathbf{p}'$. □

Theorem 5.18. *Let f be the cubic constraint on the epipole $\mathbf{p}$ obtained in Theorem 5.17. Then f contains the $\mathbf{q}_i$ for $1 \le i \le 6$. In addition, f contains a point $\mathbf{b}$ which depends on the $\mathbf{q}_i \leftrightarrow \mathbf{q}'_i$ for $1 \le i \le 5$, but which is independent of $\mathbf{q}_6 \leftrightarrow \mathbf{q}'_6$.*

Proof. The notation of Theorem 5.17 is employed. The curve f contains the fundamental points $\mathbf{q}_1$, $\mathbf{q}_2$, $\mathbf{q}_3$ of Φ. A short calculation shows that

$$\mathbf{a}_4(\Phi(\mathbf{q}_4)) = \mathbf{a}_5(\Phi(\mathbf{q}_5)) = \mathbf{a}_6(\Phi(\mathbf{q}_6)) = 0 \tag{5.78}$$

It follows from (5.77), (5.78) and the definition of k in (5.76) that f contains $\mathbf{q}_4$, $\mathbf{q}_5$, $\mathbf{q}_6$. The point $\mathbf{b}$ is obtained. Let A_4, A_5 be 3×3 matrices defined such that

$$\mathbf{a}_4(\mathbf{u}) = A_4\mathbf{u} \qquad \mathbf{a}_5(\mathbf{u}) = A_5\mathbf{u} \tag{5.79}$$

Equation (5.75) shows that A_4 depends only on $\mathbf{q}_4 \leftrightarrow \mathbf{q}'_4$ and A_5 depends only on $\mathbf{q}_5 \leftrightarrow \mathbf{q}'_5$. It follows from (5.78) that $\det(A_4) = 0$, $\det(A_5) = 0$. The generalised eigenvector equation

$$\det(\mu A_4 - \lambda A_5) = 0 \tag{5.80}$$

has three solutions for $(\lambda, \mu)^\top$, regarded as a point of $\mathbf{P}^1$. Two of these solutions are $(\lambda,\mu)^\top = (1,0)^\top$ and $(0,1)^\top$. The point $\mathbf{b}$ is defined such that $\Phi(\mathbf{b})$ is the eigenvector of $\mu A_4 - \lambda A_5$ arising from the third solution of (5.80). It follows from the definition of $\mathbf{b}$ that $\mu A_4\mathbf{b} = \lambda A_5\mathbf{b}$, where $\lambda \neq 0, \mu \neq 0$. Equation (5.79) yields

$$\mathbf{a}_4(\Phi(\mathbf{b})) \times \mathbf{a}_5(\Phi(\mathbf{b})) = 0 \tag{5.81}$$

The result $f(\mathbf{b}) = 0$ follows from (5.81) and the definitions of k and f. □

Theorems 5.17 and 5.18 are used to show that there are at most three sets of epipoles compatible with seven image correspondences.

Theorem 5.19. *Let image correspondences* $\mathbf{q}_i \leftrightarrow \mathbf{q}_i'$ *for* $1 \leq i \leq 7$ *be given in general position. Then there are at most three pairs of epipoles* $\mathbf{p}$, $\mathbf{p}'$ *such that*

$$\langle \mathbf{p}, \mathbf{q}_i \rangle \overline{\wedge} \langle \mathbf{p}', \mathbf{q}_i' \rangle \qquad (1 \leq i \leq 7) \tag{5.82}$$

Proof. It follows from Theorem 5.17 that the correspondences $\mathbf{q}_i \leftrightarrow \mathbf{q}_i'$ for $1 \leq i \leq 6$ constrain the epipole $\mathbf{p}$ to lie on a cubic plane curve f_1. Similarly, the correspondences $\mathbf{q}_i \leftrightarrow \mathbf{q}_i'$ for $1 \leq i \leq 5$ and $i = 7$ constrain $\mathbf{p}$ to lie on a second cubic plane curve f_2. The points $\mathbf{p}$ such that (5.82) holds are included amongst the nine intersections of f_1 and f_2. Six of these intersections, namely the $\mathbf{q}_i$ for $1 \leq i \leq 5$ and the point $\mathbf{b}$ found in Theorem 5.18, do not yield epipoles $\mathbf{p}$, $\mathbf{p}'$ such that (5.82) holds. It follows that there are at most $9 - 6 = 3$ possible epipoles. □

5.3.2 An Algebraic Method

In this subsection the result of Theorem 5.19 is obtained using an algebraic method. As part of the proof, an example is given in which there are exactly three epipoles compatible with seven image correspondences.

Theorem 5.20. *Let the image correspondences* $\mathbf{q}_i \leftrightarrow \mathbf{q}_i'$ *for* $1 \leq i \leq 7$, *be given in general position. Then there are exactly three sets of epipoles* $\mathbf{p}$, $\mathbf{p}'$ *such that (5.82) holds.*

Proof. It follows from Theorem 2.32 that each matrix E of rank two such that

$$\mathbf{q}_i'^{\top} E \mathbf{q}_i = 0 \qquad (1 \leq i \leq 7) \tag{5.83}$$

yields a pair of epipoles. An argument similar to that given in Theorem 5.10 establishes that the matrices E satisfying (5.83) form a line k in $\mathbb{P}^8$. The condition $\det(E) = 0$ imposes a cubic constraint on the points of k. If this cubic constraint is non-trivial then there are at most three points of k which satisfy $\det(E) = 0$. There are then at most three pairs of epipoles $\mathbf{p}$, $\mathbf{p}'$ such that (5.82) holds.

To complete the proof a single example is given. In the example the constraint $\det(E) = 0$ is non-trivial when restricted to the line k and the seven image correspondences are compatible with *exactly* three distinct pairs of epipoles. The first five image correspondences are chosen to be

$$(1,0,0)^{\top} \leftrightarrow (1,0,0)^{\top}$$

$$\begin{aligned}
(0,1,0)^\top &\leftrightarrow (0,1,0)^\top \\
(0,0,1)^\top &\leftrightarrow (0,0,1)^\top \\
(1,0,2)^\top &\leftrightarrow (1,0,2)^\top \\
(0,1,2)^\top &\leftrightarrow (0,1,2)^\top
\end{aligned} \tag{5.84}$$

The matrices E such that $\mathbf{q}'^\top E\mathbf{q} = 0$ for the five correspondences of (5.84) are given by

$$E = \begin{pmatrix} 0 & d & -h \\ g & 0 & f \\ h & -f & 0 \end{pmatrix} \tag{5.85}$$

It follows from (5.85) that

$$\det(E) = fh(d+g)$$

The three factors of $\det(E)$ each yield an epipole $\mathbf{p}$ such that $E\mathbf{p} = 0$,

$$\begin{aligned}
h &= 0 & \mathbf{p} &= (-f, 0, g)^\top \\
f &= 0 & \mathbf{p} &= (0, h, d)^\top \\
d + g &= 0 & \mathbf{p} &= (f, h, d)^\top
\end{aligned} \tag{5.86}$$

The two remaining image correspondences $\mathbf{q}_6 \leftrightarrow \mathbf{q}'_6$, $\mathbf{q}_7 \leftrightarrow \mathbf{q}'_7$ are chosen such that the equations

$$\mathbf{q}_6'^\top E\mathbf{q}_6 = 0 \qquad \mathbf{q}_7'^\top E\mathbf{q}_7 = 0$$

realise two general linear constraints on E. This is sufficient to obtain the three possible epipoles in (5.86) as functions of $\mathbf{q}_6 \leftrightarrow \mathbf{q}'_6$ and $\mathbf{q}_7 \leftrightarrow \mathbf{q}'_7$. □

Another way of showing that there are exactly three epipoles compatible with seven image correspondences is to use the critical surfaces for reconstruction up to a collineation described in Sect. 3.5. Let $\mathbf{p}$, $\mathbf{p}'$ be a pair of epipoles compatible with a general set of image correspondences $\mathbf{q}_i \leftrightarrow \mathbf{q}'_i$ for $1 \leq i \leq 7$. reconstruction up to a collineation yields the positions $\mathbf{o}$, $\mathbf{a}$ of the optical centre of the camera before and after the displacement. Let $\mathbf{p}_i$, $1 \leq i \leq 7$ be the points in space such that $\mathbf{p}_i$ projects to $\mathbf{q}_i$ when the optical centre of the camera is at $\mathbf{o}$ and to $\mathbf{q}'_i$ when the optical centre of the camera is at $\mathbf{a}$. The quadrics in $\mathbb{P}^3$ are a nine-parameter linear family, thus there is in general a unique non-singular quadric ψ containing $\mathbf{o}$, $\mathbf{a}$ and the points $\mathbf{p}_i$ for $1 \leq i \leq 7$. The three possible epipoles in the first image are the projection of the line $\langle \mathbf{o}, \mathbf{a} \rangle$ and the projections of the two generators of ψ passing through $\mathbf{o}$. Further details are given at the end of Sect. 3.5.

5.4 Reconstruction from Five Image Velocity Vectors

It is noted in Sect. 4.1 that at least five image velocity vectors are required to ensure that there are only a finite number of possible reconstructions. This is because each velocity vector yields one constraint on the velocity $\{\mathbf{v}, \mathbf{w}\}$ of the rigid surface giving rise to the image velocity vectors. In order to be certain that five image velocity vectors are sufficient to ensure only a finite number of reconstructions it is necessary to check that all five constraints are independent. Once this is done, the degree of reconstruction from image velocities is defined to be the number of reconstructions compatible with five image velocity vectors. In this section two proofs are given that five image velocity vectors are sufficient and that the degree of reconstruction is ten.

The first proof uses the quartic polynomial constraint on the translational velocity obtained in Sect. 4.3. Five image velocity vectors yield two independent quartic constraints, which intersect at 16 points. Exactly ten of these intersections yield reconstructions. The second proof uses the critical surfaces for reconstruction from image velocities. The proof begins by assuming that one of the possible camera velocities is given. After some algebra, the nine remaining reconstructions are represented as the intersection points of two cubic plane curves. The first proof is described in Sects. 5.4.1–5.4.3, and the second proof is described in Sect. 5.4.4 and Sect. 5.4.5.

5.4.1 Preliminary Results

The proof that there are, in general, exactly ten reconstructions compatible with five image velocity vectors is obtained as the culmination of a sequence of algebraic results. The strategy is to show that all the reconstructions are obtained from certain of the intersections of two quartic plane curves and then to count the number of intersections involved.

The first step is to show that the distances to points on the body surface can be eliminated without introducing any additional spurious solutions for the rigid velocity $\{\mathbf{v}, \mathbf{w}\}$. It is recalled from Sect. 4.1 that the image velocity vector $\dot{\mathbf{q}}$ with base point $\mathbf{q}$ on the unit sphere is given in terms of the translational velocity $\mathbf{v}$, the angular velocity $\mathbf{w}$ and the inverse distance K to the rigid surface from the optical centre of the camera, by the equation

$$\dot{\mathbf{q}} = (\mathbf{v} - (\mathbf{v}.\mathbf{q})\mathbf{q})K + \mathbf{w} \times \mathbf{q}$$

Let N be the number of reconstructions obtained from five general image velocity vectors. The total N includes any reconstructions for which $\mathbf{v}$ or $\mathbf{w}$ are complex. It also includes those for which some of the inverse distances K_i to points on the surface giving rise to the image velocity vectors are negative. The complex solutions are not physically realisable, but they are an inescapable part of the algebra. The number N obtained by including the complex solutions is a better measure of the difficulty of reconstruction than the (variable) number obtained by omitting the complex solutions.

Proposition 5.21. *Let* $\dot{\mathbf{q}}_i$ *for* $1 \leq i \leq 5$ *be a set of five image velocity vectors in general position, with base points* $\mathbf{q}_i$. *If* $\mathbf{v}$ *is a non-zero translational velocity* *and* $\mathbf{w}$ *is an angular velocity such that*

$$(\dot{\mathbf{q}}_i - \mathbf{w} \times \mathbf{q}_i).(\mathbf{v} \times \mathbf{q}_i) = 0 \qquad (1 \le i \le 5) \qquad (5.87)$$

then inverse distances K_i *can be found such that*

$$\dot{\mathbf{q}}_i = (\mathbf{v} - (\mathbf{v}.\mathbf{q}_i)\mathbf{q}_i)K_i + \mathbf{w} \times \mathbf{q}_i \qquad (1 \le i \le 5) \qquad (5.88)$$

Proof. The equations (5.87) are obtained by eliminating the K_i from (5.88). Suppose first that $\mathbf{v} \times \mathbf{q}_i \ne 0$ for $1 \le i \le 5$. It follows from (5.87) that there exist scalars a_i, b_i such that

$$\dot{\mathbf{q}}_i = a_i\mathbf{v} + b_i\mathbf{q}_i + \mathbf{v} \times \mathbf{q}_i \qquad (1 \le i \le 5) \qquad (5.89)$$

The scalar product of (5.89) with $\mathbf{q}_i$ yields $a_i\mathbf{v}.\mathbf{q}_i + b_i = 0$. On substituting for b_i in (5.89) the following equation is obtained:

$$\dot{\mathbf{q}}_i = [\mathbf{v} - (\mathbf{v}.\mathbf{q}_i)\mathbf{q}_i]a_i + \mathbf{v} \times \mathbf{q}_i \qquad (1 \le i \le 5) \qquad (5.90)$$

Equation (5.88) follows from (5.90) on setting $K_i = a_i$.

The proof is completed by showing that in general $\mathbf{v} \times \mathbf{q}_i \ne 0$ for $1 \le i \le 5$. Suppose, for example, that $\mathbf{v} \times \mathbf{q}_5 = 0$. Then, without loss of generality, set $\mathbf{v} = \mathbf{q}_5$. This value of $\mathbf{v}$ is a suitable translational velocity only if there exists a vector $\mathbf{w}$ such that

$$(\dot{\mathbf{q}}_i - \mathbf{w} \times \mathbf{q}_i).(\mathbf{q}_i \times \mathbf{q}_5) = 0 \qquad (1 \le i \le 4) \qquad (5.91)$$

or equivalently,

$$\dot{\mathbf{q}}_i.(\mathbf{q}_i \times \mathbf{q}_5) = (\mathbf{w} \times \mathbf{q}_i).(\mathbf{q}_i \times \mathbf{q}_5) \qquad (1 \le i \le 5) \qquad (5.92)$$

As the $\dot{\mathbf{q}}_i$ vary independently, the vector with components made from the left-hand sides of the equations (5.92),

$$(\dot{\mathbf{q}}_1.(\mathbf{q}_1 \times \mathbf{q}_5), \dot{\mathbf{q}}_2.(\mathbf{q}_2 \times \mathbf{q}_5), \dot{\mathbf{q}}_3.(\mathbf{q}_3 \times \mathbf{q}_5), \dot{\mathbf{q}}_4.(\mathbf{q}_4 \times \mathbf{q}_5))^\top$$

varies over the whole of $\mathbb{R}^4$. As $\mathbf{w}$ varies, the vector with components made from the right-hand sides of the equations (5.92),

$$\begin{pmatrix} (\mathbf{w} \times \mathbf{q}_1).(\mathbf{q}_1 \times \mathbf{q}_5) \\ (\mathbf{w} \times \mathbf{q}_2).(\mathbf{q}_2 \times \mathbf{q}_5) \\ (\mathbf{w} \times \mathbf{q}_3).(\mathbf{q}_3 \times \mathbf{q}_5) \\ (\mathbf{w} \times \mathbf{q}_4).(\mathbf{q}_4 \times \mathbf{q}_5) \end{pmatrix}$$

varies over a subspace of $\mathbb{R}^4$ of dimension three. The vector $\mathbf{v} = \mathbf{q}_5$ is real, thus $\mathbf{w}$, if it exists, is also real. It follows that for a general choice of the $\dot{\mathbf{q}}_i$, (5.91) has no solution for $\mathbf{w}$. □

The next proposition is a useful consequence of the general position of the base points.

Proposition 5.22. *If* $\mathbf{v}$ *varies with the five base points* $\mathbf{q}_i$ *fixed and in general position, then i) no three of the vectors* $(\mathbf{v}.\mathbf{q}_i)\mathbf{q}_i - \mathbf{v}$ *are ever parallel; and ii) the five vectors* $(\mathbf{v}.\mathbf{q}_i)\mathbf{q}_i - \mathbf{v}$ *always span* $\mathbb{R}^3$.

Proof. The statement (i) is proved for the three vectors given by $i = 1, 2, 3$. If

$$(\mathbf{v}.\mathbf{q}_1)\mathbf{q}_1 - \mathbf{v} \qquad \text{and} \qquad (\mathbf{v}.\mathbf{q}_2)\mathbf{q}_2 - \mathbf{v} \tag{5.93}$$

are linearly independent then the result follows. If the two vectors of (5.93) are linearly dependent, then there exist scalars λ_1, λ_2, not both zero, such that

$$\lambda_1[(\mathbf{v}.\mathbf{q}_1)\mathbf{q}_1 - \mathbf{v}] + \lambda_2[(\mathbf{v}.\mathbf{q}_2)\mathbf{q}_2 - \mathbf{v}] = 0 \tag{5.94}$$

If $\lambda_1 + \lambda_2 = 0$ then (5.94) yields

$$(\mathbf{v}.\mathbf{q}_1)\mathbf{q}_1 - (\mathbf{v}.\mathbf{q}_2)\mathbf{q}_2 = 0 \tag{5.95}$$

In general, $\mathbf{q}_1$ and $\mathbf{q}_2$ are not linearly dependent. Thus (5.95) holds only if $\mathbf{v}.\mathbf{q}_1 = \mathbf{v}.\mathbf{q}_2 = 0$. It follows that $\mathbf{v} = \mathbf{q}_1 \times \mathbf{q}_2$. Suppose next that $\lambda_1 + \lambda_2 \neq 0$. It follows from (5.94) that $\mathbf{v}$ is linearly dependent on $\mathbf{q}_1$ and $\mathbf{q}_2$. Let a_1, a_2 be scalars such that $\mathbf{v} = a_1\mathbf{q}_1 + a_2\mathbf{q}_2$. With this value of $\mathbf{v}$, (5.94) yields

$$\begin{aligned} 0 &= \lambda_1 a_2[(\mathbf{q}_2.\mathbf{q}_1)\mathbf{q}_1 - \mathbf{q}_2] + \lambda_2 a_1[(\mathbf{q}_2.\mathbf{q}_1)\mathbf{q}_2 - \mathbf{q}_1] \\ &= (\mathbf{q}_2 \times \mathbf{q}_1) \times (\lambda_1 a_2 \mathbf{q}_1 - \lambda_2 a_1 \mathbf{q}_2) \end{aligned} \tag{5.96}$$

The vectors $\mathbf{q}_1$, $\mathbf{q}_2$ are in general position, thus $(\mathbf{q}_1 \times \mathbf{q}_2) \times \mathbf{q}_1$ and $(\mathbf{q}_1 \times \mathbf{q}_2) \times \mathbf{q}_2$ are linearly independent. It follows from (5.96) that $\mathbf{v} = \mathbf{q}_1$ and $\lambda_2 = 0$ or alternatively $\mathbf{v} = \mathbf{q}_2$ and $\lambda_1 = 0$.

It is established that (5.94) holds only if $\mathbf{v} = \mathbf{q}_1$, $\mathbf{q}_2$ or $\mathbf{v} = \mathbf{q}_1 \times \mathbf{q}_2$. Thus, to prove (i) it is sufficient to observe that if $\mathbf{v} = \mathbf{q}_1$, $\mathbf{q}_2$ or $\mathbf{v} = \mathbf{q}_1 \times \mathbf{q}_2$ then $(\mathbf{v}.\mathbf{q}_3)\mathbf{q}_3 - \mathbf{v}$ is linearly independent of $(\mathbf{v}.\mathbf{q}_1)\mathbf{q}_1 - \mathbf{v}$ and $(\mathbf{v}.\mathbf{q}_2)\mathbf{q}_2 - \mathbf{v}$.

The statement (ii) is proved. It suffices to consider those values of $\mathbf{v}$ for which the vectors $(\mathbf{v}.\mathbf{q}_i)\mathbf{q}_i - \mathbf{v}$, $1 \leq i \leq 4$, do not span $\mathbb{R}^3$. The cubic polynomials f_{ijk} are defined for $1 \leq i, j, k \leq 5$ as follows:

$$f_{ijk}(\mathbf{v}) = \det \begin{pmatrix} (\mathbf{v}.\mathbf{q}_i)\mathbf{q}_i^\top - \mathbf{v}^\top \\ (\mathbf{v}.\mathbf{q}_j)\mathbf{q}_j^\top - \mathbf{v}^\top \\ (\mathbf{v}.\mathbf{q}_k)\mathbf{q}_k^\top - \mathbf{v}^\top \end{pmatrix} \tag{5.97}$$

It is shown in Theorem 4.29 that any two of the cubic plane curves f_{ijk} intersect in nine distinct points. Let $\mathbf{u}_i$, $1 \leq i \leq 9$ be the points of intersection of f_{123} and f_{124}. The values of $\mathbf{v}$ for which the vectors $(\mathbf{v}.\mathbf{q}_i)\mathbf{q}_i - \mathbf{v}$ for $1 \leq i \leq 4$ do not span $\mathbb{R}^3$ are amongst the $\mathbf{u}_i$. It follows from (i) that for each value $\mathbf{v} = \mathbf{u}_i$ the vectors $(\mathbf{v}.\mathbf{q}_i)\mathbf{q}_i - \mathbf{v}$ for $1 \leq i \leq 4$ span a plane Π_i. By hypothesis, $\mathbf{q}_5$ is in general position, thus for each i, the vector $(\mathbf{u}_i.\mathbf{q}_5)\mathbf{q}_5 - \mathbf{u}_i$ is not in Π_i. Part (ii) of the proposition follows. □

Proposition 5.23. *Each of the cubic plane curves f_{ijk} defined by (5.97) does not, in general, possess a singular point.*

Proof. The result is proved for f_{123}. The curve f_{123} has a singular point if and only if the vector equation $\nabla f_{123} = 0$ has a solution for $\mathbf{v}$. Thus f_{123} has a singular point if and only if the resultant $r(\mathbf{q}_1, \mathbf{q}_2, \mathbf{q}_3)$ obtained by eliminating $\mathbf{v}$ from $\partial f_{123}/\partial v_1$, $\partial f_{123}/\partial v_2$ and $\partial f_{123}/\partial v_3$ satisfies

$$r(\mathbf{q}_1, \mathbf{q}_2, \mathbf{q}_3) = 0 \tag{5.98}$$

Equation (5.98) is a polynomial constraint on the components of the base points $\mathbf{q}_1$, $\mathbf{q}_2$, $\mathbf{q}_3$. Either (5.98) holds for all $\mathbf{q}_1$, $\mathbf{q}_2$, $\mathbf{q}_3$ or (5.98) does not hold for a general choice of $\mathbf{q}_1$, $\mathbf{q}_2$, $\mathbf{q}_3$. To prove the proposition it suffices to find just one set $\mathbf{q}_1, \mathbf{q}_2, \mathbf{q}_3$ for which (5.98) does not hold, or equivalently, for which f_{123} does not possess a singular point. To this end, select

$$\begin{aligned}\mathbf{q}_1 &= (-1/2\sqrt{2}, \sqrt{3}/2\sqrt{2}, 1/\sqrt{2})^\top \\ \mathbf{q}_2 &= (-1/2\sqrt{2}, -\sqrt{3}/2\sqrt{2}, 1/\sqrt{2})^\top \\ \mathbf{q}_3 &= (0, 0, 1)^\top\end{aligned}$$

The substitution of the above values of $\mathbf{q}_1$, $\mathbf{q}_2$, $\mathbf{q}_3$ into (5.97) yields

$$f_{123}(\mathbf{v}) = \frac{\sqrt{3}}{4}[v_3^2 v_1 + v_3 v_2^2 + \frac{1}{4}(3v_2^2 v_1 - v_1^3)]$$

from which it follows that

$$\nabla f_{123}(\mathbf{v}) = \frac{\sqrt{3}}{4}(v_3^2 + \frac{3}{4}(v_2^2 - v_1^2), 2v_3 v_2 + \frac{3}{2}v_2 v_1, 2v_3 v_1 + v_2^2)^\top \tag{5.99}$$

It follows from (5.99) by direct calculation that $\nabla f_{123}(\mathbf{v}) \neq 0$ for all $\mathbf{v}$. Thus (5.98) is a non-trivial constraint on $\mathbf{q}_1$, $\mathbf{q}_2$ and $\mathbf{q}_3$. □

5.4.2 The Quartic Constraints

Two quartic polynomial constraints on $\mathbf{v}$ of the form defined in (4.59) are required. Define the vectors $\mathbf{r}_i$ by $\mathbf{r}_i = \dot{\mathbf{q}}_i \times \mathbf{q}_i$ for $1 \leq i \leq 5$. The first four equations of (5.90) are compatible with a single value of $\mathbf{w}$ if and only if $q_1(\mathbf{v}) = 0$, where q_1 is the quartic defined by

$$q_1(\mathbf{v}) = \det \begin{pmatrix} (\mathbf{v}.\mathbf{q}_1)\mathbf{q}_1^\top - \mathbf{v}^\top & \mathbf{r}_1.\mathbf{v} \\ (\mathbf{v}.\mathbf{q}_2)\mathbf{q}_2^\top - \mathbf{v}^\top & \mathbf{r}_2.\mathbf{v} \\ (\mathbf{v}.\mathbf{q}_3)\mathbf{q}_3^\top - \mathbf{v}^\top & \mathbf{r}_3.\mathbf{v} \\ (\mathbf{v}.\mathbf{q}_4)\mathbf{q}_4^\top - \mathbf{v}^\top & \mathbf{r}_4.\mathbf{v} \end{pmatrix} \tag{5.100}$$

Each set of four vectors drawn from the five image velocity vectors yields a quartic constraint on $\mathbf{v}$ similar in form to (5.100). Of these constraints one further is required. The following quartic q_2 is chosen:

$$q_2(\mathbf{v}) = \det \begin{pmatrix} (\mathbf{v}.\mathbf{q}_1)\mathbf{q}_1^\top - \mathbf{v}^\top & \mathbf{r}_1.\mathbf{v} \\ (\mathbf{v}.\mathbf{q}_2)\mathbf{q}_2^\top - \mathbf{v}^\top & \mathbf{r}_2.\mathbf{v} \\ (\mathbf{v}.\mathbf{q}_3)\mathbf{q}_3^\top - \mathbf{v}^\top & \mathbf{r}_3.\mathbf{v} \\ (\mathbf{v}.\mathbf{q}_5)\mathbf{q}_5^\top - \mathbf{v}^\top & \mathbf{r}_5.\mathbf{v} \end{pmatrix} \qquad (5.101)$$

The quartics q_1, q_2 are combinations of the cubic polynomials f_{ijk} defined by (5.97). It follows from (5.100) and (5.101) that any translational velocity $\mathbf{v}$ compatible with (5.90) is included amongst the 16 common zeros of q_1 and q_2.

The plane curves q_1 and q_2 have $4 \times 4 = 16$ intersections because each curve is of degree four. However, some of these intersections may be multiple. For example, if q_1 and q_2 are tangent at a point $\mathbf{p}$ then $\mathbf{p}$ is counted twice on the list of intersections. A multiple intersection can be distinguished from a simple intersection using the tangent lines of q_1 and q_2. If $q_1(\mathbf{p}) = q_2(\mathbf{p}) = 0$, but if q_1 and q_2 have distinct tangents at $\mathbf{p}$ then the intersection at $\mathbf{p}$ is a simple intersection counted once only on the list of intersections. A key property of simple intersections is that they are stable under small perturbations of q_1 and q_2. They are examples of transverse intersections, as described in Sect. 5.2.4.

Theorem 5.24. *The quartics q_1, q_2 have, in general, exactly 16 distinct common zeros.*

Proof. It is shown that there exists a pair of quartics q_1, q_2 with 16 distinct common zeros. This suffices to prove the theorem, because the condition that q_1, q_2 have one or more multiple common zeros is an algebraic constraint on the components of the $\mathbf{q}_i$, $\dot{\mathbf{q}}_i$ for $1 \le i \le 5$. A single example is enough to show that this constraint is non-trivial.

Let q_1, q_2 have a multiple common zero at $\mathbf{u}$ and n additional transverse common zeros. It is shown that there exist small perturbations δq_1, δq_2 such that the quartics $q_1 + \delta q_1$, $q_2 + \delta q_2$ have a transverse common zero at $\mathbf{u}$. If δq_1, δq_2 are sufficiently small then the n transverse common zeros of q_1 and q_2 are preserved because transverse common zeros are stable under small perturbations. As a result of the perturbation, $q_1 + \delta q_1$, $q_2 + \delta q_2$ have at least $n + 1$ transverse common zeros. On repeating this process at most 15 times the required pair of quartics is obtained.

It follows from Proposition 5.22 (part (i)) that the vectors $(\mathbf{u}.\mathbf{q}_i)\mathbf{q}_i - \mathbf{u}$, $i = 1, 2, 3$, are not all parallel. It is thus assumed without loss of generality that the vectors

$$(\mathbf{u}.\mathbf{q}_1)\mathbf{q}_1 - \mathbf{u} \qquad \text{and} \qquad (\mathbf{u}.\mathbf{q}_2)\mathbf{q}_2 - \mathbf{u} \qquad (5.102)$$

are linearly independent. It follows from Proposition 5.22 (part (ii)) that at least one of the vectors $(\mathbf{u}.\mathbf{q}_i)\mathbf{q}_i - \mathbf{u}$, $i = 3, 4, 5$, is linearly independent of the two vectors of (5.102). Thus it is the case that either $f_{123}(\mathbf{u}) \neq 0$ or $f_{124}(\mathbf{u}) \neq 0$ or $f_{125}(\mathbf{u}) \neq 0$. The proof divides into two cases.

Case 1: $f_{123}(\mathbf{u}) \neq 0$. The vectors $\mathbf{r}_4$, $\mathbf{r}_5$ of (5.100) and (5.101) are subjected to perturbations $\delta\mathbf{r}_4$, $\delta\mathbf{r}_5$ such that $\delta\mathbf{r}_4.\mathbf{u} = \delta\mathbf{r}_5.\mathbf{u} = 0$. Let δq_1 and δq_2 be the corresponding perturbations of q_1 and q_2. It follows that $\delta q_1(\mathbf{u}) = \delta q_2(\mathbf{u}) = 0$ and

$$\begin{aligned} \nabla\delta q_1(\mathbf{u}) &= f_{123}(\mathbf{u})\delta\mathbf{r}_4 \\ \nabla\delta q_2(\mathbf{u}) &= f_{123}(\mathbf{u})\delta\mathbf{r}_5 \end{aligned} \tag{5.103}$$

It follows from (5.103) that $\delta\mathbf{r}_4$ and $\delta\mathbf{r}_5$ can be chosen such that $\nabla(q_1 + \delta q_1)(\mathbf{u})$ and $\nabla(q_2 + \delta q_2)(\mathbf{u})$ are both non-zero and non-parallel.

Case 2: $f_{123}(\mathbf{u}) = 0$ and either $f_{124}(\mathbf{u}) \neq 0$ or $f_{125}(\mathbf{u}) \neq 0$. The details are given for the case $f_{124}(\mathbf{u}) \neq 0$. The case $f_{125}(\mathbf{u}) \neq 0$ is similar. The vectors $\mathbf{r}_3$, $\mathbf{r}_5$ are subjected to perturbations $\delta\mathbf{r}_3$, $\delta\mathbf{r}_5$ such that $\delta\mathbf{r}_3.\mathbf{u} = 0$, $\delta\mathbf{r}_5.\mathbf{u} \neq 0$. This yields

$$\begin{aligned} \nabla\delta q_1(\mathbf{u}) &= -f_{124}(\mathbf{u})\delta\mathbf{r}_3 \\ \nabla\delta q_2(\mathbf{u}) &= -f_{125}(\mathbf{u})\delta\mathbf{r}_3 + (\delta\mathbf{r}_5.\mathbf{u})\nabla f_{123}(\mathbf{u}) \end{aligned}$$

It follows from Proposition 5.23 that $\nabla f_{123}(\mathbf{u}) \neq 0$, because the f_{ijk} do not in general possess a singular point. Hence, $\delta\mathbf{r}_3$, $\delta\mathbf{r}_5$ can be found such that $\nabla(q_1 + \delta q_1)(\mathbf{u})$ and $\nabla(q_2 + \delta q_2)(\mathbf{u})$ are non-zero and non-parallel. □

Any translational velocity $\mathbf{v}$ compatible with (5.87) is a common zero of q_1 and q_2. However, the common zeros of q_1 and q_2 are not all compatible with (5.87).

Theorem 5.25. *Let image velocity vectors $\dot{\mathbf{q}}_i$ for $1 \leq i \leq 5$ be given in general position. Then a vector $\mathbf{u}$ is a possible translational velocity compatible with the $\dot{\mathbf{q}}_i$ for $1 \leq i \leq 5$ if and only if $\mathbf{u}$ is a common zero of q_1 and q_2 and if, in addition, the vectors*

$$((\mathbf{u}.\mathbf{q}_i)\mathbf{q}_i^\top - \mathbf{u}^\top, \mathbf{r}_i.\mathbf{u}) \qquad (i = 1, 2, 3) \tag{5.104}$$

are linearly independent.

Proof. Suppose firstly that $\mathbf{u}$ is a common zero of q_1 and q_2 such that the vectors of (5.104) are linearly independent at $\mathbf{u}$. By hypothesis, $q_1(\mathbf{u}) = q_2(\mathbf{u}) = 0$, thus there exist non-zero vectors $\mathbf{w}_1, \mathbf{w}_2$ such that

$$\begin{pmatrix} (\mathbf{u}.\mathbf{q}_1)\mathbf{q}_1^\top - \mathbf{u}^\top & \mathbf{r}_1.\mathbf{u} \\ (\mathbf{u}.\mathbf{q}_2)\mathbf{q}_2^\top - \mathbf{u}^\top & \mathbf{r}_2.\mathbf{u} \\ (\mathbf{u}.\mathbf{q}_3)\mathbf{q}_3^\top - \mathbf{u}^\top & \mathbf{r}_3.\mathbf{u} \\ (\mathbf{u}.\mathbf{q}_4)\mathbf{q}_4^\top - \mathbf{u}^\top & \mathbf{r}_4.\mathbf{u} \end{pmatrix} \mathbf{w}_1 = 0 \tag{5.105}$$

and

$$\begin{pmatrix} (\mathbf{u}.\mathbf{q}_1)\mathbf{q}_1^\top - \mathbf{u}^\top & \mathbf{r}_1.\mathbf{u} \\ (\mathbf{u}.\mathbf{q}_2)\mathbf{q}_2^\top - \mathbf{u}^\top & \mathbf{r}_2.\mathbf{u} \\ (\mathbf{u}.\mathbf{q}_3)\mathbf{q}_3^\top - \mathbf{u}^\top & \mathbf{r}_3.\mathbf{u} \\ (\mathbf{u}.\mathbf{q}_5)\mathbf{q}_5^\top - \mathbf{u}^\top & \mathbf{r}_5.\mathbf{u} \end{pmatrix} \mathbf{w}_2 = 0 \tag{5.106}$$

The vectors $\mathbf{w}_1, \mathbf{w}_2$ are both normal to the subspace of $\mathbf{R}^4$ (or $\mathbf{C}^4$ if $\mathbf{u}$ is complex) spanned by the vectors of (5.104). By hypothesis, this subspace is of dimension three in $\mathbf{R}^4$, thus $\mathbf{w}_1$ and $\mathbf{w}_2$ are parallel. The vectors $\mathbf{w}_1$ and $\mathbf{w}_2$ are scaled such that $\mathbf{w}_1 = \mathbf{w}_2$.

The component $(\mathbf{w}_i)_j$ of $\mathbf{w}_i$ is denoted by w_{ij}. If $w_{14} = 0$ then $w_{24} = 0$. It follows from (5.105) and (5.106) that the vector $(w_{11}, w_{12}, w_{13})^\top$ is normal to the vectors $(\mathbf{u}.\mathbf{q}_i)\mathbf{q}_i - \mathbf{u}$ for $1 \le i \le 5$. An application of Proposition 5.22 (part (ii)) yields $w_{11} = w_{12} = w_{13} = 0$, thus contradicting the choice of $\mathbf{w}_1$ and $\mathbf{w}_2$. It follows from this contradiction that $w_{14} \neq 0$ and $w_{24} \neq 0$. The components w_i of the three dimensional vector $\mathbf{w} = (w_1, w_2, w_3)^\top$ are defined by $w_j = -w_{1\ j+1}/w_{14}$. Equations (5.105), (5.106), and the definition of $\mathbf{w}$ yield

$$\mathbf{r}_i.\mathbf{u} = (\mathbf{u}.\mathbf{q}_i)(\mathbf{w}.\mathbf{q}_i) - \mathbf{u}.\mathbf{w} \qquad (1 \le i \le 5) \qquad (5.107)$$

It follows from (5.107) and Proposition 5.21 that $\mathbf{u}$ is a translational velocity compatible with the image velocity field.

To complete the proof, suppose that $\mathbf{u}$ is a common zero of q_1, q_2 such that the three vectors of (5.104) are linearly *dependent*. It is shown that $\mathbf{u}$ is in general not a translational velocity compatible with the image velocity field. By hypothesis, the vectors $(\mathbf{u}.\mathbf{q}_i)\mathbf{q}_i - \mathbf{u}$ for $i = 1, 2, 3$, are contained in a single plane Π. It follows from part (ii) of Proposition 5.22 that the vectors $(\mathbf{u}.\mathbf{q}_i)\mathbf{q}_i - \mathbf{u}$ for $1 \le i \le 5$ span $\mathbf{R}^3$. Hence at least one vector of the pair $(\mathbf{u}.\mathbf{q}_4)\mathbf{q}_4 - \mathbf{u}$, $(\mathbf{u}.\mathbf{q}_5)\mathbf{q}_5 - \mathbf{u}$ is not contained in Π. It is assumed, after relabelling if necessary, that $(\mathbf{u}.\mathbf{q}_4)\mathbf{q}_4 - \mathbf{u}$ is not contained in Π. It follows from Proposition 5.22 (part (i)) after relabelling if necessary that $(\mathbf{u}.\mathbf{q}_1)\mathbf{q}_1 - \mathbf{u}$ and $(\mathbf{u}.\mathbf{q}_2)\mathbf{q}_2 - \mathbf{u}$ are linearly independent. It follows that

$$\det \begin{pmatrix} (\mathbf{u}.\mathbf{q}_1)\mathbf{q}_1^\top - \mathbf{u}^\top \\ (\mathbf{u}.\mathbf{q}_2)\mathbf{q}_2^\top - \mathbf{u}^\top \\ (\mathbf{u}.\mathbf{q}_4)\mathbf{q}_4^\top - \mathbf{u}^\top \end{pmatrix} \neq 0$$

The vector $\mathbf{r}_5$, and hence $\dot{\mathbf{q}}_5$, $\mathbf{q}_5$, can thus be chosen such that

$$\det \begin{pmatrix} (\mathbf{u}.\mathbf{q}_1)\mathbf{q}_1^\top - \mathbf{u}^\top & \mathbf{r}_1.\mathbf{u} \\ (\mathbf{u}.\mathbf{q}_2)\mathbf{q}_2^\top - \mathbf{u}^\top & \mathbf{r}_2.\mathbf{u} \\ (\mathbf{u}.\mathbf{q}_4)\mathbf{q}_4^\top - \mathbf{u}^\top & \mathbf{r}_4.\mathbf{u} \\ (\mathbf{u}.\mathbf{q}_5)\mathbf{q}_5^\top - \mathbf{u}^\top & \mathbf{r}_5.\mathbf{u} \end{pmatrix} \neq 0 \qquad (5.108)$$

The condition $q_1(\mathbf{u}) = q_2(\mathbf{u}) = 0$ holds regardless of the choice of $\dot{\mathbf{q}}_5$, $\mathbf{q}_5$, because the vectors of (5.104) are, by assumption, linearly dependent. It follows from (5.108) that $\mathbf{u}$ is in general not a translational velocity compatible with the image velocity field. □

5.4.3 Counting the Solutions

The result of Theorem 5.25 is used to count the number of common zeros of q_1 and q_2 that yield reconstructions compatible with the five image velocity vectors. It follows from Theorem 5.25 that a possible translational velocity $\mathbf{v}$ is compatible with the five image velocity vectors if and only if $\mathbf{v}$ is a common zero of q_1, q_2 such that the vectors of (5.104) are linearly independent. It is shown in Theorem 5.24 that q_1 q_2 have 16 distinct common zeros. It is now shown that there are exactly six values of $\mathbf{v}$ such that the vectors of (5.104) are linearly dependent. The result that there are, in general, exactly $10 = 16 - 6$ values of $\mathbf{v}$ compatible with five image velocity vectors then follows.

It follows from elementary linear algebra that the three vectors of (5.104) are linearly dependent if and only if they form the rows of a 3×4 matrix of rank two or less. The rank of a matrix is equal to the number of linearly independent rows or to the number of linearly independent columns. The same value for the rank is obtained in both cases.

Theorem 5.26. *Let A be the matrix defined by*

$$A = \begin{pmatrix} (\mathbf{v}.\mathbf{q}_1)\mathbf{q}_1^\top - \mathbf{v}^\top & \mathbf{r}_1.\mathbf{v} \\ (\mathbf{v}.\mathbf{q}_2)\mathbf{q}_2^\top - \mathbf{v}^\top & \mathbf{r}_2.\mathbf{v} \\ (\mathbf{v}.\mathbf{q}_3)\mathbf{q}_3^\top - \mathbf{v}^\top & \mathbf{r}_3.\mathbf{v} \end{pmatrix} \tag{5.109}$$

Then there are, in general, exactly six values of $\mathbf{v}$ at which A has rank two or less.

Proof. Let the columns of A be given by $A = (A_1\mathbf{v}|A_2\mathbf{v}|A_3\mathbf{v}|A_4\mathbf{v})$, where each A_i is a 3×3 matrix independent of $\mathbf{v}$. The cubic polynomials f and g are defined by

$$f(\mathbf{v}) = \det(A_1\mathbf{v}|A_2\mathbf{v}|A_3\mathbf{v}) \qquad g(\mathbf{v}) = \det(A_2\mathbf{v}|A_3\mathbf{v}|A_4\mathbf{v})$$

The values of $\mathbf{v}$ at which A has rank two are included amongst the at most nine distinct common zeros of f and g. There are at most three values of $\mathbf{v}$ for which $A_2\mathbf{v} \times A_3\mathbf{v} = 0$; the three values are obtained from the generalised eigenvector equation

$$(A_2 - \lambda A_3)\mathbf{v} = 0 \tag{5.110}$$

Exactly three values of $\mathbf{v}$ are obtained from (5.110) in the special case given by

$$\begin{aligned} \mathbf{q}_1 &= (1,0,0)^\top \\ \mathbf{q}_2 &= (0,1,0)^\top \\ \mathbf{q}_3 &= (1/\sqrt{3}, 1/\sqrt{3}, 1/\sqrt{3})^\top \end{aligned}$$

It follows that there are, in general, exactly three values of $\mathbf{v}$ for which $A_2\mathbf{v} \times A_3\mathbf{v} = 0$. It also follows from this special case that the matrix A defined by (5.109) has rank three when $A_2\mathbf{v} \times A_3\mathbf{v} = 0$. It follows that there are *at most* six common zeros of f and g at which A has rank two. To complete the proof a single example is given in which A has rank two at six distinct values of $\mathbf{v}$. Let $\mathbf{r}_2 = \mathbf{r}_3 = 0$. The curve g then splits into the line $\mathbf{r}_1.\mathbf{v} = 0$, and a factor of degree two. The matrix A has rank two at the three values of $\mathbf{v}$ arising from the intersection of the line $\mathbf{r}_1.\mathbf{v} = 0$ with the cubic f. The matrix A also has rank two at $\mathbf{q}_2$, $\mathbf{q}_3$ and $\mathbf{q}_2 \times \mathbf{q}_3$. This gives the required total of six values of $\mathbf{v}$. □

5.4.4 Critical Surfaces

A second proof is given that the degree of reconstruction from image velocities is ten. The proof is based on the properties of critical surfaces. It requires a rigid velocity $\{\mathbf{v}, \mathbf{w}\}$ compatible with the image velocity vectors. The critical surfaces are constructed from pairs of solutions for which one member of the pair is $\{\mathbf{v}, \mathbf{w}\}$. The assumption that there is at least one solution is not a disadvantage so far as the proof goes because a suitable rigid velocity $\{\mathbf{v}, \mathbf{w}\}$ always exists for algebraic reasons. There are five unknown variables, namely the direction of $\mathbf{v}$ and the components of $\mathbf{w}$ and there are five polynomial constraints on these variables. The number of variables is equal to the number of constraints, thus at least one solution exists. The disadvantage of assuming that one solution is given is that the method of proof does not yield an algorithm for obtaining the ten reconstructions compatible with five image velocity vectors.

This second proof that the degree of reconstruction is ten is simpler than the one given in Sects. 5.4.1-5.4.3, because it makes greater use of the special properties of reconstruction. In particular, it does not involve the introduction of spurious solutions which then have to be winnowed out from the true solutions.

Theorem 5.27. *Let image velocity vectors $\dot{\mathbf{q}}_i$ for $1 \leq i \leq 5$ be given together with a compatible rigid velocity. Then a two dimensional space $\mathcal{S}^2$ of quadrics can be constructed such that any critical surface compatible with the five image velocity vectors, when viewed by a camera moving with the given velocity, is represented by a point in $\mathcal{S}^2$.*

Proof. Let the given camera velocity be $\{\mathbf{v}_1, \mathbf{w}_1\}$ and let $\mathbf{p}_i$ be the point in $\mathbb{P}^3$ moving with rigid velocity $\{\mathbf{v}_1, \mathbf{w}_1\}$ with respect to the camera, such that the projected velocity of $\mathbf{p}_i$ is $\dot{\mathbf{q}}_i$. Let $\mathbf{o}$ be the optical centre of the camera. Each critical surface, ψ, giving rise to the $\dot{\mathbf{q}}_i$ when viewed by a camera moving with velocity $\{\mathbf{v}_1, \mathbf{w}_1\}$ contains $\mathbf{o}$ and the points $\mathbf{p}_i$ for $1 \leq i \leq 5$. In addition, it follows from (4.22) that the tangent plane to ψ at $\mathbf{o}$ contains $\mathbf{v}_1$. The space of all quadric surfaces contained in $\mathbb{P}^3$ is of dimension nine. The condition that a quadric contains a known point imposes a single linear constraint, thus the quadric surfaces containing $\mathbf{o}$ and the $\mathbf{p}_i$ form a $9 - 6 = 3$ dimensional space in the space of all quadrics. The condition that the tangent plane at $\mathbf{o}$ includes $\mathbf{v}_1$ imposes an additional linear constraint. This yields the required $3 - 1 = 2$ dimensional space $\mathcal{S}^2$ of quadrics. A basis for S^2 can be calculated from $\mathbf{o}$, the $\mathbf{p}_i$ for $1 \leq i \leq 5$, and $\mathbf{v}_1$. □

With the notation of Theorem 5.27, let Cartesian coordinates be chosen such that $\mathbf{o} = (0, 0, 0, 1)^\top$. It follows from (4.22) that a quadric corresponding to a point of $\mathcal{S}^2$ has the form

$$\mathbf{x}^\top M\mathbf{x} + \mathbf{l}.\mathbf{x} = 0$$

where M is a symmetric 3×3 matrix and $\mathbf{l}$ is a vector. The space $\mathcal{S}^2$ is two dimensional (as a projective space), thus there exist symmetric matrices M_1, M_2,

M_3 and corresponding vectors $\mathbf{l}_1$, $\mathbf{l}_2$, $\mathbf{l}_3$ such that $\mathcal{S}^2$ is spanned by the quadrics

$$\psi_i = \mathbf{x}^\top M_i \mathbf{x} + \mathbf{l}_i.\mathbf{x} \tag{5.111}$$

An arbitrary quadric ψ of $\mathcal{S}^2$ has an equation

$$\psi = \lambda_1\psi_1 + \lambda_2\psi_2 + \lambda_3\psi_3 \tag{5.112}$$

where $(\lambda_1, \lambda_2, \lambda_3)^\top$ is a point of $\mathbb{P}^2$ determined uniquely by ψ.

Not every point of S^2 yields a quadric which is a critical surface. The proof that there are in general exactly ten essentially different camera velocities compatible with five image velocity vectors relies on selecting from $\mathcal{S}^2$ precisely those points corresponding to critical surfaces. As a preliminary, it is shown in the next proposition that each critical surface in $\mathcal{S}^2$ yields only one rigid velocity $\{\mathbf{v}_2, \mathbf{w}_2\}$ that is different from the given rigid velocity $\{\mathbf{v}_1, \mathbf{w}_1\}$ of Theorem 5.27, but that is at the same time compatible with the image velocities.

Proposition 5.28. *Let ψ be a critical surface for a camera with optical centre $\mathbf{o}$ and velocity $\{\mathbf{v}_1, \mathbf{w}_1\}$. Let $\dot{\mathbf{q}}$ be the resulting image velocity field. Let $\{\mathbf{v}_2, \mathbf{w}_2\}$ be the rigid velocity of a second camera with optical centre at $\mathbf{o}$ viewing a second surface which gives rise to the same image velocity field. Then $\{\mathbf{v}_2, \mathbf{w}_2\}$ is uniquely determined by ψ and $\{\mathbf{v}_1, \mathbf{w}_1\}$, up to a single unknown scale factor, the length of $\mathbf{v}_2$.*

Proof. Let Cartesian coordinates be chosen with origin $\mathbf{o}$ at the optical centre of the camera. The equation of ψ is then given by (4.22),

$$(\mathbf{w} \times \mathbf{x}).(\mathbf{v}_2 \times \mathbf{x}) + (\mathbf{v}_2 \times \mathbf{v}_1).\mathbf{x} = 0 \tag{5.113}$$

The equation (5.113) for ψ is of the form

$$\mathbf{x}^\top M\mathbf{x} + \mathbf{l}.\mathbf{x} = 0 \tag{5.114}$$

where M is a symmetric 3×3 matrix and $\mathbf{l}$ is a three dimensional vector. By hypothesis, M and $\mathbf{l}$ are known, but the individual vectors $\mathbf{w}$ and $\mathbf{v}_2$ are unknown. The vectors $\mathbf{w}$ and $\mathbf{v}_2$ are obtained as follows. The direction of $\mathbf{v}_2$ is the direction of the unique generator of ψ through $\mathbf{o}$ that contains a principal point of ψ. The direction of $\mathbf{w}$ is given by the other principal point of ψ. The magnitude of $\mathbf{w}$ is obtained by substituting any convenient value of $\mathbf{x}$ into the equation

$$(\mathbf{w} \times \mathbf{x}).(\mathbf{v}_2 \times \mathbf{x}) + (\mathbf{v}_2 \times \mathbf{v}_1).\mathbf{x} = \mathbf{x}^\top M\mathbf{x} + \mathbf{l}.\mathbf{x}$$

obtained by equating the left-hand sides of (5.113) and (5.114). The value $\mathbf{x} = \mathbf{v}_1 \times \mathbf{v}_2$ is suitable. The angular velocity $\mathbf{w}_2$ is given by $\mathbf{w}_2 = \mathbf{w} + \mathbf{w}_1$. □

5.4.5 Counting the Critical Surfaces

The critical surfaces represented by points of the space $\mathcal{S}^2$ are found.

Theorem 5.29. *Let $\dot{\mathbf{q}}_i$ for $1 \leq i \leq 5$ be a set of image velocity vectors in general position and let $\{\mathbf{v}_1, \mathbf{w}_1\}$ be a rigid velocity compatible with the $\dot{\mathbf{q}}_i$. Then there exist two cubic plane curves such that each additional camera velocity compatible with the $\dot{\mathbf{q}}_i$ arises from a common zero of the two cubic plane curves, and conversely, each common zero of the two cubic plane curves gives rise to exactly one additional camera velocity compatible with the $\dot{\mathbf{q}}_i$.*

Proof. The two cubic plane curves are contained in the space $\mathcal{S}^2$ constructed in Theorem 5.27. Let ψ be a critical surface with equation

$$\mathbf{x}^\top M\mathbf{x} + \mathbf{l}.\mathbf{x} = 0$$

where

$$\begin{aligned} M &= \lambda_1 M_1 + \lambda_2 M_2 + \lambda_3 M_3 \\ \mathbf{l} &= \lambda_1 \mathbf{l}_1 + \lambda_2 \mathbf{l}_2 + \lambda_3 \mathbf{l}_3 \end{aligned}$$

in accordance with (5.111) and (5.112). Let N be the 3×3 matrix defined by

$$N = M - (1/2)\mathrm{tr}(M)I$$

It is shown in Theorems 4.10 and 4.11 that ψ is a critical surface if and only if

$$\mathbf{l}^\top N\mathbf{l} = 0 \qquad \text{and} \qquad \det(N) = 0 \tag{5.115}$$

The two equations of (5.115) are the required cubic plane curves. The fact that each intersection of the cubic plane curves yields exactly one additional rigid velocity compatible with the $\dot{\mathbf{q}}_i$ follows from Proposition 5.28. □

The proof that the degree of reconstruction from image velocities is ten is completed by showing that the two cubic plane curves constructed in Theorem 5.29 intersect at nine distinct points.

Theorem 5.30. *Let $\dot{\mathbf{q}}_i$ for $1 \leq i \leq 5$ be a set of image velocity vectors in general position. Then there are exactly ten essentially different rigid velocities compatible with the $\dot{\mathbf{q}}_i$.*

Proof. Each rigid velocity involves five unknown parameters, three arising from the angular velocity and two arising from the direction of the translational velocity. Each velocity vector imposes one algebraic constraint (5.87) on the five parameters, thus there is at least one camera velocity $\{\mathbf{v}_1, \mathbf{w}_1\}$ compatible with the $\dot{\mathbf{q}}_i$. This is an application of the Nullstellensatz (Fulton 1969).

It follows from Proposition 5.28 and Theorem 5.29 that the rigid velocities different from $\{\mathbf{v}_1, \mathbf{w}_1\}$ but compatible with the $\dot{\mathbf{q}}_i$ arise from the intersections of the two cubic plane curves of (5.115). Thus, to prove the theorem it suffices to show that these two curves have in general the maximum number, nine, of distinct intersections. The two curves fail to have nine distinct intersections if and only if a certain algebraic constraint on the coefficients of the two curves is satisfied. If the constraint fails to hold for just one choice of five image velocity vectors then it fails to hold in general. Thus it suffices to produce a single example in which the two cubic curves of (5.115) have nine distinct common zeros.

The notation of Theorem 5.29 is employed. For the example let

$$N = \begin{pmatrix} a & d & e \\ d & b & f \\ e & f & c \end{pmatrix} \tag{5.116}$$

and let three of the reconstructed points in $\mathbf{P}^3$ be

$$\begin{aligned} \mathbf{p}_1 &= (1,0,0,0)^\top \\ \mathbf{p}_2 &= (0,1,0,0)^\top \\ \mathbf{p}_3 &= (0,0,1,0)^\top \end{aligned} \tag{5.117}$$

It follows from (5.116) and (5.117) that $a = b = c = 0$ and that $M = N$. The determinant of N splits into linear factors,

$$\det(N) = 2\,def$$

The curve $\det(N) = 0$ splits into the three lines $d = 0$, $e = 0$, $f = 0$. A method is described for finding points $\mathbf{p}_3$, $\mathbf{p}_4$ of $\mathbf{R}^3$ and a vector $\mathbf{a}$ such that

$$\begin{aligned} \mathbf{p}_4^\top M \mathbf{p}_4 + \mathbf{l}.\mathbf{p}_4 &= 0 \\ \mathbf{p}_5^\top M \mathbf{p}_5 + \mathbf{l}.\mathbf{p}_5 &= 0 \\ \mathbf{l}.\mathbf{a} &= 0 \end{aligned} \tag{5.118}$$

and such that the line $d = 0$ meets $\mathbf{l}^\top N \mathbf{l} = 0$ at three distinct points. Let p_{ij} be the ith component of $\mathbf{p}_i$, $p_{ij} = (\mathbf{p}_i)_j$. Let $\mathbf{a} = (1,1,-1)^\top$. The third equation of (5.118) yields $l_3 = l_1 + l_2$. On substituting for l_3, the first two equations of (5.118) yield

$$\begin{aligned} 2p_{43}(e\,p_{41} + f p_{42}) + l_1(p_{41} + p_{43}) + l_2(p_{42} + p_{43}) &= 0 \\ 2p_{53}(e\,p_{51} + f p_{52}) + l_1(p_{51} + p_{53}) + l_2(p_{52} + p_{53}) &= 0 \end{aligned} \tag{5.119}$$

The equation $\mathbf{l}^\top N \mathbf{l} = 0$ reduces to

$$(l_1 + l_2)(e\,l_1 + f\,l_2) = 0 \tag{5.120}$$

The component $e\,l_1 + f\,l_2$ of (5.120) is used to eliminate e from (5.119), yielding

$$\begin{aligned} 2\,p_{43} f(l_1 p_{42} - l_2 p_{41}) + l_1(p_{41} + p_{43}) + l_2(p_{42} + p_{43}) &= 0 \\ 2\,p_{53} f(l_1 p_{52} - l_2 p_{51}) + l_1(p_{51} + p_{53}) + l_2(p_{52} + p_{53}) &= 0 \end{aligned} \tag{5.121}$$

The elimination of f from (5.121) yields

$$\frac{p_{43}(l_1 p_{42} - l_2 p_{41})}{p_{53}(l_1 p_{52} - l_2 p_{51})} = \frac{l_1(p_{41} + p_{43}) + l_2(p_{42} + p_{43})}{l_1(p_{51} + p_{53}) + l_2(p_{52} + p_{53})} \tag{5.122}$$

Equation (5.122) is a homogeneous quadratic equation in l_1, l_2. For a general choice of $\mathbf{p}_4$, $\mathbf{p}_5$ this equation yields two distinct values for the ratio l_1/l_2. This gives two of the points of intersection of the line $d = 0$ with the cubic plane curve $\mathbf{l}^{\mathrm{T}} N \mathbf{l} = 0$. The third intersection arises from the first component $l_1 + l_2$ of (5.120). Similarly, the cases $e = 0$ and $f = 0$ each yield three distinct solutions to (5.115), for general choice of $\mathbf{p}_4$, $\mathbf{p}_5$. All the intersections arising from the three lines $d = 0$, $e = 0$ and $f = 0$ are distinct. This gives the required total of nine intersections. □

In Theorem 5.30 the possibility is not ruled out that the intersections of the two cubic plane curves in (5.115) yield quadrics without real generators. Such quadrics give rise to complex velocities which are algebraically compatible with the image velocity field but which are not physically acceptable. Some of the remaining intersections, although real, may yield unfeasible positions for the points $\mathbf{p}_i$ giving rise to the image velocity field. In particular, some of the points $\mathbf{p}_i$ may be behind the camera whilst other points $\mathbf{p}_i$ are in front of it.

References

Bruce J.W. & Giblin P.J. 1984 *Curves and Singularities.* Cambridge: Cambridge University Press.

Demazure M. 1988 Sur deux problèmes de reconstruction. *Technical Report No. 882, INRIA, Rocquencourt, France.*

Fulton W. 1969 *Algebraic Curves.* Reading, Massachusetts: W.A. Benjamin Inc., Mathematics Lecture Note Series (Reprinted 1974).

Kruppa E. 1913 Zur Ermittlung eines Objektes zwei Perspektiven mit innerer Orientierung. *Sitz-Ber. Akad. Wiss., Wien, math. naturw. Kl. Abt. IIa*, **122**, 1939-1948.

Hartshorne R. 1977 *Algebraic Geometry.* Graduate Texts in Mathematics **52**, Springer Verlag.

Semple J.G. & Roth R. 1949 *Introduction to Algebraic Geometry.* Oxford: Clarendon Press, reprinted 1985.

Sturm R. 1869 Das Problem der Projektivität und seine Anwendung auf die Flächen zweiten Grades. *Math. Annalen* **1**, 533-573.

6 Algorithms

Over the past twenty years numerous algorithms for reconstructing the camera motion and the shape of a scene from image motions have appeared in the computer vision literature. Useful reviews of published algorithms can be found in Adiv (1985), Aggarwal & Martin (1983) and Maybank (1987). It appears that almost every mathematical property of the equations underlying reconstruction has at one time or another been made the basis of an algorithm.

The main practical difficulty with reconstruction is instability. Reconstruction usually amplifies any errors or noise in the estimates of image motion, sometimes so much that the results of the reconstruction become wildly inaccurate. An important consequence of this instability is that it is not sufficient for an algorithm to be mathematically correct. It is possible to have two different algorithms which are mathematically equivalent, in that they compute the correct camera motion from a set of image motions that are free from errors, but which behave very differently when the image motions are subject to small errors.

Reconstruction from image velocities is particularly susceptible to noise because the image changes from which the image velocities are estimated are usually small. The advantages of reconstruction from image velocities are firstly that the underlying equations are simpler and secondly that estimates of the image velocities can be obtained quickly. Fast computation of image motions is important in applications such as collision avoidance. If a vehicle is moving at a walking pace of about 1 ms^{-1} in an indoor environment then it is necessary to detect the possibility of a collision within about 1 s of an obstacle entering the field of view.

The instability of reconstruction has been recognised in computer vision for many years, for example it is described by Fang & Huang (1984). Although the best methods for overcoming instability are not universally agreed, the following principles are clear.

- Reconstruction is unstable in that small changes in the data can cause large changes in the results. The instability is increased if the field of view is narrow, or if the camera motion produces only a small change in the image.
- The stability of reconstruction from image velocities is increased if the velocities arise from a surface showing large depth variations over small regions of the field of view.

- The stability of reconstruction is increased if prior knowledge of the camera motion or of the scene geometry is incorporated.

The accuracy and stability of reconstruction can be increased by using as many image correspondences or image velocity vectors as possible. If the camera motion is uniform over an extended time then a more accurate reconstruction can be obtained by combining the information in a sequence of images. This approach has led to the application of sophisticated filter based techniques to reconstruction (Ayache 1991). Experiments with computer generated image velocity fields suggest that the single most important factor governing the accuracy and stability of the reconstruction is the total change in the appearance of the image (Grzywacz & Hildreth 1987).

If the probability density functions describing the distributions of the errors in estimates of the image motions are known then it is possible, in principle, to estimate the camera motion in a way which in a statistical sense makes the best possible use of the available information (Bar-Shalom & Fortmann 1988). Unfortunately, the equations incorporating the information in the probability density functions are usually extremely difficult to solve. The probabilistic approach is likely to become increasingly important as the hardware available for computer vision becomes more powerful.

Four algorithms for reconstruction from image motions are described in this chapter. They are chosen because they are relatively simple in structure, they use all the available image correspondences or image velocities and they can be implemented efficiently. The algorithms do not use any information about the statistical distribution of the errors. The increased complexity of the probabilistic approach is avoided, at the cost of a greater error in the reconstruction. The full details of the implementations of the algorithms are omitted. It is usually a major undertaking to convert an algorithm, given as a mathematical 'sketch' in a paper, into a working system for estimating camera motion. Some of the issues involved in practical implementations are discussed by Adiv (1985), Lawton(1983) and Murray & Buxton (1990). Adiv describes a least squares algorithm for estimating the motions of a collection of rigid bodies. Lawton shows that the estimation of the translational velocity is a practical possibility if the angular velocity is known to be zero. Murray and Buxton describe a method based on optical flow for estimating the velocities of polyhedra.

The algorithms for reconstruction from image correspondences are described in Sect. 6.1 and the algorithms for reconstruction from image velocities are described in Sect. 6.2. One of the algorithms in Sect. 6.1 is based on the singular value decomposition of a matrix. The other algorithm of Sect. 6.1 is based on error minimisation by descent. Both of the algorithms in Sect. 6.2 seek to improve the estimates of camera velocity by exploiting irregularities in the image velocity field due to irregularities in the depth of the surface giving rise to the image velocity field. One algorithm in Sect. 6.2, the least squares algorithm, is analysed in detail in the irregular case. The algorithm is based on an error function $\mathbf{v} \mapsto \epsilon(\mathbf{v})$ defined on the unit sphere of possible directions of the translational velocity of the surface giving rise to the image velocity field. The analysis shows that if the image velocity field is irregular then $\epsilon(\mathbf{v})$ has an approximation

$$\epsilon(\mathbf{v}) \approx \tau|\mathbf{n}.\mathbf{v}| \tag{6.1}$$

where τ and $\mathbf{n}$ are independent of $\mathbf{v}$. The coefficient τ measures the amount of irregularity in the surface. Methods for estimating camera velocity based on approximations of the form (6.1) are discussed by Aisbet (1990), Heeger & Jepson (1990), Jepson & Heeger (1990) and Maybank (1987). The approximation (6.1) is closely related to the bas-relief ambiguity described in Harris (1990).

6.1 Reconstruction from Image Correspondences

Two algorithms for estimating a rigid displacement from a set of image correspondences are described. The first, taken from Toscani & Faugeras (1986), is based on the singular value decomposition (SVD) of a matrix, whilst the second, taken from Horn (1990), uses descent. The algorithms are chosen because they represent two different approaches to reconstruction. The first approach relies on the details of the mathematics of reconstruction. This approach can lead to an efficient algorithm, but the efficiency is easily lost if it is necessary to extend the algorithm, for example to incorporate knowledge of statistics of the errors in the image motions or to incorporate prior information about the camera motion. The second approach uses a general purpose method for minimising an error function defined on the space of camera displacements. Further information can be incorporated into the algorithm, at the cost of a more complicated error function.

6.1.1 An SVD Based Algorithm

In the singular value decomposition a real matrix is reduced to a standard diagonal form by pre- and postmultiplying it with orthogonal matrices. If A is a real $m \times n$ matrix, then there exists an $m \times m$ orthogonal matrix U, an $n \times n$ orthogonal matrix V and an $m \times n$ diagonal matrix Σ such that $A = U\Sigma V$. Let $k = \min\{m, n\}$. The matrices U, V are chosen such that

$$\Sigma_{11} \geq \ldots \geq \Sigma_{kk} \geq 0$$

The singular values, σ_i, of A are defined by $\sigma_i = \Sigma_{ii}$ for $1 \leq i \leq k$. Golub & Van Loan (1983) describe the theory behind the SVD and give a number of efficient algorithms for calculating it.

The SVD is closely related to the two matrix norms $\|.\|$ and $\|.\|_f$. The norm $\|.\|$ is subordinate to the Euclidean vector norm, and $\|.\|_f$ is the Frobenius norm. The definitions of the norms $\|A\|$ and $\|A\|_f$ of an $m \times n$ matrix A are

$$\|A\| = \sup\{\|A\mathbf{x}\| \mid \|\mathbf{x}\| = 1\} \tag{6.2}$$

$$\|A\|_f^2 = \sum_{i=1,j=1}^{m,n} A_{ij}^2 \tag{6.3}$$

It follows from (6.3) that

$$\|A\|_f^2 = \mathrm{tr}(A^\top A) \tag{6.4}$$

It follows from (6.2), (6.3) and the properties of the orthogonal matrices U, V used to define the SVD of A that

$$\begin{aligned} \|A\| &= \|U\Sigma V\| &= \|\Sigma\| &= \sigma_1 \\ \|A\|_f^2 &= \|U\Sigma V\|_f^2 &= \|\Sigma\|_f^2 &= \sum_{i=1}^{k}\sigma_i^2 \end{aligned}$$

Let $\mathbf{q}_i \leftrightarrow \mathbf{q}_i'$ for $1 \le i \le n$ be a set of image correspondences compatible with a camera displacement $\{R, \mathbf{a}\}$, where R is an orthogonal matrix and $\mathbf{a}$ is a non-zero translation. Let E be the essential matrix defined by $E = RT_a$. It is recalled from (2.11) that $\mathbf{q}_i'^\top E\mathbf{q}_i = 0$ for $1 \le i \le n$. An error function $E \mapsto V(E)$ is defined on the space of 3×3 matrices by

$$V(E) = \sum_{i=1}^{n}(\mathbf{q}_i'^\top E\mathbf{q}_i)^2 \tag{6.5}$$

The camera displacement $\{R, \mathbf{a}\}$ can in principle be found by searching for the minimum of V in the space of essential matrices E that satisfy $\|E\|_f = 1$. The condition $\|E\|_f = 1$ is imposed in order to exclude the trivial solution $E = 0$. No simple method for minimising V is known. An alternative computationally less expensive approach is first to find the 3×3 matrix F with unit Frobenius norm that minimises V and then to find the nearest essential matrix G to F. In the noise-free case and with a sufficiently general set of image correspondences, $F = G$ and $V(F) = 0$. There are three steps to the SVD based algorithm for estimating the camera displacement.

(1) Find the matrix F that minimises V over the space of 3×3 matrices E subject to the condition $\|E\|_f = 1$.
(2) Find the antisymmetric matrix T_a that minimises $\|F^\top F - T_a^\top T_a\|_f$.
(3) Find the orthogonal matrix R that minimises $\|F - RT_a\|_f$, where T_a is the matrix found at step (2).

The choice of the Frobenius norm and the properties of the essential matrices ensure that each step of the algorithm can be implemented efficiently. The following three theorems, one for each step, fill in some of the details of the algorithm.

Theorem 6.1. *Let E be a 3×3 matrix, let $\mathbf{q}_i \leftrightarrow \mathbf{q}_i'$ for $1 \le i \le n$ be a set of image correspondences and let V be the error function defined by (6.5). Let $\mathbf{e}$ be the nine dimensional vector defined by*

$$\mathbf{e} = (E_{11}, E_{12}, E_{13}, E_{21}, E_{22}, E_{23}, E_{31}, E_{32}, E_{33})^\top \tag{6.6}$$

There exists an $n \times 9$ matrix A such that the ith row of A depends only on the single image correspondence $\mathbf{q}_i \leftrightarrow \mathbf{q}_i'$ and such that $V(E) = \|A\mathbf{e}\|^2$. Further, *if $A = U\Sigma V$ is the singular value decomposition of A and if $\mathbf{e}_9$ is the nine dimensional vector defined by $\mathbf{e}_9 = (0,\ldots,0,1)^\top$, then $E \mapsto V(E)$ is minimised, subject to the condition $\|E\|_f = 1$, by the 3×3 matrix F obtained from the vector $V^\top \mathbf{e}_9$ by applying (6.6).*

Proof. The ith term in the sum (6.5) defining V is $(\mathbf{q}_i'^\top E\mathbf{q}_i)^2$. Let $\mathbf{a}_i$ be the nine dimensional vector depending only on $\mathbf{q}_i$, $\mathbf{q}_i'$, such that

$$(\mathbf{q}_i'^\top E\mathbf{q}_i)^2 = (\mathbf{a}_i.\mathbf{e})^2 \qquad (1 \le i \le n) \tag{6.7}$$

Then A is defined to be the $n \times 9$ matrix with ith row equal to $\mathbf{a}_i$. It follows from (6.5) and (6.7) that

$$V(E) = \sum_{i=1}^{n}(\mathbf{a}_i.\mathbf{e})^2 = \|A\mathbf{e}\|^2$$

The singular value decomposition of A yields

$$V(E) = \|A\mathbf{e}\|^2 = \|U\Sigma V\mathbf{e}\|^2 = \|\Sigma V\mathbf{e}\|^2 \ge \sigma_9^2\|V\mathbf{e}\|^2 \ge \sigma_9^2\|\mathbf{e}\|^2 \tag{6.8}$$

The condition $\|E\|_f = 1$ is identical to the condition $\|\mathbf{e}\| = 1$. On setting $\|\mathbf{e}\| = 1$ in (6.8) it follows that

$$V(E) \ge \sigma_9^2 \tag{6.9}$$

Let F be the matrix obtained from $V^\top\mathbf{e}_9$. Then $V(F) = \sigma_9^2$. The lower bound in (6.9) is thus attained. □

Theorem 6.2. *Let E be any 3×3 matrix. Then the function g defined by*

$$g(\mathbf{a}) = \|E^\top E - T_a^\top T_a\|_f^2 \tag{6.10}$$

is minimised when $\mathbf{a}$ is an eigenvector of $E^\top E$ chosen such that the eigenvalue of $\mathbf{a}$ is least among the three eigenvalues of $E^\top E$. The squared length $\|\mathbf{a}\|^2$ of $\mathbf{a}$ is equal to the average of the largest two eigenvalues of $E^\top E$.

Proof. It follows from the definition of T_a that

$$T_a^\top T_a = (\mathbf{a}.\mathbf{a})I - \mathbf{a}\otimes\mathbf{a}$$

On substituting for $T_a^\top T_a$ in (6.10) the following expression for g is obtained,

$$g(\mathbf{a}) = \|E^\top E - (\mathbf{a}.\mathbf{a})I + \mathbf{a}\otimes\mathbf{a}\|_f^2 \tag{6.11}$$

Let M be the symmetric matrix defined by $M = E^\top E$. The application of (6.4) to the right-hand side of (6.11) yields

$$\begin{aligned} g(\mathbf{a}) &= \operatorname{tr}((M - (\mathbf{a}.\mathbf{a})I + \mathbf{a}\otimes\mathbf{a})^2) \\ &= \operatorname{tr}(M^2) - 2\operatorname{tr}((\mathbf{a}.\mathbf{a})M - M\mathbf{a}\otimes\mathbf{a}) + \operatorname{tr}((\mathbf{a}.\mathbf{a}I - \mathbf{a}\otimes\mathbf{a})^2) \\ &= \operatorname{tr}(M^2) - 2(\mathbf{a}.\mathbf{a})\operatorname{tr}(M) + 2\mathbf{a}^\top M\mathbf{a} + 2(\mathbf{a}.\mathbf{a})^2 \end{aligned} \tag{6.12}$$

It follows from (6.12) that

$$\nabla g(\mathbf{a}) = -4\mathrm{tr}(M)\mathbf{a} + 4M\mathbf{a} + 8(\mathbf{a}.\mathbf{a})\mathbf{a} \tag{6.13}$$

The function g is a minimum at a point $\mathbf{a}$ if and only if $\nabla g(\mathbf{a}) = 0$. It follows from (6.13) that $\nabla g(\mathbf{a}) = 0$ only if

$$M\mathbf{a} = (\mathrm{tr}(M) - 2(\mathbf{a}.\mathbf{a}))\mathbf{a} \tag{6.14}$$

It follows from (6.14) that g is a minimum at $\mathbf{a}$ only if $\mathbf{a}$ is one of the eigenvectors of M. Let $\mathbf{a}$ be an eigenvalue of M and let λ be the eigenvalue corresponding to $\mathbf{a}$. Equation (6.12) yields

$$g(\mathbf{a}) = \mathrm{tr}(M^2) - 2(\mathbf{a}.\mathbf{a})\mathrm{tr}(M) + 2\lambda(\mathbf{a}.\mathbf{a}) + 2(\mathbf{a}.\mathbf{a})^2 \tag{6.15}$$

It follows from (6.15) that g is minimised at $\mathbf{a}$ if and only if $\mathbf{a}$ is the eigenvector of M with the least eigenvalue.

The magnitude of $\mathbf{a}$ is found. It follows from (6.14) that $\lambda = \mathrm{tr}(M) - 2(\mathbf{a}.\mathbf{a})$, thus

$$\|\mathbf{a}\|^2 = \mathbf{a}.\mathbf{a} = (\mathrm{tr}(M) - \lambda)/2 \tag{6.16}$$

The value of $\mathrm{tr}(M)$ is equal to the sum of the eigenvalues of M. Thus (6.16) shows that $\|\mathbf{a}\|^2$ is equal to the average of the two largest eigenvalues of M. □

Theorem 6.3. *Let E be any 3×3 matrix such that the least singular value of E is unique, and let $\mathbf{a}$ be the vector obtained in Theorem 6.2 that minimises g. Let $R \mapsto h(R)$ be the function defined on the space of orthogonal matrices by*

$$h(R) = \|E - RT_{\mathbf{a}}\|_f^2 \tag{6.17}$$

and let $E = U\Sigma V$ be the singular value decomposition of E. Then h has two global minima, and these are attained at the matrices R defined such that $U^\top RV^\top$ is equal to

$$\begin{pmatrix} 0 & -1 & 0 \\ 1 & 0 & 0 \\ 0 & 0 & 1 \end{pmatrix} \quad \text{or} \quad \begin{pmatrix} 0 & 1 & 0 \\ -1 & 0 & 0 \\ 0 & 0 & 1 \end{pmatrix} \tag{6.18}$$

Proof. On substituting $U\Sigma V$ for E in (6.17) it follows that

$$\begin{aligned} h(R) &= \|U\Sigma V - RT_a\|_f^2 \\ &= \|\Sigma - (U^\top RV^\top)(VT_aV^\top)\|_f^2 \end{aligned} \tag{6.19}$$

Define the antisymmetric matrix T_u by

$$T_u = VT_aV^\top \tag{6.20}$$

and define the orthogonal matrix S by $S = U^\top RV^\top$. It is an immediate consequence of (6.20) that $\mathbf{u} = V\mathbf{a}$. It is required to show that h has a global minimum if and only if S is equal to one of the two matrices of (6.18). It follows from (6.19) and the definitions of T_u and S that

$$h(R) = \|\Sigma - ST_u\|_f^2 \tag{6.21}$$

Let σ_1, σ_2, σ_3 be the singular values of E. In Theorem 6.2 it is shown that $E^\top E\mathbf{a} = \lambda\mathbf{a}$, where λ is the least eigenvalue of $E^\top E$. It follows that $\lambda = \sigma_3^2$, and that hence

$$E^\top E\mathbf{a} = \sigma_3^2\mathbf{a} \tag{6.22}$$

On substituting $U\Sigma V$ for E in (6.22) it follows that

$$(V^\top \Sigma U^\top)(U\Sigma V)\mathbf{a} = V^\top \Sigma^2 V\mathbf{a} = \sigma_3^2\mathbf{a} \tag{6.23}$$

On substituting $\mathbf{u}$ for $V\mathbf{a}$ in (6.23) it follows that

$$\Sigma^2\mathbf{u} = \sigma_3^2\mathbf{u} \tag{6.24}$$

It follows from (6.24) and the assumption that the least singular value of E is unique that

$$\mathbf{u} = \pm\|\mathbf{a}\|(0,0,1)^\top \tag{6.25}$$

The application of (6.4) to the right-hand side of (6.21) yields

$$\begin{aligned} h(R) &= \mathrm{tr}((\Sigma + T_\mathbf{u}S^\top)(\Sigma - ST_\mathbf{u})) \\ &= \mathrm{tr}(\Sigma^2) + \mathrm{tr}(T_\mathbf{u}S^\top\Sigma) - \mathrm{tr}(\Sigma ST_\mathbf{u}) - \mathrm{tr}(T_\mathbf{u}^2) \\ &= \mathrm{tr}(\Sigma^2) + 2\mathrm{tr}(T_\mathbf{u}S^\top\Sigma) - \mathrm{tr}(T_\mathbf{u}^2) \end{aligned} \tag{6.26}$$

Let the entries of S be given by $S = (s_{ij})$. It follows from (6.25) and the definition of Σ that

$$\begin{aligned} T_\mathbf{u}S^\top\Sigma &= \pm\|\mathbf{a}\| \begin{pmatrix} 0 & 1 & 0 \\ -1 & 0 & 0 \\ 0 & 0 & 0 \end{pmatrix} \begin{pmatrix} s_{11} & s_{21} & s_{31} \\ s_{12} & s_{22} & s_{31} \\ s_{13} & s_{23} & s_{33} \end{pmatrix} \begin{pmatrix} \sigma_1 & 0 & 0 \\ 0 & \sigma_2 & 0 \\ 0 & 0 & \sigma_3 \end{pmatrix} \\ &= \pm\|\mathbf{a}\| \begin{pmatrix} \sigma_1 s_{12} & \sigma_2 s_{22} & \sigma_3 s_{32} \\ -\sigma_1 s_{11} & -\sigma_2 s_{21} & \sigma_3 s_{31} \\ 0 & 0 & 0 \end{pmatrix} \end{aligned} \tag{6.27}$$

Equations (6.26) and (6.27) yield

$$h(R) = \mathrm{tr}(\Sigma^2) - \mathrm{tr}(T_\mathbf{u}^2) \pm 2\|\mathbf{a}\|(\sigma_1 s_{12} - \sigma_2 s_{21}) \tag{6.28}$$

If the plus sign is chosen on the right-hand side of (6.28) then $R \mapsto h(R)$ is minimised by setting $s_{12} = -1$, $s_{21} = 1$. The matrix S is orthogonal and $\det(S) = +1$, thus S is equal to the first matrix of (6.18). If the minus sign is chosen on the right-hand side of (6.28) then $R \mapsto h(R)$ is minimised by setting $s_{12} = 1$, $s_{21} = -1$. The matrix S is then equal to the second matrix of (6.18). □

The implementation of the SVD based algorithm for reconstructing the camera displacement is based on Theorems 6.1-6.3. To implement step (1) in the summary of the algorithm preceding Theorem 6.1 the singular value decomposition $A = U\Sigma V$ of the matrix A of Theorem 6.1 is calculated. The vector $\mathbf{e}$ (in the notation of Theorem 6.1) is obtained from the equation $\mathbf{e} = V^{\top}\mathbf{e}_9$. The vector $\mathbf{e}$ yields the 3×3 matrix F which minimises the error function V over the space of matrices E with $\|E\|_f = 1$. To implement steps (2) and (3) of the algorithm the SVD of F is calculated. Let $F = U'\Sigma'V'$. The rigid displacement $\{R, \mathbf{a}\}$ is obtained by setting $\mathbf{a} = V'\mathbf{u}$ and $R = U'SV'$, where S, $\mathbf{u}$ are as defined in Theorem 6.3. A version of this algorithm has been implemented by Toscani and Faugeras (1986), with good results.

A great advantage of the SVD is that it can be calculated using any one of a number of stable and efficient algorithms developed in numerical analysis. The algorithms are stable in that an $O(\epsilon)$ change in the entries of a matrix A produces only an $O(\epsilon)$ change in the entries of the components of the SVD of A. Golub & Van Loan (1986) describe Chan's algorithm for computing the SVD of an $m \times n$ matrix. The cost of the algorithm is approximately

$$2m^2n + 11n^3 \quad \text{flops}$$

where a flop is one floating point operation. In order to estimate the camera displacement from a set of n image correspondences two singular value decompositions are required, one for the $n \times 3$ matrix A, and one for the 3×3 matrix F. The total cost of computing the decompositions using Chan's algorithm is approximately

$$(6n^2 + 297) + 351 \quad \text{flops}$$

6.1.2 Descent Algorithms

Many algorithms for reconstruction are based on error functions. The error function usually depends on the data and on a set of unknown parameters describing the reconstruction. A typical algorithm uses an iterative search over a space of possible parameter values to find the global minimum of the error function. The reconstruction is obtained from the parameter values at which the global minimum of the error function is located. The construction of the error function usually involves a trade-off between a proper treatment of the errors present in the data and the necessity of keeping the function reasonably simple. In most approaches, including those described in this chapter, the error function is kept simple. The balance of the trade-off may change as computer hardware becomes more powerful and as the statistics of the errors in the data become better understood.

Descent is the name given to a general class of iterative methods for finding a local minimum of a function. A descent algorithm begins at an initial point in the domain of definition of the function. It searches in a neighbourhood of the initial point for a new point at which the value of the function is less. The algorithm replaces the initial point by the new point and then repeats the cycle. The algorithm terminates when no significant reduction in the value of the error function can be achieved. In most descent algorithms the search for the new point is simplified by approximating the error function in the neighbourhood of the initial point. The usual choice is a linear approximation constructed from the Taylor expansion of the error function. The descent algorithm is then first order or gradient based. A new approximation to the error function is computed in each cycle of the algorithm. There are a multitude of variations on these basic themes. Many variations involve the fine tuning of a descent algorithm to suit a particular application. Further information on gradient descent can be found in Kelley (1962).

A disadvantage of descent is that the performance of an algorithm is difficult to predict. In general the descent is not guaranteed to terminate after a fixed prespecified number of iterations. Even worse, the algorithm may become stuck in a local minimum of the error function or it may spend a long time in regions where the gradient of the error function is low. In some applications the error function has a very large number of local minima. Fortunately, the algebraic nature of reconstruction ensures that the error functions which arise generally have only a small number of local minima. The exact number of local minima depends in part on the parameterisation chosen to describe the space of possible reconstructions.

Two closely related descent algorithms for finding the rigid displacements compatible with a set of image correspondences are described. In both algorithms the descent is based on the error function $E \mapsto V(E)$ defined by (6.5). Let $\mathbf{q}_i \leftrightarrow \mathbf{q}_i'$ for $1 \le i \le n$ be a set of n image correspondences. The essential matrix E is regarded as a function of the camera displacement $\{R, \mathbf{a}\}$. The definition of $V(R, \mathbf{a}) = V(E)$ is

$$V(R,\mathbf{a}) = \sum_{i=1}^{n} ((R\mathbf{q}_i').(\mathbf{a} \times \mathbf{q}_i))^2 \tag{6.29}$$

When searching for the minimum of $V(R, \mathbf{a})$ the condition $\mathbf{a}.\mathbf{a} = 1$ is imposed in order to avoid the trivial solution $\mathbf{a} = 0$.

It is convenient to represent rotations by unit quaternions rather than orthogonal matrices. The basic properties of quaternions are recalled.

Underlying vector space. The quaternions are points of $\mathbb{R}^4$. Each quaternion $\mathbf{z}$ is an ordered pair $\mathbf{z} = (a, \mathbf{b})$, where a is the scalar part of $\mathbf{z}$ and $\mathbf{b}$ is the vector part of $\mathbf{z}$.

Addition. The addition rule for quaternions is vector space addition in $\mathbb{R}^4$. The sum $\mathbf{z}_1 + \mathbf{z}_2$ of two quaternions $\mathbf{z}_1 = (a_1, \mathbf{b}_1)$ and $\mathbf{z}_2 = (a_2, \mathbf{b}_2)$ is defined by

$$\mathbf{z}_1 + \mathbf{z}_2 = (a_1 + a_2, \mathbf{b}_1 + \mathbf{b}_2)$$

Multiplication. The product $\mathbf{z}_1 \circ \mathbf{z}_2$ of two quaternions $\mathbf{z}_1$ and $\mathbf{z}_2$ is defined by

$$\mathbf{z}_1 \circ \mathbf{z}_2 = (a_1 a_2 - \mathbf{b}_1.\mathbf{b}_2, a_1\mathbf{b}_2 + a_2\mathbf{b}_1 + \mathbf{b}_1 \times \mathbf{b}_2) \tag{6.30}$$

Quaternion multiplication is not commutative because in general $\mathbf{z}_1 \circ \mathbf{z}_2 \neq \mathbf{z}_2 \circ \mathbf{z}_1$. The multiplication distributes over addition,

$$\begin{aligned} (\mathbf{z}_1 + \mathbf{z}_2) \circ \mathbf{z}_3 &= \mathbf{z}_1 \circ \mathbf{z}_3 + \mathbf{z}_2 \circ \mathbf{z}_3 \\ \mathbf{z}_3 \circ (\mathbf{z}_1 + \mathbf{z}_2) &= \mathbf{z}_3 \circ \mathbf{z}_1 + \mathbf{z}_3 \circ \mathbf{z}_2 \end{aligned}$$

Multiplicative inverse. Each non-zero quaternion $\mathbf{z}$ has a unique inverse $\mathbf{z}^{-1}$ defined by the property

$$\mathbf{z} \circ \mathbf{z}^{-1} = \mathbf{z}^{-1} \circ \mathbf{z} = (1, 0)$$

If $\mathbf{z} = (a, \mathbf{b})$ then $\mathbf{z}^{-1} = (a^2 + \mathbf{b}.\mathbf{b})^{-1}(a, -\mathbf{b})$.

Norm. The norm of a quaternion $\mathbf{z}$ is equal to the Euclidean length of $\mathbf{z}$ as a vector in $\mathbb{R}^4$,

$$\|\mathbf{z}\| = \|(a, \mathbf{b})\| = \sqrt{a^2 + \mathbf{b}.\mathbf{b}}$$

The norm of a product of two quaternions $\mathbf{z}_1$ and $\mathbf{z}_2$ is given by

$$\|\mathbf{z}_1 \circ \mathbf{z}_2\| = \|\mathbf{z}_1\| \, \|\mathbf{z}_2\|$$

The unit quaternions are the quaternions of norm one. The product of two unit quaternions is a unit quaternion.

Quaternions and rotations. For each orthogonal matrix R there exists a unit quaternion $\mathbf{z}$ such that for all vectors $\mathbf{x}$ in $\mathbb{R}^3$,

$$R\mathbf{x} = \mathbf{z} \circ \mathbf{x} \circ \mathbf{z}^{-1} \tag{6.31}$$

In writing (6.31) $\mathbf{x}$ is regarded as the quaternion $(0, \mathbf{x})$ and R acts on $\mathbf{x}$ according to the rule $(0, \mathbf{x}) \mapsto (0, R\mathbf{x})$. If $\mathbf{z} = (a, \mathbf{b})$ then (6.31) becomes

$$(0, R\mathbf{x}) = (a, \mathbf{b}) \circ (0, \mathbf{x}) \circ (a, -\mathbf{b})$$

Further information about quaternions can be found in Fraleigh (1973).

The advantage of representing rotations by unit quaternions is that it is easier to preserve the normalisation during the descent. If the rotation is represented by a quaternion $\mathbf{z}$ and if at the next iteration $\mathbf{z}$ is replaced by $\mathbf{z} + \Delta\mathbf{z}$ then in order to preserve the normalisation it is only necessary to ensure that $\mathbf{z} + \Delta\mathbf{z}$ is a unit quaternion. It is more difficult to find a small perturbation ΔR of an orthogonal matrix R such that $R + \Delta R$ is also orthogonal.

It follows from (6.29) and (6.31) that in defining the error function V the rotation can be replaced by a quaternion,

$$V(R,\mathbf{a}) = V(\mathbf{z},\mathbf{a}) = \sum_{i=1}^{n}[(\mathbf{z}\circ\mathbf{q}_i'\circ\mathbf{z}^{-1}).(\mathbf{a}\times\mathbf{q}_i)]^2 \tag{6.32}$$

A descent algorithm is employed to find the minimum of $(\mathbf{z},\mathbf{a}) \mapsto V(\mathbf{z},\mathbf{a})$ over the five-dimensional space formed by the product of the three-dimensional space of unit quaternions $\mathbf{z}$ and the two-dimensional space of unit translation vectors, $\mathbf{a}$.

The global minimum of $(\mathbf{z},\mathbf{a}) \mapsto V(\mathbf{z},\mathbf{a})$ is not unique, even for a general choice of image correspondences. There exist multiple global minima because

$$V(\mathbf{z},\mathbf{a}) = V(\pm\mathbf{z},\pm\mathbf{a}) \tag{6.33}$$

The four choices of sign in (6.33) yield two rigid displacements $\{R,\mathbf{a}\}$, $\{R,-\mathbf{a}\}$. A change in the sign of $\mathbf{z}$ has no effect on $\{R,\mathbf{a}\}$ because of the form of the equation (6.31) relating R and $\mathbf{z}$. Four additional local minima of $(\mathbf{z},\mathbf{a}) \mapsto V(\mathbf{z},\mathbf{a})$ are obtained from the twisted pair of solutions $\{R\sigma,\mathbf{a}\}$, $\{R\sigma,-\mathbf{a}\}$. The matrix σ is a rotation of 180° about the axis $\mathbf{a}$.

The value of V tends to decrease if $\mathbf{a}$ is brought near any of the points $\mathbf{q}_i$, $R\mathbf{q}_i'$. The reduction is more pronounced if the number of image correspondences is small, or if the $\mathbf{q}_i$ or the $\mathbf{q}_i'$ are close together. The resulting bias on the estimate of $\mathbf{a}$ can be overcome by weighting each contribution to the sum defining V. Horn (1990) discusses the correct weighting.

The strategy in the descent algorithms is to begin with an estimate $(\mathbf{z},\mathbf{a})$ of the rigid displacement and to calculate the effects of small perturbations $\Delta\mathbf{z}$, $\Delta\mathbf{a}$ on the value of V. The perturbation producing the greatest decrease in the value of $V(\mathbf{z},\mathbf{a})$ is used to update $(\mathbf{z},\mathbf{a})$ according to the scheme

$$(\mathbf{z},\mathbf{a}) \mapsto (\mathbf{z}+\Delta\mathbf{z},\mathbf{a}+\Delta\mathbf{a})$$

In addition to reducing the value of V the update is chosen such that the normalisations of $\mathbf{z}$ and $\mathbf{a}$ are preserved,

$$\begin{aligned} \|\mathbf{z}+\Delta\mathbf{z}\| &= \|\mathbf{z}\| = 1 \\ \|\mathbf{a}+\Delta\mathbf{a}\| &= \|\mathbf{a}\| = 1 \end{aligned} \tag{6.34}$$

The Taylor series expansion of the error function V about the point $(\mathbf{z},\mathbf{a})$ to second order is

$$V(\mathbf{z}+\delta\mathbf{z},\mathbf{a}+\delta\mathbf{a}) = V(\mathbf{z},\mathbf{a}) + \mathbf{l}.\delta\mathbf{z} + \mathbf{m}.\delta\mathbf{a} + \delta\mathbf{z}^{\top}L\delta\mathbf{z} + 2\delta\mathbf{z}^{\top}M\delta\mathbf{a} + \delta\mathbf{a}^{\top}N\delta\mathbf{a} + R_3 \tag{6.35}$$

where $\mathbf{l}$, $\mathbf{m}$ are vectors and L, M, N are matrices. The term R_3 in (6.35) is third order in $\delta\mathbf{a}$, $\delta\mathbf{z}$. It is assumed that R_3 is small and that the first six terms on the right-hand side of (6.35) are a good approximation to $V(\mathbf{z}+\delta\mathbf{z}, \mathbf{a}+\delta\mathbf{a})$. The components of $\mathbf{l}$, $\mathbf{m}$, and L, M, N depend on $\mathbf{z}$, $\mathbf{a}$, and on the image correspondences. Explicit but not very enlightening formulae for $\mathbf{l}$, $\mathbf{m}$, L, M, N are given at the end of this subsection.

Two closely related forms of descent are described, namely first order descent and second order descent. Both forms of descent are based on (6.35). First order descent uses the terms on the right-hand side of (6.35) up to first order in $\delta\mathbf{z}$ or $\delta\mathbf{a}$ and second order descent uses the terms up to second order in $\delta\mathbf{z}$ or $\delta\mathbf{a}$.

First order descent. In first order descent the calculation of $\Delta\mathbf{z}$, $\Delta\mathbf{a}$ is based on the first three terms on the right-hand side of (6.35). Values of $\Delta\mathbf{z}$ and $\Delta\mathbf{a}$ are sought that make $\Delta\mathbf{z}.\mathbf{l}+\Delta\mathbf{a}.\mathbf{m}$ as negative as possible, subject to the constraints (6.34). An upper bound on the norms $\|\Delta\mathbf{z}\|$, $\|\Delta\mathbf{a}\|$ is imposed in order to ensure that $(\mathbf{z}+\Delta\mathbf{z}, \mathbf{a}+\Delta\mathbf{a})$ is within a region where the error in approximating $V(\mathbf{z}+\delta\mathbf{z}, \mathbf{a}+\delta\mathbf{a})$ by the first three terms on the right-hand side of (6.35) is small. The constraints on $\|\Delta\mathbf{z}\|$, $\|\Delta\mathbf{a}\|$ are $\|\Delta\mathbf{z}\| = h$ and $\|\Delta\mathbf{a}\| = h$ where h is a small fixed quantity.

The two increments $\Delta\mathbf{z}$, $\Delta\mathbf{a}$ are found separately. Let the initial point for the descent algorithm be $(\mathbf{z}_1, \mathbf{a}_1)$. Define the error function $(\lambda_1, \lambda_2, \Delta\mathbf{z}) \mapsto W(\lambda_1, \lambda_2, \Delta\mathbf{z})$ by

$$W(\lambda_1, \lambda_2, \Delta\mathbf{z}) = \mathbf{l}.\Delta\mathbf{z} + \lambda(\|\Delta\mathbf{z}\|^2 - h^2) + \lambda_2(\|\mathbf{z}_1 + \Delta\mathbf{z}\|^2 - 1) \qquad (6.36)$$

The coefficients λ_1, λ_2 on the right-hand side of (6.36) are Lagrange multipliers. Their purpose is to enforce the following two constraints on $\Delta\mathbf{z}$:

$$\|\Delta\mathbf{z}\| = h \qquad \|\mathbf{z}_1 + \Delta\mathbf{z}\| = 1$$

Further information about Lagrange multipliers can be found in Sokolnikoff & Redheffer (1966). The function W is differentiated with respect to $\Delta\mathbf{z}$, λ_1 and λ_2 in turn and the resulting derivatives set to zero to obtain

$$\begin{aligned} \partial W/\partial\Delta\mathbf{z} &= \mathbf{l} + 2\lambda_1\Delta\mathbf{z} + 2\lambda_2(\mathbf{z}_1 + \Delta\mathbf{z}) = 0 \\ \partial W/\partial\lambda_1 &= \Delta\mathbf{z}.\Delta\mathbf{z} - h^2 = 0 \\ \partial W/\partial\lambda_2 &= 2\mathbf{z}_1.\Delta\mathbf{z} + \Delta\mathbf{z}.\Delta\mathbf{z} = 0 \end{aligned} \qquad (6.37)$$

On using the first equation of (6.37) to eliminate $\Delta\mathbf{z}$ from the remaining two equations, and on replacing $\Delta\mathbf{z}.\Delta\mathbf{z}$ in the third equation by h^2 the following two equations are obtained:

$$\begin{aligned} (\mathbf{l} + 2\lambda_2\mathbf{z}_1).(\mathbf{l} + 2\lambda_2\mathbf{z}_1) - 4(\lambda_1 + \lambda_2)^2h^2 &= 0 \\ -(\mathbf{z}_1.\mathbf{l} + 2\lambda_2) + h^2(\lambda_1 + \lambda_2) &= 0 \end{aligned} \qquad (6.38)$$

The first equation of (6.38) yields

$$\mathbf{l}.\mathbf{l} + 4\lambda_2 \mathbf{l}.\mathbf{z}_1 + 4\lambda_2^2 = 4(\lambda_1 + \lambda_2)^2 h^2 \tag{6.39}$$

On using the second equation of (6.38) to substitute for $\lambda_1 + \lambda_2$ in (6.39) it follows that

$$\lambda_2^2 + (\mathbf{l}.\mathbf{z}_1)\lambda_2 + \frac{4(\mathbf{l}.\mathbf{z}_1)^2 - h^2(\mathbf{l}.\mathbf{l})}{4(4-h^2)} = 0 \tag{6.40}$$

Let $\epsilon = \pm 1$ and let α be defined by

$$\alpha = \frac{1}{2}\sqrt{\frac{\mathbf{l}.\mathbf{l} - (\mathbf{l}.\mathbf{z}_1)^2}{4-h^2}}$$

It follows from (6.40) that

$$\lambda_2 = -\frac{1}{2}(\mathbf{l}.\mathbf{z}_1) + \epsilon h \alpha \tag{6.41}$$

It follows from (6.37), (6.38) and (6.41) that

$$\begin{aligned} \Delta\mathbf{z} &= -(\mathbf{l} + 2\lambda_2 \mathbf{z}_1)/(2(\lambda_1 + \lambda_2)) \\ &= -\frac{h\epsilon}{4\alpha}(\mathbf{l} + 2\lambda_2 \mathbf{z}_1) \\ &= -\frac{h\epsilon}{4\alpha}(\mathbf{l} - (\mathbf{l}.\mathbf{z}_1)\mathbf{z}_1 + 2\epsilon h \alpha \mathbf{z}_1) \end{aligned}$$

A similar calculation yields

$$\Delta\mathbf{a} = -\frac{h\eta}{4\beta}(\mathbf{m} - (\mathbf{m}.\mathbf{z}_1)\mathbf{z}_1 + 2\eta h \alpha \mathbf{z}_1)$$

where $\eta = \pm 1$, and β is defined by

$$\beta = \frac{1}{2}\sqrt{\frac{\mathbf{m}.\mathbf{m} - (\mathbf{m}.\mathbf{a}_1)^2}{4-h^2}}$$

The updated values $(\mathbf{z}_2, \mathbf{a}_2)$ are defined by

$$(\mathbf{z}_2, \mathbf{a}_2) = (\mathbf{z}_1 + \Delta\mathbf{z}, \mathbf{a}_1 + \Delta\mathbf{a}) \tag{6.42}$$

There are four possible choices for $(\mathbf{z}_2, \mathbf{a}_2)$, depending on the values assigned to $\epsilon = \pm 1$ and $\eta = \pm 1$ on the right-hand side of (6.42). The chosen point $(\mathbf{z}_2, \mathbf{a}_2)$ is the one at which V is least.

If $V(\mathbf{z}_2, \mathbf{a}_2)$ is significantly smaller than $V(\mathbf{z}_1, \mathbf{a}_1)$ then the cycle of the descent algorithm is repeated with $(\mathbf{z}_2, \mathbf{a}_2)$ in place of $(\mathbf{z}_1, \mathbf{a}_1)$. If $V(\mathbf{z}_2, \mathbf{a}_2)$ is not significantly smaller than $V(\mathbf{z}_1, \mathbf{a}_1)$ then $\Delta\mathbf{z}$, $\Delta\mathbf{a}$ are recalculated using a smaller value of h. If the error function is not significantly decreased, even for a small value of h, then the algorithm halts and returns $(\mathbf{z}_1, \mathbf{a}_1)$ as an estimate of the position of a global minimum of V.

Second order descent. If a first order approximation to the error function V fails to produce a significant reduction in the value of V then a more complicated second order approximation can be tried. Let the initial point for the descent algorithm be $(\mathbf{z}_1, \mathbf{a}_1)$. The terms of (6.35) to second order in $\delta\mathbf{z}$, $\delta\mathbf{a}$ are used to calculate the finite increments $\Delta\mathbf{z}$, $\Delta\mathbf{a}$, subject to the constraints $\Delta\mathbf{z}.\mathbf{z}_1 = 0$, $\Delta\mathbf{a}.\mathbf{a}_1 = 0$. In order to incorporate these two constraints Lagrange multipliers λ_1, λ_2 are introduced. The function

$$(\lambda_1, \lambda_2, \Delta\mathbf{z}, \Delta\mathbf{a}) \mapsto U(\Delta\mathbf{z}, \Delta\mathbf{a}, \lambda_1, \lambda_2)$$

is minimised, where U is defined by

$$\begin{aligned} U(\lambda_1, \lambda_2, \Delta\mathbf{z}, \Delta\mathbf{a}) &= \mathbf{l}.\Delta\mathbf{z} + \mathbf{m}.\Delta\mathbf{a} + \Delta\mathbf{z}^\top L \Delta\mathbf{z} + 2\Delta\mathbf{z}^\top M \Delta\mathbf{a} + \Delta\mathbf{a}^\top N \Delta\mathbf{a} \\ &\quad + \lambda_1 \mathbf{z}.\Delta\mathbf{z} + \lambda_2 \mathbf{a}.\Delta\mathbf{a} \end{aligned} \tag{6.43}$$

On differentiating U with respect to $\Delta\mathbf{z}$ and $\Delta\mathbf{a}$ and setting the resulting expressions equal to zero the following equations constraining $\Delta\mathbf{z}$ and $\Delta\mathbf{a}$ are obtained:

$$\begin{aligned} \mathbf{l} + 2M\Delta\mathbf{a} + 2L\Delta\mathbf{z} + \lambda_1 \mathbf{z}_1 &= 0 \\ \mathbf{m} + 2M^\top \Delta\mathbf{z} + 2N\Delta\mathbf{a} + \lambda_2 \mathbf{a}_1 &= 0 \end{aligned} \tag{6.44}$$

Equations (6.44) yield a matrix equation for the vector $(\Delta\mathbf{z}, \Delta\mathbf{a})$,

$$2\begin{pmatrix} L & M \\ M^\top & N \end{pmatrix} \begin{pmatrix} \Delta\mathbf{z} \\ \Delta\mathbf{a} \end{pmatrix} = -\begin{pmatrix} \mathbf{l} + \lambda_1 \mathbf{z}_1 \\ \mathbf{m} + \lambda_2 \mathbf{a}_1 \end{pmatrix} \tag{6.45}$$

Equation (6.45) is solved for $(\Delta\mathbf{z}, \Delta\mathbf{a})$,

$$\begin{pmatrix} \Delta\mathbf{z} \\ \Delta\mathbf{a} \end{pmatrix} = -\frac{1}{2}\begin{pmatrix} L & M \\ M^\top & N \end{pmatrix}^{-1} \begin{pmatrix} \mathbf{l} + \lambda_1 \mathbf{z}_1 \\ \mathbf{m} + \lambda_2 \mathbf{a}_1 \end{pmatrix} \tag{6.46}$$

The right-hand side of (6.46) is linear in the Lagrange multipliers λ_1, λ_2. The values of λ_1, λ_2 are obtained by applying the constraints $\mathbf{z}_1.\Delta\mathbf{z} = 0$ and $\mathbf{a}_1.\Delta\mathbf{a} = 0$,

$$\begin{aligned} (\mathbf{z}_1, 0)\begin{pmatrix} L & M \\ M^\top & N \end{pmatrix}^{-1} \begin{pmatrix} \mathbf{l} + \lambda_1 \mathbf{z}_1 \\ \mathbf{m} + \lambda_2 \mathbf{a}_1 \end{pmatrix} &= 0 \\ (\mathbf{a}_1, 0)\begin{pmatrix} L & M \\ M^\top & N \end{pmatrix}^{-1} \begin{pmatrix} \mathbf{l} + \lambda_1 \mathbf{z}_1 \\ \mathbf{m} + \lambda_2 \mathbf{a}_1 \end{pmatrix} &= 0 \end{aligned} \tag{6.47}$$

The values of λ_1, λ_2 found by solving the two simultaneous equations (6.47) are substituted back into (6.46) to yield expressions for $\Delta\mathbf{z}$, $\Delta\mathbf{a}$. The new point $(\mathbf{z}_2, \mathbf{a}_2)$ is defined by

$$(\mathbf{z}_2, \mathbf{a}_2) = (\mathbf{z}_1 + \Delta\mathbf{z}, \mathbf{a}_1 + \Delta\mathbf{a})$$

If $V(\mathbf{z}_2, \mathbf{a}_2)$ is significantly smaller than $V(\mathbf{z}_1, \mathbf{a}_1)$ then the cycle of the descent algorithm is repeated with $(\mathbf{z}_2, \mathbf{a}_2)$ in place of $(\mathbf{z}_1, \mathbf{a}_1)$.

In second order descent the new point $(\mathbf{z}_2, \mathbf{a}_2)$ is obtained without introducing any bounds on $\|\Delta\mathbf{z}\|$ and $\|\Delta\mathbf{a}\|$ analogous to the bound h used in first order descent. This is because $\|\Delta\mathbf{z}\|$ and $\|\Delta\mathbf{a}\|$ are constrained by the equations governing second order descent. It is in practice advisable to place an upper bound on $\|\Delta\mathbf{z}\|$ and $\|\Delta\mathbf{a}\|$ in order to be sure of staying within the region in which the terms to second order in $\delta\mathbf{z}$ and $\delta\mathbf{a}$ on the right-hand side of (6.35) give a good approximation to $V(\mathbf{z}+\delta\mathbf{z}, \mathbf{a}+\delta\mathbf{a})$.

If the starting point for the descent algorithm is not too far from a local minimum of V then the sequence of points $(\mathbf{z}_i, \mathbf{a}_i)$, $i = 1, 2, \ldots$ generated by the descent algorithm converges rapidly to the local minimum. If the starting point is far from any local minimum then the sequence $(\mathbf{z}_i, \mathbf{a}_i)$, $i = 1, 2, \ldots$ may be very erratic. In most cases the sequence $(\mathbf{z}_i, \mathbf{a}_i)$ eventually converges to a local minimum of $V(\mathbf{z}, \mathbf{a})$, but it may be very hard to predict which one. In cases where a good estimate of the starting point is not available the descent algorithm can be run many times starting from different points scattered over the space $S^4 \times S^2$ of values of $(\mathbf{z}, \mathbf{a})$. If the starting points are scattered widely enough and densely enough then all the local minima of V will be located. Further details are given in Horn (1990).

The Taylor expansion of $V(\mathbf{z}, \mathbf{a})$. Expressions are obtained for the vectors $\mathbf{l}$, $\mathbf{m}$ and for the matrices L, M, N appearing in (6.35). In order to simplify the notation functions $(\mathbf{z}, \mathbf{a}) \mapsto f_i(\mathbf{z}, \mathbf{a})$ are defined by

$$\begin{aligned} f_i(\mathbf{z}, \mathbf{a}) &= (\mathbf{z} \circ \mathbf{q}_i' \circ \mathbf{z}).(\mathbf{a} \times \mathbf{q}_i) \\ &= M_{ijkl} z_j z_k a_l \qquad (1 \le i \le n) \end{aligned} \tag{6.48}$$

The four-dimensional array M_{ijkl} depends on $\mathbf{z}$, $\mathbf{a}$ and on the ith pair of corresponding points, $\mathbf{q}_i \leftrightarrow \mathbf{q}_i'$. The usual convention of summing over repeated indices is adopted in (6.48) and the formulae which follow. The indices i, j, k, l have ranges $1 \le i \le n$, $1 \le j, k \le 4$, $1 \le l \le 3$. The array M_{ijkl} is symmetric in j and k, in that $M_{ijkl} = M_{ikjl}$. It follows from (6.32) and (6.48) that

$$V(\mathbf{z}, \mathbf{a}) = \sum_{i=1}^{n} f_i(\mathbf{z}, \mathbf{a})^2 = \sum_{i=1}^{n} (M_{ijkl} z_j z_k a_l)^2$$

Let l_s be the sth component of $\mathbf{l}$ and let m_s be the sth component of $\mathbf{m}$. The first order derivatives of V yield

$$\begin{aligned} l_s &= \frac{\partial V}{\partial z_s} &&= 2\sum_{i=1}^{n} f_i \frac{\partial f_i}{\partial z_s} \\ & &&= 4\sum_{i=1}^{n} f_i M_{iskl} z_k a_l \\ m_s &= \frac{\partial V}{\partial a_s} &&= 2\sum_{i=1}^{n} f_i \frac{\partial f_i}{\partial a_s} \\ & &&= 2\sum_{i=1}^{n} f_i M_{ijks} z_j z_k \end{aligned}$$

The matrices L, M, N are obtained by differentiating V twice,

$$\begin{aligned}
L_{uv} &= \frac{\partial^2 V}{\partial z_u \partial z_v} = 2\sum_{i=1}^{n}[\frac{\partial f_i}{\partial z_u}\frac{\partial f_i}{\partial z_v} + f_i \frac{\partial^2 f_i}{\partial z_u \partial z_v}] \\
&= 4\sum_{i=1}^{n}[2M_{iukl}M_{ivst}z_k z_s a_l a_t + f_i M_{iuvl}a_l] \\
M_{uv} &= \frac{\partial^2 V}{\partial z_u \partial a_v} = 4\sum_{i=1}^{n}[M_{iukl}M_{ijsv}z_j z_s z_k a_l + f_i M_{iukv}z_k] \\
N_{uv} &= \frac{\partial^2 V}{\partial a_u \partial a_v} = 2\sum_{i=1}^{n}\frac{\partial f_i}{\partial a_u}\frac{\partial f_i}{\partial a_v} \\
&= 2\sum_{i=1}^{n} M_{ijku}M_{istv}z_j z_k z_s z_t
\end{aligned}$$

6.2 Reconstruction from Image Velocities

Two algorithms for reconstruction from image velocities are described. The first algorithm uses a least squares approach to define an error function $\mathbf{v} \mapsto \epsilon(\mathbf{v})$ on the unit sphere of possible directions for the translational velocity of the rigid surface giving rise to the image velocity field. The second algorithm uses estimates of the derivatives of the image velocity field to first order; two sets of derivatives are required, obtained from different parts of the field of view. The algorithms are called respectively the least squares algorithm and the first order algorithm.

Both algorithms illustrate the importance of using any irregularities in the image velocity field to improve estimates of the camera velocity. In the case of the least squares algorithm the effects of irregularity are described in detail. If the image velocity field is irregular then

$$\epsilon(\mathbf{v}) \sim \tau|\mathbf{n}.\mathbf{v}| \tag{6.49}$$

where τ and $\mathbf{n}$ are independent of $\mathbf{v}$. The coefficient τ measures the irregularity of the surface giving rise to the image velocity field and $\mathbf{n}$ is normal to the true translational velocity $\hat{\mathbf{v}}$. A consequence of the approximation (6.49) is that one component of $\hat{\mathbf{v}}$ and one component of the true angular velocity $\hat{\mathbf{w}}$ can be estimated more accurately than the other components. The least squares algorithm can be applied to any image velocity field arising from rigid motion. If the surface is smooth rather than irregular then the simple approximation (6.49) no longer applies.

The first order algorithm applies only to image velocity fields that are differentiable in two different parts of the field of view. If the two sets of derivatives differ significantly then the stability of the first order algorithm is increased. Differentiation amplifies any errors in the estimates of the image velocity vectors.

The first order algorithm is thus likely to perform better than algorithms which rely on derivatives of the image velocity field to second order.

The least squares algorithm is defined in Sect. 6.2.1 and its properties investigated in Sects. 6.2.2-6.2.4. The first order algorithm is defined in Sect. 6.2.5 and its properties investigated in Sect. 6.2.5 and Sect. 6.2.6.

6.2.1 A Least Squares Algorithm

In the least squares algorithm an error function, $\mathbf{v} \mapsto \epsilon(\mathbf{v})$, is defined on the unit sphere of possible directions of the translational velocity $\mathbf{v}$ of the rigid surface relative to the camera. Let the image velocity field consist of n velocity vectors $\dot{\mathbf{q}}_i$ with base points $\mathbf{q}_i$ on the unit sphere. Coordinates are chosen in $\mathbb{R}^3$ such that the centre of the sphere is at the origin. The centre of the sphere is, by definition, the optical centre of the camera. Let the vectors $\mathbf{r}_i$ be defined by

$$\mathbf{r}_i = \dot{\mathbf{q}}_i \times \mathbf{q}_i \qquad (1 \leq i \leq n)$$

Let $\mathbf{w}$ be the angular velocity of the rigid surface giving rise to the image velocity field, taken about the optical centre of the camera. Each image velocity vector yields a scalar constraint on the rigid velocity $\{\mathbf{v}, \mathbf{w}\}$. This constraint is given as equation (4.8) in Sect. 4.1. For each i it has the form

$$\mathbf{r}_i.\mathbf{v} = [(\mathbf{v}.\mathbf{q}_i)\mathbf{q}_i - \mathbf{v}].\mathbf{w} \qquad (1 \leq i \leq n) \tag{6.50}$$

Let A be the $n \times 3$ matrix with ith row equal to

$$(\mathbf{v}.\mathbf{q}_i)\mathbf{q}_i^{\mathsf{T}} - \mathbf{v}^{\mathsf{T}}$$

and let $\mathbf{l}$ be the n dimensional vector with ith entry equal to $\mathbf{r}_i.\mathbf{v}$. The n equations of (6.50) form the rows of the matrix equation $A\mathbf{w} = \mathbf{l}$. The value $\epsilon(\mathbf{v})$ is defined to be the least squares error in solving $A\mathbf{w} = \mathbf{l}$ for $\mathbf{w}$ at a fixed value of $\mathbf{v}$. In detail,

$$\epsilon(\mathbf{v}) = \min\left\{\|A\mathbf{w} - \mathbf{l}\| \;\middle|\; \mathbf{w} \in \mathbb{R}^3\right\} \tag{6.51}$$

If it is necessary to specify the image velocity field used in (6.51) then the notation $\epsilon(\langle\dot{\mathbf{q}}\rangle, \mathbf{v})$ is used. It is assumed that $\mathbf{v} \neq 0$ and that $n \geq 4$. This is to avoid the trivial case in which ϵ is zero for almost all $\mathbf{v}$.

Some additional notation is required. Let l_i be the ith component of $\mathbf{l}$. The quantities $\epsilon_i(\mathbf{v}, \mathbf{w})$ are defined by

$$\epsilon_i(\mathbf{v}, \mathbf{w}) = (\mathbf{v}.\mathbf{q}_i)(\mathbf{w}.\mathbf{q}_i) - \mathbf{w}.\mathbf{v} - l_i \qquad (1 \leq i \leq n) \tag{6.52}$$

It follows from (6.50) and (6.52) that

$$\epsilon(\mathbf{v}) = \min\left\{\sqrt{\sum_{i=1}^{n} \epsilon_i^2} \;\middle|\; \mathbf{w} \in \mathbb{R}^3\right\} \tag{6.53}$$

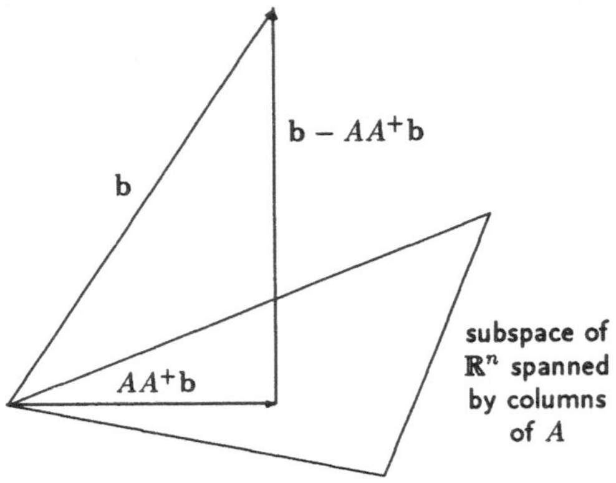

Fig. 6.1. The Moore-Penrose pseudo-inverse

The separate contributions ϵ_i to $\epsilon(\mathbf{v})$ are not weighted to take account of their varying degrees of reliability. This is because a weighting would lead to an error function too complicated to analyse.

A closed form expression for $\epsilon(\mathbf{v})$ is obtained using the Moore-Penrose pseudo-inverse A^+ of the matrix A. If the matrix $A^\top A$ is invertible then A^+ is defined by

$$A^+ = (A^\top A)^{-1} A^\top$$

If A is invertible then A^+ is equal to the usual inverse A^{-1}. The matrix AA^+ has a geometric interpretation, illustrated in Fig. 6.1. Let L be the subspace of $\mathbb{R}^n$ spanned by the columns of the $m \times n$ matrix A and let $\mathbf{b}$ be any vector in $\mathbb{R}^n$. Then $AA^+\mathbf{b}$ is the orthogonal projection of $\mathbf{b}$ onto L. Thus $\|AA^+\mathbf{b} - \mathbf{b}\|$ is the minimum distance from $\mathbf{b}$ to L. It follows that $\|A\mathbf{x} - \mathbf{b}\|$ is minimised over all $\mathbf{x}$ in $\mathbb{R}^n$ when $\mathbf{x} = A^+\mathbf{b}$. It follows from (6.51) and the definition of A^+ that

$$\epsilon(\mathbf{v}) = \|AA^+\mathbf{l} - \mathbf{l}\| \tag{6.54}$$

The value of the angular velocity $\mathbf{w}$ at which the minimum of (6.51) is attained is $\mathbf{w} = A^+\mathbf{l}$. Information about the Moore-Penrose pseudo-inverse can be found in Golub & Van Loan (1983).

The least squares algorithm involves a search over the unit sphere for the value of $\mathbf{v}$ at which ϵ is a minimum. In practice the search can be carried out quickly because the unit sphere is only of dimension two, and it has a finite area. There is no loss of generality in assuming $\|\mathbf{v}\| = 1$, because $\mathbf{v}$ is determined by the image velocity field only up to a non-zero scalar multiple.

6.2.2 Properties of the Least Squares Error Function

It is shown that the least squares error function $\mathbf{v} \mapsto \epsilon(\mathbf{v})$ is independent of the component of the image velocity field arising from the angular velocity. It is also shown that ϵ has a continuity property: if $\epsilon(\langle\dot{\mathbf{q}}\rangle, \mathbf{v})$ is small then there exists a rigid surface moving with the velocity $\{\mathbf{v}, \mathbf{w}(\mathbf{v})\}$ that produces an image velocity field close to $\langle\dot{\mathbf{q}}\rangle$.

Proposition 6.4. *If an image velocity field due to angular velocity only is added to the given image velocity field then the values of ϵ are unchanged for all $\mathbf{v}$.*

Proof. Let $\dot{\mathbf{q}}_i$ for $1 \leq i \leq n$ be the given image velocity field and let $\mathbf{w}'$ be an additional angular velocity. Let $\dot{\mathbf{q}}_i'$ be the image velocity vectors defined by

$$\dot{\mathbf{q}}_i' = \dot{\mathbf{q}}_i + (\mathbf{w}' \times \mathbf{q}_i) \times \mathbf{q}_i \qquad (1 \leq i \leq n)$$

Let $\mathbf{l}$, $\mathbf{l}'$ be the n dimensional vectors with ith components given by

$$\begin{aligned} l_i &= (\dot{\mathbf{q}}_i \times \mathbf{q}_i).\mathbf{v} \\ l_i' &= (\dot{\mathbf{q}}_i' \times \mathbf{q}_i).\mathbf{v} \qquad (1 \leq i \leq n) \end{aligned}$$

It follows from the definition (6.51) of $\epsilon(\mathbf{v})$ that

$$\begin{aligned} \epsilon(\langle\dot{\mathbf{q}}\rangle, \mathbf{v}) &= \min\{\|A\mathbf{w} - \mathbf{l}\| \mid \mathbf{w} \in \mathbb{R}^3\} \\ &= \min\{\|A(\mathbf{w} - \mathbf{w}') - \mathbf{l}\| \mid \mathbf{w} \in \mathbb{R}^3\} \\ &= \min\{\|A\mathbf{w} - \mathbf{l}'\| \mid \mathbf{w} \in \mathbb{R}^3\} \\ &= \epsilon(\langle\dot{\mathbf{q}}'\rangle, \mathbf{v}) \end{aligned}$$

as required. □

Corollary. The values of ϵ depend only on the component of the image velocity field arising from the translational velocity.

In the case $n = 4$ the function ϵ is closely related to the quartic polynomial constraint q defined by (4.58) in Sect. 4.3.1.

Proposition 6.5. *If the number of image velocity vectors is four then*

$$|q(\mathbf{v})| = \sqrt{\det(A^\top A)}\, \epsilon(\mathbf{v}) \qquad (6.55)$$

where A is the matrix of (6.53).

Proof. Let V be an orthogonal matrix chosen such that $\det(V) = 1$ and such that $V^\top A$ is upper triangular. If A has rank three then (6.51) yields

$$\begin{aligned} \epsilon(\mathbf{v}) &= \min\{\|A\mathbf{w} - \mathbf{l}\| \mid \mathbf{w} \in \mathbb{R}^3\} \\ &= \min\{\|V^\top A\mathbf{w} - V^\top\mathbf{l}\| \mid \mathbf{w} \in \mathbb{R}^3\} \\ &= |(V^\top\mathbf{l})_4| \qquad (6.56) \end{aligned}$$

In the case of q, (4.58) yields

$$\begin{aligned} q(\mathbf{v}) &= \det(A|\mathbf{l}) \\ &= \det(V^\top A|V^\top\mathbf{l}) \\ &= \pm\sqrt{\det(A^\top A)}\,(V^\top\mathbf{l})_4 \qquad (6.57) \end{aligned}$$

Equation (6.55) follows from (6.56) and (6.57). If A has rank two or less then (6.55) holds because $q(\mathbf{v}) = 0$ and $\det(A^{\mathsf{T}}A) = 0$. □

In the next theorem it is shown that ϵ has a continuity property: if ϵ is small for some $\mathbf{v}$, then the image velocity field is nearly compatible with a rigid velocity $\{\mathbf{v}, \mathbf{w}\}$, where $\mathbf{w}$ is an angular velocity depending on $\mathbf{v}$. In the noise-free case, $\epsilon(\mathbf{v}) = 0$ if and only if $\mathbf{v}$ is the translational velocity of a rigid surface giving rise to the image velocity field.

Theorem 6.6. *Let ϵ be the least squares error function constructed from the image velocity vectors $\dot{\mathbf{q}}_i$, $1 \le i \le n$. Let $\mathbf{u}$ be a unit vector such that $\epsilon(\mathbf{u}) \le \delta$ for some $\delta > 0$. Then there exists an image velocity field $\dot{\mathbf{q}}_i'$ for $1 \le i \le n$ with the same set of base points $\mathbf{q}_i$, such that the $\dot{\mathbf{q}}_i'$ arise from a surface moving with translational velocity $\mathbf{u}$ and such that*

$$\sum_{i=1}^{n} \|\dot{\mathbf{q}}_i - \dot{\mathbf{q}}_i'\|^2 \le r^{-2}\delta^2 \tag{6.58}$$

where r is defined by

$$r = \min\{\|\mathbf{u} \times \mathbf{q}_i\| \mid 1 \le i \le n\} \tag{6.59}$$

Proof. Let the ϵ_i be defined as in (6.52) and let $\mathbf{w} = \mathbf{w}(\mathbf{u})$ be the angular velocity such that

$$\epsilon(\mathbf{u})^2 = \sum_{i=1}^{n} \epsilon_i(\mathbf{u}, \mathbf{w})^2$$

The hypothesis is that

$$\sum_{i=1}^{n} \epsilon_i^2 = \epsilon(\mathbf{u})^2 < \delta^2$$

If $\mathbf{u} \times \mathbf{q}_i = 0$ for some i then r^{-2} is infinite and (6.58) holds trivially. It is thus assumed that $\mathbf{u} \times \mathbf{q}_i \ne 0$ for $1 \le i \le n$. The substitution of $l_i = (\dot{\mathbf{q}}_i \times \mathbf{q}_i).\mathbf{u}$ into (6.52) yields

$$\begin{aligned} \epsilon_i &= -(\mathbf{w} \times \mathbf{q}_i).(\mathbf{u} \times \mathbf{q}_i) - \dot{\mathbf{q}}_i.(\mathbf{q}_i \times \mathbf{u}) \\ &= (\mathbf{u} \times \mathbf{q}_i).(\dot{\mathbf{q}}_i - \mathbf{w} \times \mathbf{q}_i) \qquad (1 \le i \le n) \end{aligned} \tag{6.60}$$

For each value of i, the vectors $\mathbf{u}$, $\mathbf{q}_i$, $\mathbf{u} \times \mathbf{q}_i$ span $\mathbb{R}^3$. Let a_i, b_i, c_i be scalars such that

$$\dot{\mathbf{q}}_i - \mathbf{w} \times \mathbf{q}_i = a_i\mathbf{u} + b_i\mathbf{q}_i + c_i\mathbf{u} \times \mathbf{q}_i \qquad (1 \le i \le n) \tag{6.61}$$

The scalar product of (6.61) with $\mathbf{q}_i$ yields

$$\dot{\mathbf{q}}_i.\mathbf{q}_i = a_i\mathbf{u}.\mathbf{q}_i + b_i = 0 \tag{6.62}$$

On using (6.62) to substitute for b_i in (6.61) the following equation is obtained:

$$\dot{\mathbf{q}}_i = a_i[\mathbf{u} - (\mathbf{u}.\mathbf{q}_i)\mathbf{q}_i] + c_i\mathbf{u} \times \mathbf{q}_i + \mathbf{w} \times \mathbf{q}_i \qquad (1 \le i \le n) \qquad (6.63)$$

The image velocity field $\dot{\mathbf{q}}_i'$ is defined by

$$\dot{\mathbf{q}}_i' = a_i[\mathbf{u} - (\mathbf{u}.\mathbf{q}_i)\mathbf{q}_i] + \mathbf{w} \times \mathbf{q}_i \qquad (1 \le i \le n) \qquad (6.64)$$

It follows from (6.63) and (6.64) that $\dot{\mathbf{q}}_i - \dot{\mathbf{q}}_i' = c_i\mathbf{u} \times \mathbf{q}_i'$, thus

$$\|\dot{\mathbf{q}}_i - \dot{\mathbf{q}}_i'\| = |c_i|\,\|\mathbf{u} \times \mathbf{q}_i\| \qquad (1 \le i \le n) \qquad (6.65)$$

It follows from (6.60) and (6.63) that

$$\begin{aligned} \epsilon_i &= (\dot{\mathbf{q}}_i - \mathbf{w} \times \mathbf{q}_i).(\mathbf{u} \times \mathbf{q}_i) \\ &= [a_i(\mathbf{u} - (\mathbf{u}.\mathbf{q}_i)\mathbf{q}_i) + c_i\mathbf{u} \times \mathbf{q}_i].(\mathbf{u} \times \mathbf{q}_i) \\ &= c_i(\mathbf{u} \times \mathbf{q}_i).(\mathbf{u} \times \mathbf{q}_i) \\ &= c_i\|\mathbf{u} \times \mathbf{q}_i\|^2 \qquad (1 \le i \le n) \qquad (6.66) \end{aligned}$$

The elimination of c_i from (6.65) and (6.66) yields

$$\|\dot{\mathbf{q}}_i - \dot{\mathbf{q}}_i'\| = \epsilon_i\|\mathbf{u} \times \mathbf{q}_i\|^{-1} \qquad (1 \le i \le n) \qquad (6.67)$$

It follows from (6.67) and the definition (6.59) of r that

$$\begin{aligned} \sum_{i=1}^{n} \|\dot{\mathbf{q}}_i - \dot{\mathbf{q}}_i'\|^2 &= \sum_{i=1}^{n} \epsilon_i^2\,\|\mathbf{u} \times \mathbf{q}_i\|^{-2} \\ &\le r^{-2}\delta^2 \end{aligned}$$

as required. □

6.2.3 Irregular Image Velocity Fields

An irregular image velocity field is one in which nearby velocity vectors differ significantly in length or orientation, because they arise from points on the rigid surface that are at different distances from the camera. In this subsection a number of results are obtained with the same general form: if the image velocity field is irregular then the values $\epsilon(\mathbf{v})$ are approximated by

$$\epsilon(\mathbf{v}) \sim \tau|\mathbf{n}.\mathbf{v}| \qquad (6.68)$$

where τ, $\mathbf{n}$ are independent of $\mathbf{v}$ and where $\mathbf{n}$ is normal to the true translational velocity. The factor τ is a measure of the irregularity of the rigid surface giving rise to the image velocity field. The results differ in the precise assumptions made about the irregularity in the image velocity field. It seems difficult to obtain (6.68) in a very general case, especially if a bound on the error $|\epsilon(\mathbf{v}) - \tau|\mathbf{n}.\mathbf{v}||$

in the approximation (6.68)) is required. An alternative analysis of the approximation (6.68) is described in Heeger & Jepson (1990) and Jepson & Heeger (1990).

The approximation (6.68) is one of the most important properties of the least squares error function. It illustrates the way in which reconstruction from an irregular image velocity field is simplified as compared with reconstruction from a smooth image velocity field. Experiments with computer generated image velocity fields reported by Maybank (1987) and by Heeger & Jepson (1990) show that (6.68) is usually an extremely good approximation to $\epsilon(\mathbf{v})$. The approximation is useful even when the irregularity in the image velocity field is slight. The accuracy of the approximation increases as the measure τ of the irregularity increases.

Throughout this subsection $\hat{\mathbf{v}}$ and $\hat{\mathbf{w}}$ are the true (veridical) translational and angular velocities of the rigid surface giving rise to the image velocity field and $\mathbf{v}$, $\mathbf{w}$ are trial velocities, which are, in general, not equal to $\hat{\mathbf{v}}$, $\hat{\mathbf{w}}$.

Theorem 6.7. *Let $\dot{\mathbf{q}}_i$ for $1 \le i \le n$ be a set of image velocity vectors such that the base points $\mathbf{q}_i$ are coplanar with the true translational velocity $\hat{\mathbf{v}}$. Let $\mathbf{n}$ be the unit vector normal to this plane. Then the least squares error function is given by*

$$\epsilon(\mathbf{v}) = \tau \mid \mathbf{n}.\mathbf{v} \mid \tag{6.69}$$

where τ is independent of $\mathbf{v}$.

Proof. Without loss of generality coordinates are chosen such that $\mathbf{n} = (1/\sqrt{2}, 0, 1/\sqrt{2})^\top$. In view of Proposition 6.4, it is assumed that the true angular velocity $\hat{\mathbf{w}}$ is zero. Let $\mathbf{l}$ be the vector with ith component $l_i = (\dot{\mathbf{q}}_i \times \mathbf{q}_i).\mathbf{v}$ and let K_i be the inverse distance to the rigid surface in the direction $\mathbf{q}_i$. Equation (4.7) yields

$$\dot{\mathbf{q}}_i \times \mathbf{q}_i = (\hat{\mathbf{v}} \times \mathbf{q}_i)K_i \qquad (1 \le i \le n)$$

It follows from the hypotheses of the theorem that $\hat{\mathbf{v}} \times \mathbf{q}_i$ is parallel to $\mathbf{n}$. This yields

$$\begin{aligned} l_i &= [(\hat{\mathbf{v}} \times \mathbf{q}_i).\mathbf{v}]K_i \\ &= \|\hat{\mathbf{v}} \times \mathbf{q}_i\|(\mathbf{n}.\mathbf{v})K_i \qquad (1 \le i \le n) \end{aligned} \tag{6.70}$$

Let $\mathbf{m}$ be the n-dimensional vector with ith component $m_i = \|\mathbf{v} \times \mathbf{q}_i\|K_i$. It follows from (6.70) that $\mathbf{l} = (\mathbf{n}.\mathbf{v})\mathbf{m}$. The definition (6.51) of ϵ is applied to obtain

$$\begin{aligned} \epsilon(\mathbf{v}) &= \min\{\|A\mathbf{w} - \mathbf{l}\| \mid \mathbf{w} \in \mathbb{R}^3\} \\ &= \min\{\|A\mathbf{w} - (\mathbf{n}.\mathbf{v})\mathbf{m}\| \mid \mathbf{w} \in \mathbb{R}^3\} \\ &= |\mathbf{n}.\mathbf{v}| \min\{\|A\mathbf{w} - \mathbf{m}\| \mid \mathbf{w} \in \mathbb{R}^3\} \end{aligned} \tag{6.71}$$

The coordinates of $\mathbf{q}_i$ are given by $\mathbf{q}_i = (x_i, 0, z_i)^\top$ for $1 \le i \le n$. The ith component of $A\mathbf{w}$ is obtained:

$$(A\mathbf{w})_i = (\mathbf{v}.\mathbf{q}_i)(\mathbf{w}.\mathbf{q}_i) - \mathbf{v}.\mathbf{w}$$

$$\begin{aligned} &= (v_1x_i + v_3z_i)(w_1x_i + w_3z_i) - (\mathbf{v}.\mathbf{w})(x_i^2 + z_i^2) \\ &= -(v_2w_2 + v_3w_3)x_i^2 + (v_3w_1 + v_1w_3)x_iz_i - (v_1w_1 + v_2w_2)z_i^2 \end{aligned} \quad (6.72)$$

The components a_1, a_2, a_3 of the three dimensional vector $\mathbf{a}$ are defined by

$$\begin{aligned} a_1 &= -(v_2w_2 + v_3w_3) \\ a_2 &= v_3w_1 + v_1w_3 \\ a_3 &= -(v_1w_1 + v_2w_2) \end{aligned} \quad (6.73)$$

It follows from (6.72) and (6.73) that

$$(A\mathbf{w})_i = a_1x_i^2 + a_2x_iz_i + a_3z_i^2 \quad (6.74)$$

It follows from (6.74) that $A\mathbf{w} = B\mathbf{a}$, where B is the $n \times 3$ matrix with ith row equal to (x_i^2, x_iz_i, z_i^2). The matrix B is independent of $\mathbf{v}$. The substitution of $B\mathbf{a}$ for $A\mathbf{w}$ in (6.71) yields

$$\epsilon(\mathbf{v}) = |\mathbf{n}.\mathbf{v}| \min\{\|B\mathbf{a} - \mathbf{m}\| \mid \mathbf{a} \in \mathbb{R}^3\} \quad (6.75)$$

Let τ be the scalar defined by

$$\tau = \min\{\|B\mathbf{a} - \mathbf{m}\| \mid \mathbf{a} \in \mathbb{R}^3\} \quad (6.76)$$

Equation (6.69) follows from (6.75) and (6.76). □

In the next theorem the condition that the base points be coplanar with $\hat{\mathbf{v}}$ is relaxed. Instead it is required that two of the base points be coincident. The number of base points is restricted to four to ensure that effects arising from the coincidence of just two of the base points are not swamped by the influence of a large number of other base points. As in Theorem 6.7, $\epsilon(\mathbf{v})$ is given exactly by an expression of the form (6.68). This case of coincident base points is similar to an example considered by Longuet-Higgins & Prazdny (1980), in which two different image velocity fields are obtained in the same area of the image. Image velocity fields of this type are physically possible. For example, they arise when a scene is viewed through a glass with marks or stains on it.

Theorem 6.8. *Let the $\dot{\mathbf{q}}_i$ for $1 \leq i \leq 4$ be image velocity vectors such that $\mathbf{q}_1 = \mathbf{q}_2 = \mathbf{q}$, but $\dot{\mathbf{q}}_1 \neq \dot{\mathbf{q}}_2$. Let K_1, K_2 be the inverse distances to the rigid body surface at $\mathbf{q}_1$, $\mathbf{q}_2$, respectively and let $\mathbf{n}$ be a unit vector parallel to $\mathbf{q} \times \hat{\mathbf{v}}$. If the matrix A used to define the least squares error function ϵ in (6.51) has rank three then*

$$\epsilon(\mathbf{v}) = \frac{1}{\sqrt{2}}\|\hat{\mathbf{v}} \times \mathbf{q}\| \; |K_1 - K_2| \; |\mathbf{n}.\mathbf{v}|$$

Proof. As in Theorem 6.7 it is assumed without loss of generality that the angular velocity $\hat{\mathbf{w}}$ of the rigid surface giving rise to the image velocity field is zero. Let

H be the 3×3 matrix formed from the last three rows of A. By hypothesis, H has rank three, thus H is invertible. The application of H^{-1} to the right of A yields

$$AH^{-1} = \begin{pmatrix} 1 & 0 & 0 \\ 1 & 0 & 0 \\ 0 & 1 & 0 \\ 0 & 0 & 1 \end{pmatrix} \tag{6.77}$$

The expression (6.54) for ϵ is used. It follows from the properties of the Moore-Penrose pseudo-inverse A^+ of A that

$$AA^+ = (AH^{-1})(AH^{-1})^+$$

Equation (6.77) thus yields

$$AA^+ = \begin{pmatrix} 1/2 & 1/2 & 0 & 0 \\ 1/2 & 1/2 & 0 & 0 \\ 0 & 0 & 1 & 0 \\ 0 & 0 & 0 & 1 \end{pmatrix} \tag{6.78}$$

On using (6.78) to substitute for AA^+ in (6.107) it follows that

$$\epsilon(\mathbf{v}) = |l_1 - l_2|/\sqrt{2} \tag{6.79}$$

It follows from the definition $l_i = (\dot{\mathbf{q}}_i \times \mathbf{q}_i).\mathbf{v}$ and the assumption $\hat{\mathbf{w}} = 0$, that

$$\begin{aligned} l_1 - l_2 &= [(\hat{\mathbf{v}} \times \mathbf{q}).\mathbf{v}](K_1 - K_2) \\ &= \|\hat{\mathbf{v}} \times \mathbf{q}\|(\mathbf{n}.\mathbf{v})(K_1 - K_2) \end{aligned} \tag{6.80}$$

The result follows from (6.79) and (6.80). □

In practical applications of the least squares algorithm a more general version of Theorem 6.8 is required, in which the base points $\mathbf{q}_1$, $\mathbf{q}_2$ are close together rather than coincident. The expression (6.69) for $\epsilon(\mathbf{v})$ then ceases to be exact and the problem arises of calculating an upper bound for $|\epsilon(\mathbf{v}) - \tau|\mathbf{n}.\mathbf{v}||$ in terms of the motion and shape of the rigid surface giving rise to the image velocity vectors. The calculation appears to be complicated. The results obtained so far (Maybank 1987) are unsatisfactory in that the upper bounds are up to ten times larger than the values of $|\epsilon(\mathbf{v}) - \tau|\mathbf{n}.\mathbf{v}||$.

The next two theorems illustrate the difficulties involved in obtaining an upper bound to $|\epsilon(\mathbf{v}) - \tau|\mathbf{n}.\mathbf{v}||$ that is simultaneously simple and accurate.

Theorem 6.9. *Let image velocity vectors $\dot{\mathbf{q}}_i$ be given with base points $\mathbf{q}_i$ for $1 \le i \le 4$. As in the proof of Theorem 6.8, let H be the 3×3 matrix formed by the last three rows of the matrix A used in the definition (6.51) of ϵ,*

$$H = \begin{pmatrix} (\mathbf{v}.\mathbf{q}_2)\mathbf{q}_2^\top - \mathbf{v}^\top \\ (\mathbf{v}.\mathbf{q}_3)\mathbf{q}_3^\top - \mathbf{v}^\top \\ (\mathbf{v}.\mathbf{q}_4)\mathbf{q}_4^\top - \mathbf{v}^\top \end{pmatrix}$$

Let r_1, r_2, r_3 be defined by the equation

$$(1 + r_1, r_2, r_3) = [(\mathbf{v}.\mathbf{q}_1)\mathbf{q}_1 - \mathbf{v}]^\top H^{-1} \tag{6.81}$$

Let $\mathbf{l}$ be the four dimensional vector with ith component $l_i = (\dot{\mathbf{q}}_i \times \mathbf{q}_i).\mathbf{v}$. Let the base points $\mathbf{q}_1$, $\mathbf{q}_2$ be sufficiently close together and let $\dot{\mathbf{q}}_1$, $\dot{\mathbf{q}}_2$ be sufficiently far apart to ensure that $|l_1 - l_2| \gg |r_i|$ for $i = 1, 2, 3$. Let $\eta = sign(l_2 - l_1)$. Then $\epsilon(\mathbf{v})$ has an approximation of the form

$$\epsilon(\mathbf{v}) = \frac{1}{\sqrt{2}}|l_1 - l_2| + \frac{\eta}{2\sqrt{2}}[r_1(l_1 + l_2) + r_2 l_3 + r_3 l_4] + O(r_i^2) \tag{6.82}$$

Proof. Let s be defined by $s = 1 + r_1$ and let the matrix B be defined by $B = AH^{-1}$. It follows from the definition of H that

$$B = AH^{-1} = \begin{pmatrix} s & r_2 & r_3 \\ 1 & 0 & 0 \\ 0 & 1 & 0 \\ 0 & 0 & 1 \end{pmatrix} \tag{6.83}$$

It follows from the properties of the Moore-Penrose pseudo-inverses, A^+, B^+, that $AA^+ = BB^+$. The product BB^+ is evaluated using the equation $BB^+ = B(B^\top B)^{-1}B^\top$. Equation (6.83) yields

$$B^\top B = \begin{pmatrix} 1 + s^2 & r_2 s & r_3 s \\ r_2 s & 1 + r_2^2 & r_2 r_3 \\ r_3 s & r_2 r_3 & 1 + r_3^2 \end{pmatrix}$$

and

$$\det(B^\top B) = 1 + r_2^2 + r_3^2 + s^2$$

The inverse of $B^\top B$ is given by

$$(B^\top B)^{-1} = (1 + r_2^2 + r_3^2 + s^2)^{-1} \begin{pmatrix} 1 + r_2^2 + r_3^2 & -r_2 s & -r_3 s \\ -r_2 s & 1 + s^2 + r_3^2 & -r_2 r_3 \\ -r_3 s & -r_2 r_3 & 1 + s^2 + r_2^2 \end{pmatrix} \tag{6.84}$$

Equations (6.83) and (6.84) yield

$$(1 + r_2^2 + r_3^2 + s^2)BB^+ = \begin{pmatrix} s^2 + r_2^2 + r_3^2 & s & r_2 & r_3 \\ s & 1 + r_2^2 + r_3^2 & -r_2 s & -r_3 s \\ r_2 & -r_2 s & 1 + s^2 + r_3^2 & -r_2 r_3 \\ r_3 & -r_3 s & -r_2 r_3 & 1 + s^2 + r_2^2 \end{pmatrix} \tag{6.85}$$

The equality $AA^+ = BB^+$, (6.85) and the equality $s = 1 + r_1$ together yield

$$\begin{aligned} AA^+ &= \frac{1}{2}(1+r_1)^{-1}\begin{pmatrix} 1+2r_1 & 1+r_1 & r_2 & r_3 \\ 1+r_1 & 1 & -r_2 & -r_3 \\ r_2 & -r_2 & 2(1+r_1) & 0 \\ r_3 & -r_3 & 0 & 2(1+r_1) \end{pmatrix} + O(r_i^2) \\ &= \begin{pmatrix} (1+r_1)/2 & 1/2 & r_2/2 & r_3/2 \\ 1/2 & (1-r_1)/2 & -r_2/2 & -r_3/2 \\ r_2/2 & -r_2/2 & 1 & 0 \\ r_3/2 & -r_3/2 & 0 & 1 \end{pmatrix} + O(r_i^2) \end{aligned} \tag{6.86}$$

Let $\mathbf{m}$ be the four dimensional vector defined by $\mathbf{m} = AA^+\mathbf{l} - \mathbf{l}$. It follows from (6.86) that the components $(m_1, m_2, m_3, m_4)^\top$ of $\mathbf{m}$ are given by

$$\begin{aligned} m_1 &= \frac{1}{2}(1+r_1)l_1 + \frac{1}{2}l_2 + \frac{1}{2}(r_2l_3 + r_3l_4) - l_1 + O(r_i^2) \\ &= \frac{1}{2}(l_2 - l_1) + \frac{1}{2}(r_1l_1 + r_2l_3 + r_3l_4) + O(r_i^2) \\ m_2 &= \frac{1}{2}l_1 + \frac{1}{2}(1-r_1)l_2 - \frac{1}{2}(r_2l_3 + r_3l_4) - l_2 + O(r_i^2) \\ &= \frac{1}{2}(l_1 - l_2) - \frac{1}{2}(r_1l_2 + r_2l_3 + r_3l_4) + O(r_i^2) \\ m_3 &= r_2(l_1 - l_2)/2 + O(r_i^2) \\ m_4 &= r_3(l_1 - l_2)/2 + O(r_i^2) \end{aligned} \tag{6.87}$$

Equation (6.54) yields $\epsilon(\mathbf{v}) = \|\mathbf{m}\|$. On using (6.87) to expand $\|\mathbf{m}\|^2$ it follows that

$$\begin{aligned} \epsilon(\mathbf{v})^2 &= m_1^2 + m_2^2 + O(r_i^2) \\ &= \frac{1}{2}(l_2 - l_1)^2 + \frac{1}{2}r_1(l_2 - l_1)(l_1 + l_2) + (l_2 - l_1)(r_2l_3 + r_3l_4) + O(r_i^2) \\ &= \frac{1}{2}(l_2 - l_1)^2\left[1 + r_1\left(\frac{l_2 + l_1}{l_2 - l_1}\right) + \left(\frac{r_2l_3 + r_3l_4}{l_2 - l_1}\right)\right] + O(r_i^2) \end{aligned} \tag{6.88}$$

The square root of (6.88) yields

$$\epsilon(\mathbf{v}) = \frac{1}{\sqrt{2}}|l_2 - l_1|\left[1 + \frac{r_1}{2}\left(\frac{l_2 + l_1}{l_2 - l_1}\right) + \frac{1}{2}\left(\frac{r_2l_3 + r_3l_4}{l_2 - l_1}\right)\right] + O(r_i^2) \tag{6.89}$$

The result, (6.82), follows from (6.89). □

Let $\mathbf{q}$ be the unit vector in the direction $\mathbf{q}_1 + \mathbf{q}_2$. The leading order term, $|l_2 - l_1|/\sqrt{2}$, on the right-hand side of (6.82) is approximated as follows:

$$\begin{aligned} |l_2 - l_1|/\sqrt{2} &= \frac{1}{\sqrt{2}}|(\dot{\mathbf{q}}_2 \times \mathbf{q}_2).\mathbf{v} - (\dot{\mathbf{q}}_1 \times \mathbf{q}_1).\mathbf{v}| \\ &= \frac{1}{\sqrt{2}}|K_2(\hat{\mathbf{v}} \times \mathbf{q}_2).\mathbf{v} - K_1(\hat{\mathbf{v}} \times \mathbf{q}_1).\mathbf{v}| \\ &= \frac{1}{\sqrt{2}}|K_2 - K_1|\,|(\hat{\mathbf{v}} \times \mathbf{q}).\mathbf{v}| + O(\|\mathbf{q}_2 - \mathbf{q}_1\|) \end{aligned}$$

The approximation (6.82) to $\epsilon(\mathbf{v})$ is unsatisfactory because the terms on the right-hand side of (6.82) are not given explicitly in terms of the velocity and shape of the surface giving rise to the image velocity field. If the velocity and shape are introduced then the right-hand side of (6.82) becomes more complicated. A reasonably simple approximation to the right-hand side of (6.82) is sought such that the translational velocity $\hat{\mathbf{v}}$ and the inverse distances K_i appear explicitly. In order to make this approximation further assumptions about the positions of the base points are introduced. It is assumed that two of the base points are close together in comparison with the distances between the remaining base points and at the same time it is assumed that the maximum distance between any two base points is small when compared with the size of the projection sphere. More formally, let $\mathbf{q}_1$, $\mathbf{q}_2$ be the two base points that are exceptionally close together, and let δ_1, δ_2 be defined by

$$\begin{aligned} \delta_1 &= \|\mathbf{q}_1 - \mathbf{q}_2\| \\ \delta_2 &= \max\{\|\mathbf{q}_i - \mathbf{q}_j\| \mid i, j = 2, 3, 4\} \end{aligned}$$

The assumptions are

$$\delta_1 \ll \delta_2 \ll 1 \tag{6.90}$$

Let ρ be the ratio defined by

$$\rho = \frac{\delta_1}{\delta_2} \ll 1 \tag{6.91}$$

and let the $\mathbf{a}_i$ be vectors defined by

$$\mathbf{a}_i = (\mathbf{v}.\mathbf{q}_i)\mathbf{q}_i - \mathbf{v} \qquad (1 \le i \le 4)$$

Let H be the matrix defined in the statement of Theorem 6.9,

$$H = (\mathbf{a}_2, \mathbf{a}_3, \mathbf{a}_4)^{\top}$$

The inverse matrix, H^{-1}, is given by

$$H^{-1} = \frac{1}{\mathbf{a}_2.(\mathbf{a}_3 \times \mathbf{a}_4)} (\mathbf{a}_3 \times \mathbf{a}_4 \,|\, \mathbf{a}_4 \times \mathbf{a}_2 \,|\, \mathbf{a}_2 \times \mathbf{a}_3) \tag{6.92}$$

Coordinates are chosen such that the centroid of the base points $\mathbf{q}_i$ is on the line joining the centre of the projection sphere to the point $\mathbf{q} = (0,0,1)^{\top}$. Let t be the scalar defined such that $t\delta_2^2$ is the area of the triangle with vertices $\mathbf{q}_2$, $\mathbf{q}_3$, $\mathbf{q}_4$. If the distances between the vertices of the triangle are comparable in magnitude then the order of magnitude of t is one. For example, if the triangle is equilateral then $t = \sqrt{3}/4$. The term $\det(H)$ is identical to the polynomial f_{234} described in Theorem 4.16 of Sect. 4.3.2. It is shown in Theorem 4.16 that

$$\det(H) = \mathbf{a}_2.(\mathbf{a}_3 \times \mathbf{a}_4) = \pm 2\, t\, \delta_2^2 |v_3| (v_1^2 + v_2^2) + O(\delta_2^3) \tag{6.93}$$

The right-hand side of (6.82) is approximated as follows.

Theorem 6.10. *With the notation and hypotheses of Theorem 6.9 and with the additional hypothesis (6.90), the least squares error function ϵ has an approximation of the form*

$$\epsilon(\mathbf{v}) = \frac{1}{\sqrt{2}}|(\mathbf{v} \times \hat{\mathbf{v}}).\mathbf{q}|\,|K_1 - K_2| + \frac{\xi\rho|\overline{K}(\mathbf{v} \times \hat{\mathbf{v}}).\mathbf{q}|}{\sqrt{2}t|v_3|(v_1^2 + v_2^2)} + O(\delta_1) \qquad (6.94)$$

where $\overline{K}$ is the average of the four inverse distances K_i, ξ is a number in the interval $[-1,1]$ and ρ is defined by (6.91).

Proof. Let $\mathbf{\Delta}$ be defined by $\mathbf{\Delta} = \mathbf{a}_1 - \mathbf{a}_2$. It follows from (6.81) that

$$\begin{aligned}(1 + r_1, r_2, r_3) &= \mathbf{a}_1^\top H^{-1} \\ &= (\mathbf{a}_2 + \mathbf{\Delta})^\top H^{-1} \\ &= (1,0,0) + \mathbf{\Delta}^\top H^{-1} \end{aligned} \qquad (6.95)$$

It follows from (6.92) and (6.95) that

$$\begin{aligned}(r_1, r_2, r_3) &= \mathbf{\Delta}^\top H^{-1} \\ &= (\mathbf{a}_2.(\mathbf{a}_3 \times \mathbf{a}_4))^{-1}\mathbf{\Delta}^\top(\mathbf{a}_3 \times \mathbf{a}_4|\mathbf{a}_4 \times \mathbf{a}_2|\mathbf{a}_2 \times \mathbf{a}_3)\end{aligned}$$

The application of (6.93) yields

$$\begin{aligned}(r_1, r_2, r_3) &= (\mathbf{a}_2.(\mathbf{a}_3 \times \mathbf{a}_4))^{-1}\mathbf{\Delta}^\top(\mathbf{a}_3 \times \mathbf{a}_4|\mathbf{a}_4 \times \mathbf{a}_2|\mathbf{a}_2 \times \mathbf{a}_3) \\ &= \pm\frac{\mathbf{\Delta}^\top(\mathbf{a}_3 \times \mathbf{a}_4 \,|\, \mathbf{a}_4 \times \mathbf{a}_2 \,|\, \mathbf{a}_2 \times \mathbf{a}_3)}{2t\,\delta_2^2|v_3|(v_1^2 + v_2^2) + O(\delta_2^3)}\end{aligned} \qquad (6.96)$$

The norm of each of the vectors $\mathbf{a}_3 \times \mathbf{a}_4$, $\mathbf{a}_4 \times \mathbf{a}_2$, $\mathbf{a}_2 \times \mathbf{a}_3$ is no greater that δ_2. The norm of $\mathbf{\Delta}$ is no greater that δ_1. It thus follows from (6.96) that

$$\begin{aligned}\max\{|r_1|, |r_2|, |r_3|\} &= \frac{\max\{|\mathbf{\Delta}.(\mathbf{a}_3 \times \mathbf{a}_4)|, |\mathbf{\Delta}.(\mathbf{a}_4 \times \mathbf{a}_2)|, |\mathbf{\Delta}.(\mathbf{a}_2 \times \mathbf{a}_3)|\}}{2t\,\delta_2^2|v_3|(v_1^2 + v_2^2) + O(\delta_2^3)} \\ &\leq \frac{\|\mathbf{\Delta}\| \max\{\|\mathbf{a}_3 \times \mathbf{a}_4\|, \|\mathbf{a}_4 \times \mathbf{a}_2\|, \|\mathbf{a}_2 \times \mathbf{a}_3\|\}}{2t\delta_2^2|v_3|(v_1^2 + v_2^2) + O(\delta_2^3)} \\ &\leq \frac{\delta_1}{2t\delta_2|v_3|(v_1^2 + v_2^2) + O(\delta_2^2)} \\ &\leq \frac{\rho}{2t|v_3|(v_1^2 + v_2^2)} + O(\rho\delta_2)\end{aligned} \qquad (6.97)$$

The product $\rho\delta_2$ on the right-hand side of (6.97) is equal to δ_1. It follows from (6.97) that there exists a number ξ' in the interval $[0,1]$ such that

$$\max\{|r_1|, |r_2|, |r_3|\} = \frac{\xi'\rho}{2t|v_3|(v_1^2 + v_2^2)} + O(\delta_1)$$

It is assumed without loss of generality that the true angular velocity $\hat{\mathbf{w}}$ is equal to zero. It follows from the remarks preceding the statement of this theorem that

$$|l_2 - l_1| = |K_2 - K_1|\,|(\mathbf{v} \times \hat{\mathbf{v}}).\mathbf{q}| + O(\delta_1)$$

Let $\bar{l}$ be the average value of the $|l_i|$. It follows that

$$\begin{aligned} |r_1(l_1+l_2)+r_2l_3+r_3l_4| &\le 4\max\{|r_1|,|r_2|,|r_3|\}\bar{l} \\ &\le \frac{2\xi'\rho\bar{l}}{2t|v_3|(v_1^2+v_2^2)}+O(\delta_1) \end{aligned} \tag{6.98}$$

The average $\bar{l}$ is given by

$$\bar{l} = |\overline{K}(\hat{\mathbf{v}}\times\mathbf{v}).\mathbf{q}| + O(\delta_2) \tag{6.99}$$

Equation (6.94) follows from (6.82), (6.98) and (6.99). The scalar ξ is defined by $\xi = \eta\xi'$. □

Theorems 6.8-6.10 describe a range of cases in which the image velocity field is irregular and the leading order term of ϵ has the form $\tau|\mathbf{n}.\mathbf{v}|$, where τ depends only on the degree of irregularity of the surface giving rise to the image velocities and $\mathbf{n}$ is a vector independent of $\mathbf{v}$. If the surface is highly irregular then τ is large. Experiments with computer generated image velocity fields indicate that the difference $|\epsilon(\mathbf{v})-\tau|\mathbf{n}.\mathbf{v}||$ is very small, especially if the irregularity of the surface is pronounced or if the base points are close together. The results of a typical experiment to test the accuracy of the approximation $\epsilon(\mathbf{v})\approx\tau|\mathbf{n}.\mathbf{v}|$ are displayed in Fig. 6.2.

The graph in Fig. 6.2 shows the values of ϵ on the circle defined by

$$\mathbf{v} = \left(\frac{\sqrt{2}}{\sqrt{3}}\cos(\phi), \frac{\sqrt{2}}{\sqrt{3}}\sin(\phi), \frac{1}{\sqrt{3}}\right)^\top \qquad (0\le\phi\le 2\pi)$$

The image velocity field arises from a rigid moving surface consisting of two planar facets with outward normals

$$(\cos(\zeta), 0, -\sin(\zeta))^\top \quad \text{and} \quad (-\cos(\zeta), 0, -\sin(\zeta))^\top$$

respectively. The angle ζ is set equal to 45°. The surface forms a 'peak' pointing towards the projection sphere. The true translational velocity $\hat{\mathbf{v}}$ is equal to $(1/\sqrt{3}, 1/\sqrt{3}, 1/\sqrt{3})^\top$ and the true angular velocity, $\hat{\mathbf{w}}$, is zero. The image velocity vectors are scaled to have average length $d/5$, where d is the radius of the field of view. The base points are

$$\begin{aligned} \mathbf{q}_1 &= (d/\sqrt{2}, d/\sqrt{2}, \sqrt{1-d^2})^\top \\ \mathbf{q}_2 &= (-d/\sqrt{2}, d/\sqrt{2}, \sqrt{1-d^2})^\top \\ \mathbf{q}_3 &= (-d/\sqrt{2}, -d/\sqrt{2}, \sqrt{1-d^2})^\top \\ \mathbf{q}_4 &= (d/\sqrt{2}, -d/\sqrt{2}, \sqrt{1-d^2})^\top \\ \mathbf{q}_5 &= (0,0,1)^\top \end{aligned}$$

The first four base points are at the vertices of a square and the fifth base point is over the centre of the square. The value of d is set at $d = 0.1$.

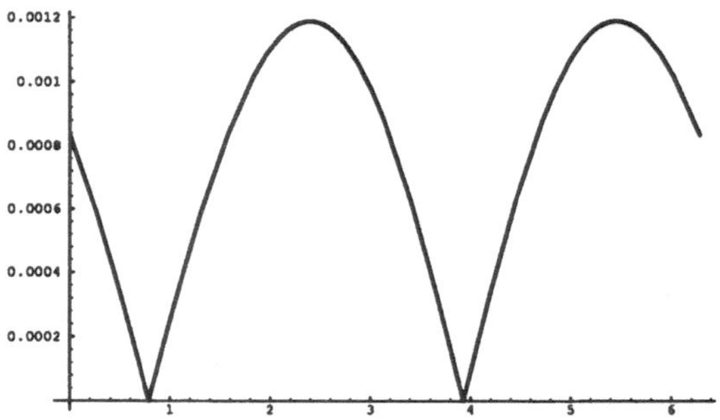

Fig. 6.2. Graph of ϵ for $v_3 = 3^{-1/2}$

It follows from the approximation $\epsilon(\mathbf{v}) \sim \tau|\mathbf{n}.\mathbf{v}|$ that one component of the true translational velocity $\hat{\mathbf{v}}$ and one component of the true angular velocity $\hat{\mathbf{w}}$ can be found accurately. The function ϵ takes low values on or near to the great circle defined by $\mathbf{n}.\mathbf{v} = 0$, where $\mathbf{n}$ is a vector which is independent of $\mathbf{v}$ and normal to $\hat{\mathbf{v}}$. This great circle passes near to the true translational velocity $\hat{\mathbf{v}}$ and it also passes near to the base points. Let coordinate axes be chosen such that the base points are close to $(0,0,1)^\top$. The position of the great circle yields an accurate estimate of $\tan^{-1}(\hat{v}_1/\hat{v}_2)$, provided $\hat{\mathbf{v}}$ is not close to the base points.

The ratio $\hat{v}_1/\hat{v}_2$ and the zero order approximation (f_1, g_1) to the image velocity field at $(0,0,1)^\top$ together determine the component of the angular velocity $\hat{\mathbf{w}}$ parallel to $(\hat{v}_1, \hat{v}_2, 0)^\top$. The expressions for f_1 and g_1 are obtained from (4.149),

$$\begin{aligned} f_1 &= \hat{v}_1 k_1 + \hat{w}_2 \\ g_1 &= \hat{v}_2 k_1 - \hat{w}_1 \end{aligned} \qquad (6.100)$$

Let $\hat{w}_{||}$ be the component of $\hat{\mathbf{w}}$ parallel to $(\hat{v}_1, \hat{v}_2, 0)^\top$. It follows from (6.100) and the definition of $\hat{w}_{||}$ that

$$\hat{w}_{||} = \frac{\hat{v}_1\hat{w}_1 + \hat{v}_2\hat{w}_2}{\sqrt{\hat{v}_1^2 + \hat{v}_2^2}} = \frac{\hat{v}_2 f_1 - \hat{v}_1 g_1}{\sqrt{\hat{v}_1^2 + \hat{v}_2^2}}$$

The component $\hat{w}_{||}$ is thus determined by the ratio $\hat{v}_1/\hat{v}_2$.

6.2.4 The First Order Algorithm

In the first order algorithm the velocity of a rigid moving surface is reconstructed from an image velocity field which is differentiable to first order in two different parts of the field of view. It is assumed that both parts of the image velocity field arise from the same moving rigid surface. The first order algorithm obtains stable estimates of velocity by combining the information obtained from the two sets of derivatives. It avoids estimating the derivatives of the image velocity field to second order. This is an advantage because such second order estimates are severely affected by small errors in the image velocity vectors. Another advantage of the first order algorithm is that the translational velocity is obtained from one of the roots of a quadratic equation in one variable. This is in contrast with algorithms which use the derivatives of the image velocity field to second order. Such algorithms usually involve the solution of an equation of degree three (Waxman *et al.* 1987). Suitable image velocity fields for the first order algorithm occur in practice, especially if the camera is moving through a man-made environment. For example, if the camera is moving indoors then the part of the image velocity field arising from the floor could yield one set of derivatives and the part arising from one of the walls could yield the other set of derivatives.

The equations linking the rigid velocity $\{\mathbf{v}, \mathbf{w}\}$ of the surface to the derivatives of the image velocity field are obtained in this subsection. In Sect. 6.2.5 two homogeneous polynomial constraints on $\mathbf{v}$ are obtained from these equations. One constraint is of degree one and the other constraint is of degree two.

Coordinate axes are chosen such that the points at which the derivatives of the image velocity field are obtained are $(0,0,1)^\top$ and $(\sin(\theta), 0, \cos(\theta))^\top$. The image is projected onto the unit sphere centred at the origin. Let $\dot{\mathbf{q}} = (\dot{x}, \dot{y}, \dot{z})^\top$ be the image velocity field, and let f_i, g_i, $1 \le i \le 3$, be the spatial derivatives of the image velocity field to first order at $\mathbf{q}_0 = (0,0,1)^\top$. The projection sphere is tangent to the projection plane at $(0,0,1)^\top$, thus the derivatives f_i, g_i, $1 \le i \le 3$, agree with the derivatives to first order on the projection plane, as given in Sect. 4.5.1. The derivatives f_i, g_i, $i = 1,2,3$ are defined by

$$
\begin{array}{lll}
f_1 = \dot{x}(\mathbf{q}_0) & f_2 = \left.\dfrac{\partial \dot{x}}{\partial x}\right|_{\mathbf{q}_0} & f_3 = \left.\dfrac{\partial \dot{x}}{\partial y}\right|_{\mathbf{q}_0} \\[2ex]
g_1 = \dot{y}(\mathbf{q}_0) & g_2 = \left.\dfrac{\partial \dot{y}}{\partial x}\right|_{\mathbf{q}_0} & g_3 = \left.\dfrac{\partial \dot{y}}{\partial y}\right|_{\mathbf{q}_0}
\end{array}
\qquad (6.101)
$$

It follows from (6.101) and the condition $\dot{\mathbf{q}}.\mathbf{q} = 0$ that in a neighbourhood of $\mathbf{q}_0$,

$$
\begin{aligned}
\dot{\mathbf{q}} &= (\dot{x}, \dot{y}, \dot{z})^\top \\
&= (f_1 + f_2 x + f_3 y, g_1 + g_2 x + g_3 y, -f_1 x - g_1 y)^\top + R_2
\end{aligned}
\qquad (6.102)
$$

The term R_2 on the right-hand side of (6.102) is second order in x, y. The symbol R_2 is used for all such second order terms throughout this subsection. It follows from (6.102) that

$$
\dot{\mathbf{q}} \times \mathbf{q} = (g_1 + g_2 x + g_3 y, -f_1 - f_2 x - f_3 y, f_1 x - g_1 y)^\top + R_2 \qquad (6.103)
$$

It is recalled from (4.8) that

$$
(\dot{\mathbf{q}} \times \mathbf{q}).\mathbf{v} = (\mathbf{w}.\mathbf{q})(\mathbf{v}.\mathbf{q}) - \mathbf{w}.\mathbf{v} \qquad (6.104)
$$

On using (6.103) to substitute for $\dot{\mathbf{q}} \times \mathbf{q}$ in (6.104) and expanding the resulting equation it follows that

$$
(g_1 + g_2 x + g_3 y)v_1 - (f_1 + f_2 x + f_3 y)v_2 + (f_1 y - g_1 x)v_3
$$

$$= (w_1x + w_2y + w_3)(v_1x + v_2y + v_3) - \mathbf{w}.\mathbf{v} + R_2 \tag{6.105}$$

On equating the coefficients of the constant terms and the coefficients of the terms linear in x or y on the left- and right-hand sides of (6.105) three equations are obtained,

$$\begin{aligned} w_1v_1 + w_2v_2 &= -g_1v_1 + f_1v_2 \\ w_1v_3 + w_3v_1 &= g_2v_1 - f_2v_2 - g_1v_3 \\ w_2v_3 + w_3v_2 &= g_3v_1 - f_3v_2 + f_1v_3 \end{aligned} \tag{6.106}$$

Let r_1, r_2 and r_3 be linear forms in $\mathbf{v}$ defined by

$$\begin{aligned} r_1(\mathbf{v}) &= g_1v_1 - f_1v_2 \\ r_2(\mathbf{v}) &= g_2v_1 - f_2v_2 - g_1v_3 \\ r_3(\mathbf{v}) &= g_3v_1 - f_3v_2 + f_1v_3 \end{aligned}$$

The signs of the r_i are chosen such that

$$(\dot{\mathbf{q}} \times \mathbf{q}).\mathbf{v} = r_1(\mathbf{v}) + r_2(\mathbf{v})x + r_3(\mathbf{v})y + R_2$$

The coefficients of the r_i are known because they are the derivatives of the image velocity field, evaluated at $\mathbf{q}_0$. The equations of (6.106) are

$$\begin{aligned} r_1(\mathbf{v}) &= -w_1v_1 - w_2v_2 \\ r_2(\mathbf{v}) &= w_1v_3 + w_3v_1 \\ r_3(\mathbf{v}) &= w_2v_3 + w_3v_2 \end{aligned} \tag{6.107}$$

A second set of equations similar to (6.106) is obtained from the derivatives of $(\dot{\mathbf{q}} \times \mathbf{q}).\mathbf{v}$ evaluated at

$$\mathbf{q}_\theta = (\sin(\theta), 0, \cos(\theta))^\top$$

The method for obtaining these equations is similar to that used to obtain (6.107), but the details are more complicated because the derivatives of the image velocity field are obtained at $\mathbf{q}_\theta$ rather than at $(0,0,1)^\top$. The first step is to examine the derivatives of $\dot{\mathbf{q}}$ at $\mathbf{q}_\theta$. Let $(\delta x, \delta y, \delta z)^\top$ be a small perturbation such that the point

$$\mathbf{q} = (\sin(\theta) + \delta x, \delta y, \cos(\theta) + \delta z)^\top$$

is on the unit sphere and close to $\mathbf{q}_\theta$. The condition $\|\mathbf{q}\| = 1$ yields

$$\delta z = -\tan(\theta)\delta x + R_2$$

from which it follows that

$$\mathbf{q} = (\sin(\theta) + \delta x, \delta y, \cos(\theta) - \tan(\theta)\delta x)^\top + R_2 \tag{6.108}$$

Let r_4, r_5, r_6 be linear forms in $\mathbf{v}$ defined such that the following equation holds near to $(\sin(\theta), 0, \cos(\theta))^\top$:

$$(\dot{\mathbf{q}} \times \mathbf{q}).\mathbf{v} = r_4(\mathbf{v}) + r_5(\mathbf{v})\delta x + r_6(\mathbf{v})\delta y + R_2 \tag{6.109}$$

The Taylor series for $\dot{\mathbf{q}} \times \mathbf{q}$ in the neighbourhood of $\mathbf{q}_\theta$ is

$$\dot{\mathbf{q}} \times \mathbf{q} = (f_1' + f_2'\delta x + f_3'\delta y, g_1' + g_2'\delta x + g_3'\delta y, h_1' + h_2'\delta x + h_3'\delta y)^\top + R_2$$

where the coefficients f_i', g_i', h_i', $1 \leq i \leq 3$, are spatial derivatives of $\dot{\mathbf{q}} \times \mathbf{q}$ evaluated at $\mathbf{q}_\theta$. The condition $(\dot{\mathbf{q}} \times \mathbf{q}).\mathbf{q} = 0$ is used to obtain the h_i' as functions of the f_i' and the g_i', as follows:

$$\begin{aligned} 0 &= (\dot{\mathbf{q}} \times \mathbf{q}).\mathbf{q} \\ &= (f_1' + f_2'\delta x + f_3'\delta y)(\sin(\theta) + \delta x) + (g_1' + g_2'\delta x + g_3'\delta y)\delta y \\ &\quad + (h_1' + h_2'\delta x + h_3'\delta y)(\cos(\theta) - \tan(\theta)\delta x) + R_2 \end{aligned} \tag{6.110}$$

Equation (6.111) holds identically to first order in δx, δy thus

$$\begin{aligned} f_1' \sin(\theta) + h_1' \cos(\theta) &= 0 \\ f_1' + f_2' \sin(\theta) + h_2' \cos(\theta) - h_1' \tan(\theta) &= 0 \\ f_3' \sin(\theta) + g_1' + h_3' \cos(\theta) &= 0 \end{aligned} \tag{6.111}$$

Equations (6.111) are solved for the h_i',

$$\begin{aligned} h_1' &= -f_1' \tan(\theta) \\ h_2' &= -f_1' \sec^3(\theta) - f_2' \tan(\theta) \\ h_3' &= -f_3' \tan(\theta) - g_1' \sec(\theta) \end{aligned}$$

The next step is to express the coefficients r_4, r_5, r_6 defined by (6.109) as functions of the f_i', g_i', h_i'. The scalar product of $\dot{\mathbf{q}} \times \mathbf{q}$ with $\mathbf{v}$ yields

$$\begin{aligned} (\dot{\mathbf{q}} \times \mathbf{q}).\mathbf{v} &= v_1 f_1' + v_2 g_1' + v_3 h_1' + \delta x(v_1 f_2' + v_2 g_2' + v_3 h_2') \\ &\quad + \delta y(v_1 f_3' + v_2 g_3' + v_3 h_3') \end{aligned} \tag{6.112}$$

It follows from (6.109) and (6.112) that

$$\begin{aligned} r_4 &= v_1 f_1' + v_2 g_1' + v_3 h_1' \\ r_5 &= v_1 f_2' + v_2 g_2' + v_3 h_2' \\ r_6 &= v_1 f_3' + v_2 g_3' + v_3 h_3' \end{aligned} \tag{6.113}$$

From this point on the derivatives f_i', g_i', h_i' of $\dot{\mathbf{q}} \times \mathbf{q}$ do not appear explicitly in the calculations. They are replaced by the linear forms r_4, r_5, r_6 defined in (6.109).

It follows from (6.108) and (6.109) that

$$(\mathbf{w}.\mathbf{q})(\mathbf{v}.\mathbf{q}) - \mathbf{w}.\mathbf{v} = r_4(\mathbf{v}) + r_5(\mathbf{v})\delta x + r_6(\mathbf{v})\delta y + R_2 \tag{6.114}$$

On using (6.108) to substitute for $\mathbf{q}$ on the left-hand side of (6.114) it follows that

$$\begin{aligned}(w_1\sin(\theta)+w_3\cos(\theta))(v_1\sin(\theta)+v_3\cos(\theta)+v_1\delta x+v_2\delta y-v_3\tan(\theta)\delta x)\\ +(w_1\delta x+w_2\delta y-w_3\tan(\theta)\delta x)(v_1\sin(\theta)+v_3\cos(\theta))-\mathbf{w}.\mathbf{v}\\ =r_4(\mathbf{v})+r_5(\mathbf{v})\delta x+r_6(\mathbf{v})\delta y+R_2\end{aligned}\tag{6.115}$$

The terms of (6.115) independent of δx and δy yield

$$\begin{aligned}r_4(\mathbf{v}) &= (w_1\sin(\theta)+w_3\cos(\theta))(v_1\sin(\theta)+v_3\cos(\theta))-\mathbf{w}.\mathbf{v}\\ &= -w_1v_1-w_2v_2+(w_3v_1+w_1v_3)\sin(\theta)\cos(\theta)\\ &\quad +(w_1v_1-w_3v_3)\sin^2(\theta)\end{aligned}\tag{6.116}$$

The terms of (6.115) linear in δx yield

$$\begin{aligned}r_5(\mathbf{v}) &= (w_1\sin(\theta)+w_3\cos(\theta))(v_1-v_3\tan(\theta))\\ &\quad +(w_1-w_3\tan\theta)(v_1\sin(\theta)+v_3\cos(\theta))\\ &= (w_3v_1+w_1v_3)\cos(\theta)+2(w_1v_1-w_3v_3)\sin(\theta)\\ &\quad -(w_1v_3+w_3v_1)\sin(\theta)\tan(\theta)\end{aligned}\tag{6.117}$$

The terms of (6.115) linear in δy yield

$$\begin{aligned}r_6(\mathbf{v}) &= (w_1\sin(\theta)+w_3\cos(\theta))v_2+w_2(v_1\sin(\theta)+v_3\cos(\theta))\\ &= (w_3v_2+w_2v_3)\cos(\theta)+(w_1v_2+w_2v_1)\sin(\theta)\end{aligned}\tag{6.118}$$

The six equations constraining $\mathbf{v}$, $\mathbf{w}$ are the three equations of (6.107), and the equations (6.116), (6.117) and (6.118).

6.2.5 Constraints on the Translational Velocity

In Sect. 6.2.4 six equations constraining the rigid velocity $\{\mathbf{v},\mathbf{w}\}$ are obtained. In this subsection the components of $\mathbf{w}$ are eliminated to give three equations in $\mathbf{v}$, the coefficients of which are functions of the derivatives of the image velocity field. The aim is to obtain equations constraining $\mathbf{v}$ which are of as low a degree as possible. A straightforward calculation with determinants yields three equations in $\mathbf{v}$, each of degree four. However, a more careful examination of the six equations yields one equation in $\mathbf{v}$ of degree one and a second equation in $\mathbf{v}$ of degree two. It is thus possible to obtain the value of $\mathbf{v}$, up to a two way ambiguity, by solving a quadratic equation. The correct value of $\mathbf{v}$ is determined from a third equation of degree four.

It is assumed that the angle θ in the six equations constraining $\{\mathbf{v},\mathbf{w}\}$ is non-zero. The case $\theta=0$ is straightforward because it leads directly to the following three linear constraints on $\mathbf{v}$,

$$r_1(\mathbf{v})=r_4(\mathbf{v})\qquad r_2(\mathbf{v})=r_5(\mathbf{v})\qquad r_3(\mathbf{v})=r_6(\mathbf{v})$$

To return to the case $\theta \neq 0$, the six equations obtained in Sect. 6.2.4 are collected together,

$$\begin{aligned}
r_1(\mathbf{v}) &= -w_1v_1 - w_2v_2 \\
r_2(\mathbf{v}) &= w_1v_3 + w_3v_1 \\
r_3(\mathbf{v}) &= w_2v_3 + w_3v_2 \\
r_4(\mathbf{v}) &= -w_1v_1 - w_2v_2 + (w_3v_1 + w_1v_3)\sin(\theta)\cos(\theta) + (w_1v_1 - w_3v_3)\sin^2(\theta) \\
r_5(\mathbf{v}) &= (w_3v_1 + w_1v_3)\cos(\theta) + 2(w_1v_1 - w_3v_3)\sin(\theta) \\
&\quad - (w_1v_3 + w_3v_1)\sin(\theta)\tan(\theta) \\
r_6(\mathbf{v}) &= (w_3v_2 + w_2v_3)\cos(\theta) + (w_1v_2 + w_2v_1)\sin(\theta)
\end{aligned} \tag{6.119}$$

The first three equations of (6.119) are used to eliminate the terms

$$v_1w_1 + v_2w_2 \qquad v_1w_3 + v_3w_1 \qquad v_2w_3 + v_3w_2$$

from the right-hand sides of the last three equations of (6.119). This yields three equations

$$\begin{aligned}
r_4(\mathbf{v}) - r_1(\mathbf{v}) &= (w_3v_1 + w_1v_3)\sin(\theta)\cos(\theta) + (w_1v_1 - w_3v_3)\sin^2(\theta) \\
r_5(\mathbf{v}) - r_2(\mathbf{v})\cos(\theta) &= 2(w_1v_1 - w_3v_3)\sin(\theta) - (w_1v_3 + w_3v_1)\sin(\theta)\tan(\theta) \\
r_6(\mathbf{v}) - r_3(\mathbf{v})\cos(\theta) &= (w_1v_2 + w_2v_1)\sin(\theta)
\end{aligned} \tag{6.120}$$

The term $w_3v_1 + w_1v_3$ on the right-hand side of the first two equations of (6.120) is replaced by $r_2(\mathbf{v})$. A rearrangement of the resulting equations yields

$$\begin{aligned}
v_1w_1 - v_3w_3 &= [r_4(\mathbf{v}) - r_1(\mathbf{v})]\mathrm{cosec}^2(\theta) - r_2(\mathbf{v})\cot(\theta) \\
v_1w_1 - v_3w_3 &= \frac{1}{2}r_5(\mathbf{v})\mathrm{cosec}(\theta) - r_2(\mathbf{v})\cot(2\theta) \\
v_1w_2 + v_2w_1 &= r_6(\mathbf{v})\mathrm{cosec}(\theta) - r_3(\mathbf{v})\cot(\theta)
\end{aligned} \tag{6.121}$$

The elimination of the term $v_1w_1 - v_3w_3$ from the first two equations of (6.121) yields a linear constraint on $\mathbf{v}$,

$$2r_4(\mathbf{v}) - 2r_1(\mathbf{v}) - r_2(\mathbf{v})\tan(\theta) - r_5(\mathbf{v})\sin(\theta) = 0 \tag{6.122}$$

An additional algebraic constraint on $\mathbf{v}$ of degree two is obtained as follows.

Theorem 6.11. *The equations of (6.119) yield a homogeneous quadratic constraint on* $\mathbf{v}$.

Proof. Let a, b, c be the coefficients of (6.122). Equation (6.122) is written in the form

$$av_1 + bv_2 + cv_3 = 0 \tag{6.123}$$

After some algebraic manipulation, the first three equations of (6.119) and the last two equations of (6.121) yield

$$\begin{aligned}
(aw_1 + bw_2 + cw_3)v_1 - 2aw_1v_1 &= cr_2 + b[r_6\mathrm{cosec}(\theta) - r_3\cot(\theta)] \\
(aw_1 + bw_2 + cw_3)v_2 + 2bw_1v_1 &= -2br_1 + cr_3 + a[r_6\mathrm{cosec}(\theta) - r_3\cot(\theta)] \\
(aw_1 + bw_2 + cw_3)v_3 - 2cw_1v_1 &= ar_2 + br_3 + c[2r_2\cot(2\theta) \\
&\quad - r_5\mathrm{cosec}(\theta)]
\end{aligned} \tag{6.124}$$

The equations (6.124) are regarded as a set of three equations in the two unknowns w_1 and $a\,w_1 + b\,w_2 + c\,w_3$. The three equations are linear, thus a solution exists if and only if

$$\det\begin{pmatrix} v_1 & -a & cr_2 + b[r_6\mathrm{cosec}(\theta) - r_3\cot(\theta)] \\ v_2 & b & -2br_1 + cr_3 + a[r_6\mathrm{cosec}(\theta) - r_3\cot(\theta)] \\ v_3 & -c & ar_2 + br_3 + c[2r_2\cot(2\theta) - r_5\mathrm{cosec}(\theta)] \end{pmatrix} = 0 \qquad (6.125)$$

Equation (6.125) is the required quadratic constraint on $\mathbf{v}$. □

There are, in general, at most two values of $\mathbf{v}$ satisfying both the linear constraint (6.123) and the quadratic constraint (6.125). The two values can be found by solving a second order polynomial equation in one variable. In general only one of these two values is a possible translational velocity. The correct value can be identified by the condition that it leads to a solution of all six equations (6.119). If the image velocity field is ambiguous then both values of $\mathbf{v}$ are possible translational velocities.

References

Adiv G. 1985 Determining three-dimensional motion and structure from optical flow generated by several moving objects. *IEEE Trans. Pattern Analysis and Machine Intelligence* **7**, 384-401.

Aggarwal J.K. & Martin W.N. 1983 Dynamic scene analysis. *Image Sequence Processing and Dynamic Scene Analysis* (Ed. Huang T.S.). NATO ASI Series F: Computer and Systems Sciences No. 2, 40-73. Berlin, Heidelberg and New York: Springer-Verlag.

Aisbet J. 1990 An iterated estimation of the motion parameters of a rigid body from noisy displacement vectors. *IEEE Trans. Pattern Analysis and Machine Intelligence* **12**, 1092-1098.

Ayache N. 1991 *Artificial Vision for Mobile Robots: stereo vision and multisensory perception.* Cambridge, Massachusetts: The MIT Press.

Bar-Shalom Y. & Fortmann T.E. 1988 *Tracking and Data Association.* Mathematics in Science and Engineering Series, vol. **179**, Orlando, Florida: Academic Press.

Fang J.-Q. & Huang T.S. 1984 Some experiments on estimating the 3-D motion parameters of a rigid body from two consecutive images. *IEEE Trans. Pattern Analysis and Machine Intelligence* **6** 545-554.

Fraleigh J.B. 1973 *A First Course in Abstract Algebra.* Reading, Massachusetts: Addison Wesley.

Golub G.H. & Van Loan C.F. 1983 *Matrix Computations.* Oxford: North Oxford Publishing Co. Ltd.

Grzywacz N.M. & Hildreth E.C. 1987 Incremental rigidity scheme for recovering structure from motion: position-based versus velocity-based methods. *J. Optical Soc. America*, Series A **4**, 503-518.

Harris C. 1990 Resolution of the bas-relief ambiguity in structure-from-motion under orthographic projection. *BMVC90 Proc. British Machine Vision Conference, Oxford*, 67-77.

Heeger D.J. & Jepson A.D. 1990 Subspace methods for recovering rigid motion I: algorithm and implementation. *University of Toronto Report on Research in Biological and Computational Vision, RBCV-TR-90-35.*

Horn B.K.P. 1990 Relative orientation. *International J. Computer Vision* **4**, 59-78.

Jepson A.D. & Heeger D.J. 1990 Subspace methods for recovering rigid motion, part II: theory. *University of Toronto Report on Research in Biological and Computational Vision, RBCV-TR-90-36.*

Kelley H.J. 1962 Methods of gradients. *Optimization Techniques with Applications to Aerospace Systems.* Ed. Leitmann G., Series: Mathematics in Science and Engineering, vol 5. New York: Academic Press.

Lawton D. 1983 Processing translational motion sequences. *Computer Vision, Graphics, and Image Processing* **22**, 116-144.

Longuet-Higgins H.C. & Prazdny K. 1980 The interpretation of a moving retinal image. *Proc. Royal Soc. London, Series B* **208**, 385-397.

Maybank S.J. 1987 *A theoretical study of optical flow.* PhD thesis, University of London, Birkbeck College.

Murray D.W. & Buxton B.F. 1990 *Experiments in the Machine Interpretation of Visual Motion.* Cambridge, Massachusetts: The MIT Press.

Sokolnikoff I.S. & Redheffer R.M. 1966 *Mathematics of Physics and Modern Engineering.* New York: McGraw-Hill.

Toscani G. & Faugeras O. 1986 Structure and motion from two noisy perspective images. *Proc. IEEE Conf. on Robotics and Automation, Raleigh, N. Carolina, USA*, 221-227.

Waxman A.M., Kamgar-Parsi B. & Subbarao M. 1987 Closed form solutions to image flow equations for 3D structure and motion. *International J. Computer Vision* **1**, 239-258.

Subject Index